Protein Phosphorylation and Meat Quality

Dequan Zhang • Xin Li • Li Chen •
Chengli Hou • Zhenyu Wang

Protein Phosphorylation and Meat Quality

Springer

Dequan Zhang
Institute of Food Science and Technology, Chinese Academy of Agricultural Sciences
Beijing, China

Xin Li
Institute of Food Science and Technology, Chinese Academy of Agricultural Sciences
Beijing, China

Li Chen
Institute of Food Science and Technology, Chinese Academy of Agricultural Sciences
Beijing, China

Chengli Hou
Institute of Food Science and Technology, Chinese Academy of Agricultural Sciences
Beijing, China

Zhenyu Wang
Institute of Food Science and Technology, Chinese Academy of Agricultural Sciences
Beijing, China

ISBN 978-981-15-9440-3 ISBN 978-981-15-9441-0 (eBook)
https://doi.org/10.1007/978-981-15-9441-0

This Springer imprint is published by the registered company Springer Nature Singapore Pte Ltd.
The registered company address is: 152 Beach Road, #21-01/04 Gateway East, Singapore 189721, Singapore

Preface

Meat quality is an important characteristic that determines the value of meat products and economic benefits. Although research has been performed on different aspects of meat quality, meat quality issues still exist because meat quality is complex and controlled by many factors and mechanisms. Protein phosphorylation is regarded as one of the most common protein post-translational modifications that is considered universal across all the domains of life. In the field of meat science, the role of protein phosphorylation in meat quality characteristics is gaining great interest and more studies are being conducted. This book is based on the achievements of our team research focusing on protein phosphorylation and meat quality. Our aim was to provide new ideas for exploring the mechanism of meat quality and possible solutions for improving meat quality.

The book is divided into three parts preceded by an introduction and background chapter providing an overview of meat quality mechanism, protein post-translational modification on meat quality, and current research of protein phosphorylation in live and postmortem. The first section (Chaps. 2–5) gave a brief introduction of the methods used for protein phosphorylation detection and discussed the relationship between protein phosphorylation and meat quality attributes of color, tenderness, and water holding capacity. The second section (Chaps. 6–9) elucidated possible mechanisms of the effect of protein phosphorylation on meat quality from the aspects of glycolysis, myofibril protein degradation, calpain activity, and myoglobin. The third section (Chaps. 10–12) focused on the improvement of meat quality by regulating protein phosphorylation, which included the effects of temperature and ionic strength on protein phosphorylation and three novel technologies.

We wish to express our gratitude to all the contributors to this book, including Meng Li, Zheng Li, Lijuan Chen and Fan He for their contributions on the studies related with Part I, Ying Wang, Zheng Li, Manting Du and Meng Li for their contributions on the studies related with Part II, Chi Ren, Caixia Zhang, Chunhui Zhang, Yan Zhang and Weiwei Xu for their contributions on the studies related with

Part III, Manting Du, Ying Wang, Chi Ren, Lichuang Cao, Muawuz Ijaz, Yingxin Zhao, Caiyan Huang, Yejun Zhang, Xiangru Wei, Xu Wang, Yujun Xu, Yue Ge, Dan Jiao and Tongjing Yan for their contributions on preparation of this book, Caiwei Liu for her contributions on proof of the language. We also want to thank the production team at Springer for their kind assistance in this book.

Beijing, China Dequan Zhang

Contents

Chapter 1
Introduction

1.1 Meat Quality and Mechanism from Muscle to Meat

1.1.1 Importance of Meat Quality

Meat is a kind of important agricultural product and how to maximize meat quality is an urgent issue as it is of great concern of every person involved with the meat industry, including farmers, producers, consumers, and scientists. Interestingly, the definition of meat quality changes from person to person. For producers, meat quality is mainly determined by not only the parameters demanded by the consumer but also the objective factors representing the industrial meat characteristics, including meat weight losses. For the supply chain, pH, color, water holding capacity (WHC), cooking loss, and tenderness all are important for integrated quality and homogenous products. Consumers mainly focus on color and tenderness and some are conscious of the fat contents. Above these, for scientists, all these parameters are of equal importance. All the meat quality parameters are interconnected and dependent on many intrinsic (innate to the animals, such as breed, genetics, and age) and extrinsic (related with the pre- and postmortem management and environment) factors.

With the increased great demand of meat, the awareness of consumer from all of the world has transformed into meat quality and safety (Taheri-Garavand et al. 2019). Generally, meat quality includes five aspects: eating quality, nutritional quality, technological quality, safety quality, and humane quality (Zhou et al. 2007). Eating quality mainly involves color, tenderness, flavor, juiciness, water holding capacity, etc. The nutrition quality consists of contents of water, protein, fat, etc. The different kinds of fatty acid profiles lead to various sensory traits of meat. The status of shortening, rigor, the content of connective tissue, the degree of protein denaturation, the ability of oxidation and/or pH value affect the technological quality of meat. Microbial spoilage can evaluate the freshness of meat by using the figure of total volatile basic nitrogen (TVB-N) or aerobic bacterial count. The

D. Zhang et al., *Protein Phosphorylation and Meat Quality*,
https://doi.org/10.1007/978-981-15-9441-0_1

residual veterinary drugs, pesticides and heavy metals are also factors that have large negative effects on meat safety quality. With the increasing awareness of protecting animals and environment, humane quality of meat has become a new aspect (Warriss 2000). As a result, the system and condition of animal feeding are friendlier and of less stress. However, the total meat quality is affected not only by these five aspects but also innate traits like breed and age. Dhanda et al. (2003) studied six goat genotypes and the results showed that the subcutaneous fat, shear force, juiciness, and other sensory characteristics were different in these goats. Therefore, meat quality is mainly affected by the animal species, age, sex, physiological state, and postmortem factors including postmortem muscle biochemistry and conversion of muscle into meat.

1.1.2 Mechanism of Meat Quality Formation During Postmortem

Before slaughtering, energy for muscle contraction and relaxation derives from the adenosine triphosphate (ATP) through mitochondrial oxidative metabolism, phosphocreatine, and myokinase (Scheffler and Gerrard 2007). But after animal slaughtering, the supply of oxygen to the muscles stops and oxidative decarboxylation and phosphorylation no longer operate. At this point, phosphocreatine is the main supplier of ATP. After most of the phosphocreatine are depleted, the level of ATP content decreases quickly. The glycogen in the muscle is degraded and a small number of ATP are produced through glycolysis. As a result, lactic acid generates and pH value declines, consequently acidifying the environment, leading to the conversion of muscles into the meat. This acid declines the pH from neutral muscle pH (7.0) to the acidic meat pH (5.5). When muscle is stored at high temperature, it is easy to become pale, soft, exudative (PSE). After electrical stimulation, pH value declines rapidly and PSE-like traits are generated (Bowker et al. 1999). With ATP consumption, the actomyosin cross-bridge starts to form, which is the signal of rigor mortis. When ATP content is extremely consumed, the process of rigor finishes. With storage time increasing, muscle tenderness becomes tender because of the degradation of structural myofibrillar proteins.

During the process of muscle to meat, meat tenderness, color, and WHC change a lot. Tenderness is an important feature of meat quality, having a close relationship with WHC. When actomyosin cross-bridge cannot be destructed, it reaches the largest tension of muscle and the least WHC. With the release of enough Ca^{2+}, calpain proteolytic enzymes improve some structural myofibrillar proteins degrade, leading to tenderization (Ertbjerg and Puolanne 2017). Meat color is another important sensory feature and myoglobin (Mb) is the main protein which controls the meat color. Many factors such as glycogen storage, chilling rate, antioxidant accumulation, genetics, diet, etc. affect meat color. The physical, chemical, and biological reactions are complex in postmortem muscle, many quality traits are impressionable, so keeping the best meat quality is a meaningful subject.

1.2 Protein Post-Translational Modification on Meat Quality

1.2.1 Definition and the Role of Protein Post-Translational Modification

Post-translational modifications, referring to the covalent enzymatic modifications of proteins, are mainly responsible for all the biological and biochemical mechanisms by regulating proteins function via modulating their stability and shape. To understand the biochemistry of any mechanism, analysis of these protein modifications is essential. The main post-translational modifications are protein phosphorylation, acetylation, ubiquitination, nitrosylation, succinylation, etc.

The protein phosphorylation is a process, where the transfer of phosphate of ATP or guanosine triphosphate (GTP) to the hydroxyl groups of its proteins under protein kinases and reversion by the counterpart phosphatases. The role of phosphatases is to catalyze the transfer of phosphate of phosphoprotein to the water molecule (Ubersax and Ferrell 2007). Although both protein kinase and phosphatase are phosphotransferases, they catalyze opposite processes to regulate the functions of many cellular proteins ranging from the fate of cell to regulation of metabolism. Protein phosphorylation is the most common type of post-translational modification and essentially affects every basic cellular process.

Acetylation, controlled by acetyltransferases enzymes which transfer the acetyl group from a specific amino acid to either the α-amino group of amino-terminal residues or the ε-amino group of lysine residues at various positions, changes the structure, molecular weight, and ultimately the function of targeted molecule (Polevoda and Sherman 2002). Most of the enzymes involved in anabolism and catabolism have been shown to be acetylated (Guan and Xiong 2011).

S-nitrosylation is a nitric oxide (NO) signaling pathway independent of guanosine 3′, 5′-cyclic phosphate (cGMP), and NO plays an important role as a signaling molecule and covalently binds to free mercaptocysteine residues of certain proteins to form S-nitrosothiol (SNO) (Foster et al. 2009). The products of SNO have been observed to regulate protein conformation, activity, and function (Hess et al. 2005). The NO and NO-induced S-nitrosylation take part in glucose uptake, muscle contraction, and muscular dystrophies in skeletal muscles by managing cGMP and ryanodine receptor 1 (Stamler and Meissner 2001), and subsets of S-nitrosylated proteins play a major role in the regulation of tissue homeostasis (Furuta 2017).

Ubiquitination means the addition of ubiquitin to a substrate protein, and ubiquitin is a small regulatory protein (8.6 kDa) found in most tissues of eukaryotic organisms (Majetschak 2011). Ubiquitination is crucial for protein localization, metabolism, function, regulation, and degradation (Schreiweis et al. 2005).

Succinylation is a recently discovered novel post-translational modification in which succinyl-CoA modifies protein lysine groups (Yang and Gibson 2019). This modification is found in many proteins, including histones (Xie et al. 2012). The potential role of succinylation is still under investigation, but as the addition of succinyl group changes lysine's charge from +1 to −1 and introduces a relatively large structural moiety (100 Da), bigger than acetylation (42 Da), it is expected that succinylation can lead to more significant changes in protein structure and function than that of acetylation (Zhang et al. 2011).

1.2.2 Role of Protein Post-Translational Modification on Meat Quality

Meat quality is an important aspect that determines the value and economic benefit of meat products. To obtain high quality meat, it is necessary to clarify the changes and mechanisms in the process of meat quality formation. To better understand the mechanism of meat quality regulation, proteomics has been widely used to describe the mechanism of meat quality formation and to explore its biomarkers (Lametsch and Bendixen 2001; Lametsch et al. 2003; Mekchay et al. 2010). The mentioned protein modifications above have a significant effect on the development of ultimate meat quality. Recently, it was found out that protein phosphorylation showed a significant effect on the color, tenderness, and WHC of meat. Li et al. (Li et al. 2017a, b, c) measured the phosphorylation level of myofibrillar proteins in different tenderness groups and examined the phosphoproteomes in ovine muscle with different degrees of tenderness over time. Most of the different phosphoproteins maintained sarcomeric functions or were involved in glycometabolism. Allison et al. (2003) believed that protein phosphorylation could improve muscle glycolysis and reduce postmortem proteolysis, which had adverse effects on meat color, WHC, and tenderness. Li et al. (2017a, b, c) obtained the sarcoplasmic proteins with different phosphorylation levels by adding phosphates and protein kinase inhibitors to ground lamb meat and investigated their effects on meat color stability. They discovered that the global phosphorylation level of sarcoplasmic protein was inversely related to lamb color stability and proposed that protein phosphorylation can influence meat color by regulating the glycolytic enzymes and redox form of myoglobin. Scheffler and Gerrard (2007) found out that the phosphorylation levels of rate-limiting enzymes, such as glycogen phosphorylase (GP), phosphofructokinase (PFK), pyruvate kinase (PK), etc., mainly controlled the glycolysis, protein degradation, and ultimately the meat fluid losses. Likewise, protein acetylation regulated the postmortem muscle glycolysis (Li et al. 2017a, b, c) and it suggested that 5′ adenosine monophosphate-activated protein kinase (AMPK) can be targeted to control the postmortem metabolism. Therefore, by doing so, pale, soft, exudative (PSE) meat in poultry can be prevented. Li et al. (2017a, b, c) studied the effect of AMPK on protein acetylation and glycolysis in postmortem muscles and reported that AMPK controlled the glycolysis through protein acetylation in pre-slaughter stressed animals, whereas Mora et al. (2015) derived 68 peptides from ubiquitin-60S ribosomal protein which contributed to the better knowledge of protein degradation and the responsible intrinsic enzymes and suggested that these peptides can be used as biomarkers to track the processing time. In postmortem muscles, ubiquitination, directly related to tenderness, can be an indicator of the extent and rate of meat tenderization (Picard and Gagaoua 2017). Recently, it was also speculated that nitric oxide and protein S-nitrosylation were also involved in the conversion of muscle into meat by regulating the biochemical processes including glycolysis, release of calcium, and breakdown of myofibrillar proteins (Liu et al. 2018).

1.3 Current Research of Protein Phosphorylation (Live and Postmortem)

Among all post-translation modifications, protein phosphorylation is one of the most important and common modifications which has been extensively investigated. Protein phosphorylation is the most important cellular regulatory event, as so many enzymes and receptors are regulated through the phosphorylation and dephosphorylation, and it controls the protein synthesis, cell growth and division, signal transduction, and process of aging. Phosphorylation is a reversible mechanism, which is mainly activated by the actions of kinases. During the process, the phosphate group is added to the polar side of several amino acids to modify the structure and allow them to interact with other molecules. On the contrary, dephosphorylation, mediated by phosphatases, antagonizes the actions of kinases. Different internet websites, including PhosphoSitePlus (www.phosphosite.org) and PhosphoNET (www.phosphonet.ca), have listed around 200,000 known phosphosites.

Besides controlling the live animal processes, phosphorylation also has a significant influence on postmortem muscle biochemistry. By using proteomic tools, many studies have explored the role of protein phosphorylation in postmortem meat quality development. Meat color and tenderness, which are the most common criteria used to predict the meat quality, are affected by protein phosphorylation. The global phosphorylation level of tough meat is higher than tender meat. Meanwhile, the phosphorylation status of actin and myosin light chain 2 is also critical in determining the ultimate meat tenderness (Chen et al. 2016; Li et al. 2017a). The degradation of myofibrillar proteins by μ-calpain, which tenderizes the meat, is also decreased by its phosphorylation (Li et al. 2017a, b, c). Furthermore, dephosphorylation and protein kinase A (PKA) phosphorylation positively regulate the functions of μ-calpain (Du et al. 2018). Similarly, protein phosphorylation also significantly affects the meat color stability, by regulating the glycolysis and redox form of myoglobin (Li et al. 2017c, 2018). Moreover, different patterns of sarcoplasmic proteins phosphorylation, including pyruvate kinase and triosephosphate isomerase-1, are related to the postmortem muscle pH decline (Huang et al. 2011). Season also significantly affects the phosphorylation level of post-slaughter glycolytic enzymes and plays its role in mediating the effects of pre-slaughter environmental temperature on post-slaughter glycolysis and meat quality development (Li et al. 2015a). It was also reported that electrical stimulation affected the postmortem rate of pH decline by regulating the phosphorylation of sarcoplasmic and myofibrillar proteins (Li et al. 2015b).

References

Allison, C. P., Bates, R. O., Booren, A. M., Johnson, R. C., & DoumIT, M. E. (2003). Pork quality variation is not explained by glycolytic enzyme capacity. *Meat Science, 63*(1), 17–22.

Bowker, B. C., Wynveen, E. J., Grant, A. L., & Gerrard, D. E. (1999). Effects of electrical stimulation on early postmortem muscle pH and temperature declines in pigs from different genetic lines and halothane genotypes. *Meat Science, 53*(2), 125–133.

Chen, L. J., Li, X., Ni, N., Liu, Y., Chen, L., Wang, Z. Y., et al. (2016). Phosphorylation of myofibrillar proteins in post-mortem ovine muscle with different tenderness. *Journal of the Science of Food and Agriculture, 96*(5), 1474–1483.

Dhanda, J. S., Taylor, D. G., & Murray, P. J. (2003). Part 1. Growth, carcass and meat quality parameters of male goats: Effects of genotype and liveweight at slaughter. *Small Ruminant Res, 50*, 57–66.

Du, M. T., Li, X., Li, Z., Shen, Q. W., Wang, Y., Li, G. X., et al. (2018). Phosphorylation regulated by protein kinase A and alkaline phosphatase play positive roles in μ-calpain activity. *Food Chemistry, 252*, 33–39.

Ertbjerg, P., & Puolanne, E. (2017). Muscle structure, sarcomere length and influences on meat quality: A review. *Meat Science, 132*, 139–152.

Foster, M. W., Hess, D. T., & Stamler, J. S. (2009). Protein S-nitrosylation in health and disease: A current perspective. *Trends in Molecular Medicine, 15*(9), 391–404.

Furuta, S. (2017). Basal S-nitrosylation is the guardian of tissue homeostasis. *Trends Cancer, 3*(11), 744–748.

Guan, K. L., & Xiong, Y. (2011). Regulation of intermediary metabolism by protein acetylation. *Trends in Biochemical Sciences, 36*(2), 108–116.

Hess, D. T., Matsumoto, A., Kim, S., Marshall, H. E., & Stamler, J. S. (2005). Protein S-nitrosylation: Purview and parameters. *Nature Reviews Molecular Cell Biology, 6*, 150–166.

Huang, H. G., Larsen, M. R., Karlsson, A. H., Pomponio, L., Costa, L. N., & Lametsch, R. (2011). Gel-based phosphoproteomics analysis of sarcoplasmic proteins in postmortem porcine muscle with pH decline rate and time differences. *Proteomics, 11*, 4063–4076.

Lametsch, R., & Bendixen, E. (2001). Proteome analysis applied to meat science: Characterizing postmortem changes in porcine muscle. *Journal of Agricultural and Food Chemistry, 49*(10), 4531–4537.

Lametsch, R., Karlsson, A., Rosenvold, K., Andersen, H. J., Roepstorff, P., & Bendixen, E. (2003). Postmortem proteome changes of porcine muscle related to tenderness. *Journal of Agricultural and Food Chemistry, 51*(24), 6992–6997.

Li, C. B., Zhou, G. H., Xu, X. L., Lundström, K., Karlsson, A., & Lametsch, R. (2015b). Phosphoproteome analysis of sarcoplasmic and myofibrillar proteins in bovine longissimus muscle in response to postmortem electrical stimulation. *Food Chemistry, 175*, 197–202.

Li, M., Li, X., Xin, J. Z., Li, Z., Li, G. X., Zhang, Y., et al. (2017c). Effects of protein phosphorylation on color stability of ground meat. *Food Chemistry, 219*, 304–310.

Li, M., Li, Z., Li, X., Xin, J. Z., Wang, Y., Li, G. X., et al. (2018). Comparative profiling of sarcoplasmic phosphoproteins in ovine muscle with different color stability. *Food Chemistry, 240*, 104–111.

Li, Q., Li, Z., Lou, A., Wang, Z., Zhang, D. Q., & Shen, Q. W. (2017b). Histone acetyltransferase inhibitors antagonize AMP-activated protein kinase in postmortem glycolysis. *Asian-Australasian Journal of Animal Sciences, 30*(6), 857–864.

Li, X., Chen, L. J., He, F., Li, M., Shen, Q. W., & Zhang, D. Q. (2017a). A comparative analysis of phosphoproteome in ovine muscle at early postmortem in relationship to tenderness. *Journal of the Science of Food and Agriculture, 97*(13), 4571–4579.
Li, X., Fang, T., Zong, M. H., Shi, X. Q., Xu, X. L., Dai, C., et al. (2015a). Phosphorproteome changes of myofibrillar proteins at early post-mortem time in relation to pork quality as affected by season. *Journal of Agricultural and Food Chemistry, 63*, 10287–10294.
Liu, R., Warner, R. D., Zhou, G. H., & Zhang, W. G. (2018). Contribution of nitric oxide and protein S-nitrosylation to variation in fresh meat quality. *Meat Science, 144*, 135–148.
Majetschak, M. (2011). Extracellular ubiquitin: Immune modulator and endogenous opponent of damage-associated molecular pattern molecules. *Journal of Leukocyte Biology, 89*(2), 205–219.
Mekchay, S., Teltathum, T., Nakasathien, S., & Pongpaichan, P. (2010). Proteomic analysis of tenderness trait in Thai native and commercial broiler chicken muscles. *The Journal of Poultry Science, 47*(1), 8–12.
Mora, L., Gallego, M., Aristoy, M. C., Fraser, P. D., & Toldrá, F. (2015). Peptides naturally generated from ubiquitin-60S ribosomal protein as potential biomarkers of dry-cured ham processing time. *Food Control, 48*, 102–107.
Picard, B., & Gagaoua, M. (2017). Proteomic investigations of beef tenderness. In *Proteomics in food science* (pp. 177–197). Oakville, ON: Delve Publishing.
Polevoda, B., & Sherman, F. (2002). The diversity of acetylated proteins. *Genome Biology, 3*(5), 1–6.
Scheffler, T. L., & Gerrard, D. E. (2007). Mechanisms controlling pork quality development: The biochemistry controlling postmortem energy metabolism. *Meat Science, 77*(1), 7–16.
Schreiweis, M. A., Hester, P. Y., & Moody, D. E. (2005). Identification of quantitative trait loci associated with bone traits and body weight in an F2 resource population of chickens. *Genetics Selection Evolution, 37*(6), 677–698.
Stamler, J. S., & Meissner, G. (2001). Physiology of nitric oxide in skeletal muscle. *Physiological Reviews, 81*(1), 209–237.
Taheri-Garavand, A., Fatahi, S., Omid, M., & Makino, Y. (2019). Meat quality evaluation based on computer vision technique: A review. *Meat Science, 156*, 183–195.
Ubersax, J. A., & Ferrell, J. E. (2007). Mechanisms of specificity in protein phosphorylation. *Nature Reviews Molecular Cell Biology, 8*(7), 530–541.
Warriss, P. D. (2000). *Meat science: An introductory text*. Wallingford, UK: CABI Publishing.
Xie, Z., Dai, J., Dai, L., Tan, M., Cheng, Z., Wu, Y., et al. (2012). Lysine succinylation and lysine malonylation in histones. *Molecular & Cellular Proteomics, 11*(5), 100–107.
Yang, Y., & Gibson, G. E. (2019). Succinylation links metabolism to protein functions. *Neurochemical Research, 44*(10), 2346–2359.
Zhang, Z., Tan, M., Xie, Z., Dai, L., Chen, Y., & Zhao, Y. (2011). Identification of lysine succinylation as a new post-translational modification. *Nature Chemical Biology, 7*(1), 58.
Zhou, G. H., Li, C. B., & Xu, X. L. (2007). Advances in methods for evaluating meat palatability. *Science, 2*(2), 75–82.

Part I
Relationship Between Protein Phosphorylation and Meat Quality

Chapter 2
Protein Phosphorylation Detection Method

Abstract With the advancement of technical approaches, signaling pathways have been continuously discovered in numerous biological processes. Majority of these signaling pathways include the protein phosphorylation, which is a main post-translational modification affecting the structure and function of proteins within the cell. Consequently, determining the locations and extent of the protein phosphorylation of any specific protein or set of proteins, either qualitatively or quantitatively, has become a routine and extremely vital step in many fields of life sciences, and recently in meat science as well. However, when choosing the method to detect protein phosphorylation, it can be different depending on several aspects, particularly under consideration of the availability of the related reagents and equipment. Although different methods have their own advantages, but this is also a matter of fact that every method is associated with several drawbacks such as poor specificity, low sensitivity, safety issues, or intensive labor. In this chapter, we outlined the routine methods used to detect the protein phosphorylation and discussed their advantages and disadvantages, especially focusing on the usage of gel and liquid chromatography coupled with tandem mass spectrometry (LC-MS/MS) techniques to identify phosphorylated proteins in meat science field.

Keywords Protein phosphorylation · Methods · 2DE-gel · Pro-Q Diamond · TiO_2 enrichment · LC-MS/MS

2.1 Introduction

Proteins can be modified through several modifications with several functional groups, such as phosphate, ubiquitin, nitric oxide, sulfate, glycan, ribose, lipid, etc. Protein phosphorylation involving the phosphate group is an important post-translational modification that affects the structure and function of proteins. Kinases can add phosphate group to threonine, serine, or tyrosine residues. On the contrary, phosphatases have the ability to remove phosphate group from these residues. These amino acids enclose a nucleophilic group that can react with the adenosine triphosphate (ATP), replace the oxygen present on terminal phosphorous, and eject the

D. Zhang et al., *Protein Phosphorylation and Meat Quality*,
https://doi.org/10.1007/978-981-15-9441-0_2

adenosine diphosphate (ADP) from the amino acid. In eukaryotic cells, kinases and phosphatases are mainly in their active form, which makes phosphorylation a vital post-translational modification for biological processes. It is estimated that 30% of eukaryotic cell proteins are involved with phosphorylation, which makes it the most interesting protein modification for the scientists to explore its many biological functions (Scorsone et al. 1987).

In the past decade, significance of protein phosphorylation in medical field has urged the meat scientists to study the protein phosphorylation of postmortem muscles in order to explore the mechanisms of meat quality formation. In this context, different techniques have been used in this field. However, the principal approaches to evaluate the protein phosphorylation are gel-based methods to stain specific phosphoproteins, for instance, Pro-Q Diamond staining (Steinberg et al. 2003). In comparison with traditional antibody-based and radioactive labeling approaches, Pro-Q Diamond dye has significant benefits such as evasion of radioactivity, no sequence specific binding to phosphorylated amino acids, and no need of time-consuming, electro-blotting, or expensive antibodies. Pro-Q Diamond dye allows the rapid determination of phosphor-serine, phosphor-threonine, and phosphor-tyrosine proteins, as well as the identification of the phosphoproteins and their phosphorylated peptides by mass spectrometry (Schulenberg et al. 2003). In the same way, mass spectrometry is considered as a major tool to study protein post-translational modifications on large scale, and it also helps to explore the unique phosphorylated proteins and phosphorylation sites (Mann et al. 2002).

The increase in demand and improvement in methodologies bring researchers to a closer understanding of the complex processes, which ultimately control cellular function and meat quality formation. However, to select the best method under different circumstances, it is important to pick the method that best fits the objective and experimental design of the study. This chapter will provide a brief overview of the most commonly used techniques in field of meat science, with special focus of different gels and LC-MS/MS related techniques to measure protein phosphorylation.

2.2 Principle of Phosphorylated Protein Detection

Analysis of the phosphorylated protein is necessary for interpreting the fundamental biological processes and signaling networks at the molecular level. Several methods have been described for the analysis of phosphorylated proteins according to different principles. However, there are several limitations with those methods, so that none of them could be used to detect all kinds of protein phosphorylation.

One of the methods that used at early stage for detecting protein phosphorylation is the labeling of proteins with the radiolabeled 32P-orthophosphate (Erikson and Erikson 1980). In this method, kinase transfers radiolabel to its substrate that allows to detect its activity. For which, the sample is incubated with the radiolabel for a set amount of time prior to harvesting and preparing the sample, and isolate the proteins by gel electrophoresis (± an immunoprecipitation step for a particular protein). Then make the gel dry and expose it to film on a phosphorimager screen to detect labeled bands. Radiolabeling works well if we want to determine the protein phosphorylation without knowing the site of phosphorylation or the kinase involved.

On the other hand, antibodies have turned out to be a great tool to detect protein phosphorylation at a particular site. Antibodies can be fixed to specific site, thus identifying related protein phosphorylation. The antibodies are named as phospho-specific antibodies and are becoming a critical method to identify protein phosphorylation at basic research and commercial diagnosis levels.

Post-translational modifications including protein phosphorylation can be easily identified on two-dimensional electrophoresis (2DE) gels. The phosphorylation replaces neutral hydroxyl groups on threonines, serines, or tyrosines with negatively charged phosphates. Therefore, under pH 5.5, the phosphates added a single negative charge and adjacent to pH 6.5, phosphates added the 1.5 negative charges, moreover, beyond pH 7.5, phosphates added 2 negative charges. Thus, relative amount of each isoform including protein phosphorylation can be easily and rapidly determined from staining intensity on 2DE gels.

It is possible to detect phosphorylation based on the shift of protein electrophoretic mobility just on simple 1-dimensional SDS-PAGE gels. SDS-PAGE is a very versatile technique to separate complex protein mixture and is the most commonly and widely used technique in different research areas. Radioactive labeling of phosphorylated proteins combined with gel separation is a very sensitive and quantitative approach to separate the phosphorylated from the non-phosphorylated proteins. Detection of antibodies bonded with specific protein or phosphorylated amino acid could be used to analyze small-scale protein phosphorylation. A major limitation of this technique is the lack of specificity of the antibody in binding with specific phosphorylated protein. Another commonly used method is the application of commercially available Pro-Q Diamond phosphoprotein stain. This is a simple technique that selectively marks the phosphoproteins of phosphorylated spots on tyrosine, serine, or threonine residues.

In summary, all the electrochemical methods to detect protein phosphorylation including SDS-PAGE are usually based on one of the following two principles: (1) the addition of negative charges to the protein with the transfer of phosphoryl groups; (2) the release of protons in the reaction buffer upon phosphorylated protein (Bhalla et al. 2014). There have been also attempts to detect the changes in the protein charge after phosphorylation by measuring the alterations on the surface charge of an electrode in contact with the protein, which is recorded in the form of current as a function of time.

2.3 Gel-Based Phosphorylated Protein Detecting Methods

2.3.1 Separation of Proteins by 2DE Gel

Two-dimensional electrophoresis of proteins, firstly introduced by O'Farrell (1975) and Klose (1975), was the sign of beginning of proteomics. The 2DE is a commonly used technique to study the innovative protein mixes extracted from the cell, tissue, or any other biological sample. According to the name, this technique is mainly comprised of two steps; isoelectric focusing as first dimension, which separates the proteins based on their isoelectric point by immobilized pH gradient; and SDS-PAGE as second dimension, which segregates the proteins based on their molecular weight (Fig. 2.1). On the two-dimensional array, every spot represents one kind of protein specie of the tested sample. Then targeted spots are removed and digested to identify the proteins. Thus, thousands of proteins could be separated and obtained from the sample by the two mentioned physical techniques of this method.

Although it is an old and no longer the only method of proteomics, it still has many salient features and advantages. Using 2DE, thousands of proteins can be identified by a single run, and resolved the intact full length protein on single gel. The method of 2DE includes the visualized detection of isoelectric point and molecular weight by quantification based on the spot intensity. Moreover, 2DE is a compatible technique for further analysis of proteins with many other biochemical methods, such as polypeptides can also be further identified with antibodies and tested for post-translational modifications like protein phosphorylation. Furthermore, continuous improvement in its methodologies also increases the robustness of 2DE. The possible limitations of these techniques are the big amount of sample handling, smaller dynamic with limited reproducibility. Some of the proteins, with

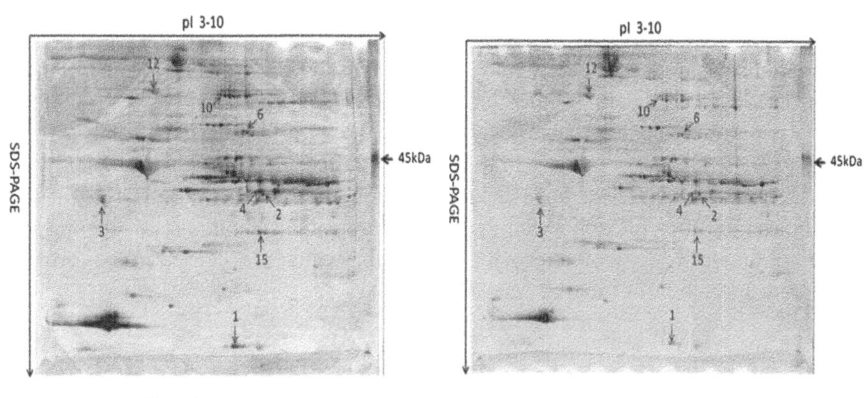

Fig. 2.1 Images of 2DE gels of the differentially phosphorylated proteins at 4 h postmortem between groups with different tenderness. The figures on the image and in the parenthesis represent spot number (Li et al. 2017)

low quantity, acidic, basic, or hydrophobic behavior, of very small or very large size, are difficult to be separated through this technique.

2.3.2 Detection of Phosphorylated Protein by Pro-Q Diamond Gel Stain

Fluorescent stains are developed recently to identify the proteins, including their post-translational modifications, with high sensitivity and linearity. Among these stains, Pro-Q Diamond phosphoprotein stain is a frequently used technique, commercially manufactured by the Invitrogen™. Pro-Q Diamond is an advanced method that offers a simple and direct way to bind with the phosphate moiety of phosphoproteins. It can work without involvement of numerous steps or sample pretreatment for staining the phosphoproteins, which phosphorylated on threonine tyrosine, or serine residues. Pro-Q Diamond is compatible with standard SDS-PAGE or with 2DE gels, and there is no need of specific antibody or Western blotting (Fig. 2.2). This technique can also be combined with mass spectrometry, therefore, allowing meaningful examination of phosphorylation status of entire proteome for the first time.

Although the specificity and sensitivity of Pro-Q Diamond gel stain are good, however, high cost hinders its usage to high-throughput phosphoproteomics. In this context, recently Wang et al. (2014) introduced an alternative quercetin stain, which could selectively detect the 16–32 ng of phosphoproteins. Quercetin stain requires relatively lower cost and readily available chemicals and is more time saving.

2.3.3 Detection of Phosphorylated Protein by Phos-Tag Method

A dinuclear metal complex of 1, 3-bis[bis(pyridin-2-ylmethyl) amino] propan-2-olate is a phosphate-binding tag (Phos-tag) aqueous solution, firstly described by Kinoshita-Kikuta et al. (2007). Manganese homolog (Mn^{2+}-Phos-tag) can be bonded with phosphate entity, such as phosphotyrosine and phosphoserine at the alkaline pH (pH = 9.0). The Mn^{2+}-Phos-tag can be polymerized with the separating gel of SDS-PAGE and used for discovering phosphoproteins. This method does not involve with radioactive or other chemical labels and phosphate-binding specificity is also independent from the amino acid sequence, therefore, it is widely used for determining the phosphorylation state of many proteins.

Phos-tag method had certain constraint in finding the mobility shift with phosphorylated forms of some of the proteins. Some proteins treated with different protein kinases failed to present any change in the band shift using this technique.

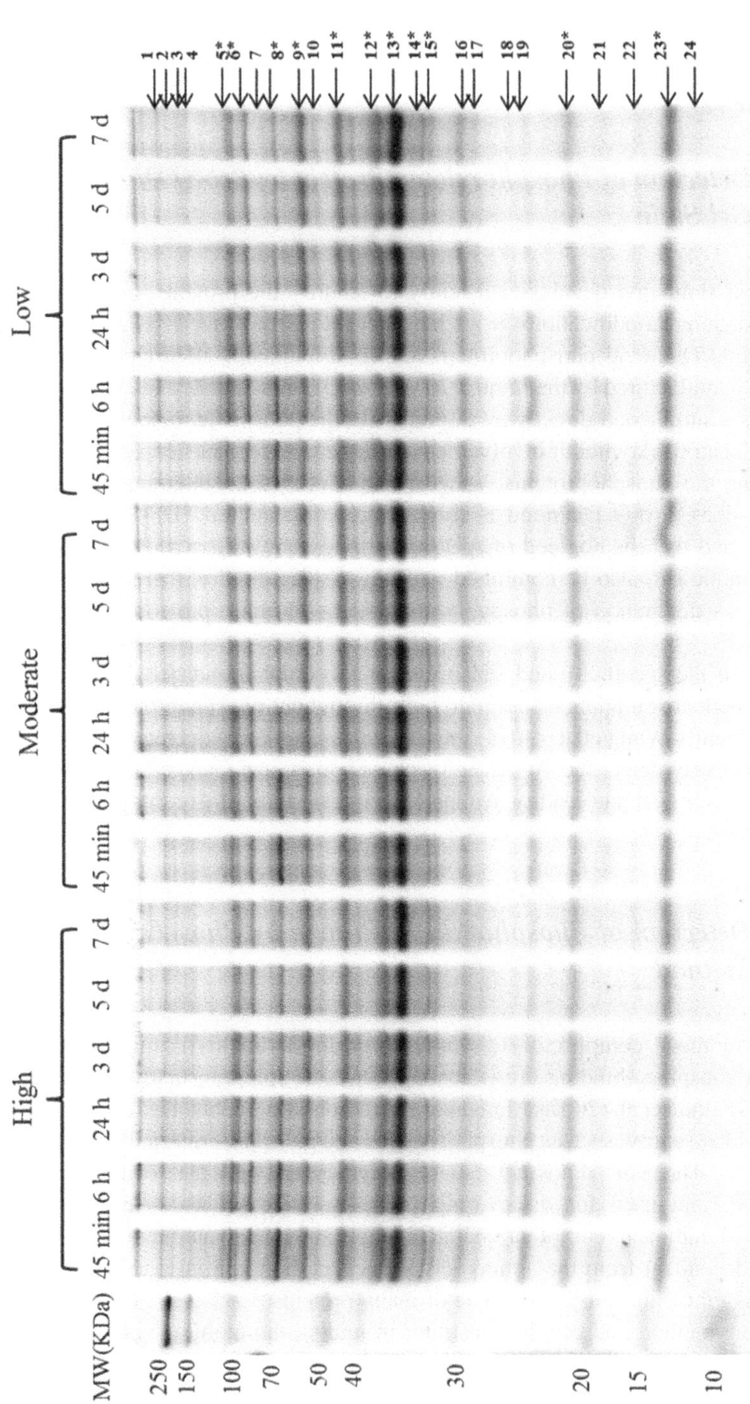

Fig. 2.2 Images of gels stained with Pro-Q Diamond. Bands marked with * were selected for protein identification by LC-MS/MS. High, moderate, and low represent high color stability group, moderate color stability group, and low color stability group, respectively. Adapted from (Li et al. 2018)

Moreover, the alkaline buffer used for the SDS-PAGE gel gave short-term stability for the separated phosphoproteins by this method.

To deal with limitations of Mn^{2+}-Phos-tag assay, Kinoshita-Kikuta et al. (2007) modified the protocol and introduced Zn^{2+}-Phos-tag SDS-PAGE. This new method used the neutral pH buffer and Zinc (II) complex (Zn^{2+}-Phos-tag acrylamide) instead of using alkaline pH and Mn^{2+} of the Mn^{2+}-Phos-tag assay. The application of neutral pH was the main development and it made it possible to store the gels for nearly 6 months without any deterioration compared to alkaline pH. Additionally, use of Zn^{2+}-Phos-tag gave a better mobility shift for some of the proteins which failed to show the shift or showed the smaller shift on the gel with Mn^{2+}-Phos-tag assay.

2.4 Detection of Phosphorylated Proteins by LC-MS/MS

2.4.1 Labeling Treatment of Samples

Over the past few years, several approaches have been developed for the incorporation of specific tags into the peptides or proteins. These mainly include the introduction of isotopic or chemical tags at specific function group on polypeptides, use of heavy amino acids for metabolic isotope labeling and different methods of introducing the stable isotopes by enzymatic reactions. Every method has its own specific strengths and weaknesses. For example, incorporation of stable isotopes by chemical reactions provides a good selectivity and specificity for tagging the reactive groups on proteins or peptides. Because it occurs after isolation, avoiding possible side reactions is important for application of this procedure. While in metabolic isotope labeling, cell culturing without any chemical reaction is used to label the proteins, so it is easy to perform. However, this technique is limited to in vitro labeling of cells through cell culturing. Incorporation of stable isotopes by enzymatic reactions is straightforward, however, difference created by this technique is sometimes 4 or 2 Da, which is difficult for quantitative analysis without using the algorithm for deconvolution of isotope pattern.

Among these techniques, isotope coded affinity tag (ICAT) is the most commonly used method, where proteome of two samples are labeled on the side chains of reduced cysteinyl residues using one chemically same but isotopically different reagent. Important feature of this technique is the introduction of biotin loving tag in the ICAT reagents, which helps in selective isolation and purification of analytes, thus enabling a substantial reduction in sample complexity.

2.4.2 Label Free Treatment of Samples

Label free treatment of samples for LC-MS/MS is a smart method for high-throughput quantitative proteomics that aims to detect the relative abundance of proteins of two or more biological samples. Unlike other techniques, label free quantification does not use a stable isotopes and is comprised of two main approaches: (1) measurement of peptide peak intensity, which measures and compares the peaks of chromatographic peptide precursors ions of a particular proteins; and (2) spectral counting, which involves the calculating and matching of the number of spectra of the tested protein peptides in order to know the relative quantity of that specific protein. Both of these approaches are simple and economical with high reproducibility at both protein and peptide levels. However, application of these procedures requires more computational power accomplished with the handling of chromatographic peaks, peptide intensity measurement, and spectral counting for examining the small significant changes, such as protein phosphorylation, that are biologically meaningful. The main steps involved in both mentioned procedures are: (1) preparation of sample involving extraction, reduction, alkylation, and digestion of proteins; (2) sample separation and analysis by LC-MS/MS; (3) data analysis including identification, quantification, and statistical analysis; and at the end (4) interpretation of results.

In recent time, an increasing number of publications show that label free quantification is a viable and appropriate method to identify biomarkers within given samples. Besides its breakthrough in biological sciences, there are several issues regarding this method. These issues mainly include the problems related with chromatographic alignment, peptide qualification for quantification and normalization. However, advancements in label free protein quantification methods will further assist to fulfill the demands of advanced proteomics.

2.4.3 Enrichment of Phosphorylated Peptides Using the TiO_2 Beads

Since past few years, adsorption of proteins with titanium dioxide (TiO_2) has been explored extensively with the goal of finding a technique for studying the bioelectrochemical functions of proteins. Interestingly for the study of phosphoproteomics, TiO_2 is shown to have affinity with phosphate ions, and TiO_2 chromatography is used as an effective method for phosphopeptides enrichment. Several scientists introduced the enrichment of phosphorylated peptides using the TiO_2 material. For example, Pinkse et al. (2004) explored the capability of TiO_2 to bind with the selective phosphopeptides by using the online two-dimensional LCMS setup with the particles of TiO_2 as the first dimension and reversed phase material as second dimension. Thingholm et al. (2006) introduced another offline setup TiO_2 chromatography with stronger buffers, including the use of 2, 5-dihydrobenzoic acid and comparatively higher concentrations of trifluoroacetic acid, which reduced the non-specific binding of TiO_2 with compounds other than phosphorylated peptides.

The higher selectivity of TiO_2 with the phosphopeptides makes it a stronger method for phosphoproteomics study. Furthermore, this technique is exceptionally acceptable towards most of the salts and buffers used in biochemistry and cell biology studies. Because the whole procedure only requires approximately 15 min per sample with prepared buffers, the application of the TiO_2 beads for enrichment of phosphorylated peptides is also a highly robust technique and has become the first choice in laboratory as well as in large-scale phosphoproteomic studies. Another benefit of this technique is the highly efficiency of phosphopeptides purification, which is good to study both in vitro and in vivo phosphoproteins when combined with mass spectrometry.

2.4.4 LC-MS/MS Analysis

Mass spectrometry (MS) is the most modern approach for the determination of phosphorylation and it significantly advances the research in protein phosphorylation. MS is not only used for detection of phosphorylation but also helps to identify the phosphorylation sites. Determination of phosphorylation by mass spectrometry is based on the spectrum generated by the peptides which are previously digested by trypsin. Liquid chromatography with tandem mass spectrometry is the most advance form of MS and has been extensively used for phosphoproteomic studies as a powerful analytical approach that uses the separating ability of liquid chromatography along with the enhanced sensitivity and selectivity MS capability of triple quadrupole MS.

The LC-MS/MS has many advantages over other phosphoproteomic techniques, as liquid chromatography offers a vast range of separation options. It provides high specificity detectors and there is no need for confirmatory detection. Similarly, it also shows high sensitivity and can identify the compounds lower than 1 part per trillion. LC-MS/MS has great reproducibility if internal standards are stable labeled. It also has restrictions, such as, when the internal standards are not stable labeled, the reproducibility is usually not as good as other LC detectors. It is also limited for large molecules (>4000 m/z), and the matrix co-extractants can suppress or boost the ionization potential and may needed to apply matrix matched calibrants for quantification.

2.5 Conclusions

Protein phosphorylation is an extraordinarily important component of various signal transduction pathways underlying postmortem biochemical processes. In this chapter, we summarized the different approaches commonly used in meat science field.

However, despite all these given and other available techniques, detection and quantification of protein phosphorylation in any given biological samples is still hampered and challenging. The study of protein phosphorylation on a large scale has been limited due to intensive labor, high time consuming, and low cost effective. Therefore, it is urgently needed to discover the simple, reliable, and rapid techniques for wider application of protein phosphorylation to effectively explore the hidden part of meat quality formation.

Acknowledgments Parts of this chapter are reprinted from Journal of the Science of Food and Agriculture, 97, Li, X., et al., A comparative analysis of phosphoproteome in ovine muscle at early postmortem in relationship to tenderness, 4571-4579; Food Chemistry, 240, Li, M., et al., Comparative profiling of sarcoplasmic phosphoproteins in ovine muscle with different color stability, 104-111. Copyright (2020), with permission from Elsevier.

References

Bhalla, N., Di Lorenzo, M., Pula, G., & Estrela, P. (2014). Protein phosphorylation analysis based on proton release detection: Potential tools for drug discovery. *Biosensors and Bioelectronics, 54*, 109–114.

Erikson, E., & Erikson, R. L. (1980). Identification of a cellular protein substrate phosphorylated by the avian sarcoma virus-transforming gene product. *Cell, 21*(3), 829–836.

Kinoshita-Kikuta, E., Aoki, Y., Kinoshita, E., & Koike, T. (2007). Label-free kinase profiling using phosphate affinity polyacrylamide gel electrophoresis. *Molecular & Cellular Proteomics, 6*(2), 356–366.

Klose, J. (1975). Protein mapping by combined isoelectric focusing and electrophoresis of mouse tissues. *Humangenetik, 26*(3), 231–243.

Li, M., Li, Z., Li, X., Xin, J., Wang, Y., Li, G., et al. (2018). Comparative profiling of sarcoplasmic phosphoproteins in ovine muscle with different color stability. *Food Chemistry, 240*, 104–111.

Li, X., Chen, L., He, F., Li, M., Shen, Q., & Zhang, D. (2017). A comparative analysis of phosphoproteome in ovine muscle at early postmortem in relationship to tenderness. *Journal of the Science of Food and Agriculture, 97*(13), 4571–4579.

Mann, M., Ong, S. E., Grønborg, M., Steen, H., Jensen, O. N., & Pandey, A. (2002). Analysis of protein phosphorylation using mass spectrometry: Deciphering the phosphoproteome. *Trends in Biotechnology, 20*(6), 261–268.

O'Farrell, P. H. (1975). High resolution two-dimensional electrophoresis of proteins. *Journal of Biological Chemistry, 250*(10), 4007–4021.

Pinkse, M. W., Uitto, P. M., Hilhorst, M. J., Ooms, B., & Heck, A. J. (2004). Selective isolation at the femtomole level of phosphopeptides from proteolytic digests using 2D-NanoLC-ESI-MS/MS and titanium oxide precolumns. *Analytical Chemistry, 76*(14), 3935–3943.

Schulenberg, B., Aggeler, R., Beechem, J. M., Capaldi, R. A., & Patton, W. F. (2003). Analysis of steady-state protein phosphorylation in mitochondria using a novel fluorescent phosphosensor dye. *Journal of Biological Chemistry, 278*(29), 27251–27255.

Scorsone, K. A., Panniers, R., Rowlands, A. G., & Henshaw, E. C. (1987). Phosphorylation of eukaryotic initiation factor 2 during physiological stresses which affect protein synthesis. *Journal of Biological Chemistry, 262*(30), 14538–14543.

Steinberg, T. H., Agnew, B. J., Gee, K. R., Leung, W. Y., Goodman, T., Schulenberg, B., et al. (2003). Global quantitative phosphoprotein analysis using multiplexed proteomics technology. *Proteomics, 3*(7), 1128–1144.

Thingholm, T. E., Jørgensen, T. J., Jensen, O. N., & Larsen, M. R. (2006). Highly selective enrichment of phosphorylated peptides using titanium dioxide. *Nature Protocols, 1*(4), 1929.

Wang, X., Ni, M., Niu, C., Zhu, X., Zhao, T., Zhu, Z., et al. (2014). Simple detection of phosphoproteins in SDS-PAGE by quercetin. *EuPA Open Proteomics, 4*, 156–164.

Chapter 3
Protein Phosphorylation Affects Meat Color

Abstract The relationship between meat color stability and protein phosphorylation was studied. By adding phosphatase and protein kinase inhibitors into sarcoplasmic proteins from minced ovine *longissimus thoracis* et *lumborum* (LTL) muscle, we found out that protein phosphorylation had a significant relationship with lactate accumulation, as well as pH decline rate and extent. Also, meat color stability had a negative correlation with sarcoplasmic protein phosphorylation level. Therefore, how protein phosphorylation regulated glycolysis and myoglobin redox forms, and furthermore meat color stability was investigated in the second step. Results showed that phosphorylation differences existed in some individual protein bands, but not in whole sarcoplasmic proteins, from muscle groups with different meat color stability. Glycolytic enzymes turned out to be the most color stability related proteins, and myoglobin phosphorylation level was negatively related to meat color stability, which meant protein phosphorylation might regulate meat color stability via glycolysis and myoglobin redox stability. At the third step, the protein phosphorylation level of muscles with different color stability was examined with quantitative analysis. We successfully identified 3412 phosphopeptides from 1070 phosphoproteins, and among which 243 had a significant correlation between phosphorylation level and meat color stability. With further study into these phosphopeptides through informatics analysis, 27 of them were highly related to meat color stability, and glycolytic enzymes were the biggest group. Moreover, there was a negative correlation between the Ser133 phosphorylation level in myoglobin and meat color stability. In conclusion, our results revealed that protein phosphorylation may regulate meat color stability through glycolytic enzymes and myoglobin redox forms, and Ser133 played a key role in the process.

Keywords Meat color · Protein phosphorylation · Sarcoplasmic proteins · Glycolysis · Glycolytic enzyme · Phosphoproteomics

---

Electronic Supplementary Material The online version of this chapter (https://doi.org/10.1007/978-981-15-9441-0_3) contains supplementary material, which is available to authorized users.

D. Zhang et al., *Protein Phosphorylation and Meat Quality*,
https://doi.org/10.1007/978-981-15-9441-0_3

3.1 Introduction

With the improvement of residents' living standards, consumers' demand for food has changed from quantity to quality, safety, and nutrition. Meat and meat products are indispensable parts of human diet. The evaluation indexes of meat quality mainly include tenderness, flavor, meat color, water resistance, and juiciness. Among them, meat color is the most intuitive impression of consumers when they purchase meat or meat products (Yin et al. 2011). Meat and meat products with poor color give consumers the impression of unsanitary and low quality, resulting in huge economic losses caused by the discount sale of meat due to the deterioration of meat color every year (Jeyamkondan et al. 2000; Canto et al. 2015). Therefore, meat color is one of the most important factors influencing consumers' purchase decisions (Clydesdale 1978; Mancini and Hunt 2005).

Protein phosphorylation is one of the most common and important post-translational modifications in biology system. It refers to the process of transferring γ-phosphorylation group from GTP or ATP to the amino acid substrate protein residue under the catalysis of kinase, which mainly occurs on the side chains of serine, tyrosine, and threonine. Phosphatase can dephosphorylate proteins, which is a dynamic balance, and the result of its comprehensive action determines the phosphorylation state of proteins (Fischer 1993; Krebs 1993). The phosphorylation and dephosphorylation of proteins play an important role in the process of cell signal transmission, as a switch that controls many cell biological functions, which involve almost all vital activities, including cell signal transduction, gene expression, apoptosis, muscle contraction, metabolism, tumor generation, etc. (Cohen 2002).

3.2 Changes of Meat Color After Protein Phosphorylation Regulation

Protein phosphorylation in postmortem muscle played essential roles in meat quality. The influence of protein phosphorylation on meat color stability was investigated in the minced ovine *longissimus thoracis et lumborum*. After trimming off visible fat and connective tissue, the LTL muscle was divided into four treatment groups (150 g for each group): phosphatase inhibitor group (high phosphorylation level), saline control group (C1), muscle was added with the vehicle of saline, kinase inhibitor group (low phosphorylation level) and DMSO control group (C2), muscle was added with dimethyl sulfoxide (DMSO) and saline. The lactate and pH values were measured to check the protein phosphorylation while the protein band was studied to evaluate the intensity of phosphoproteins. Moreover, the analysis of meat color and the relative content of myoglobin redox forms calculations were used to check the meat color stability related to the phosphorylation of sarcoplasmic proteins. Thus, the protein phosphorylation may be involved in meat color development by regulating glycolysis and the redox stability of myoglobin.

3.2.1 Global Level of Protein Phosphorylation After Regulation of Protein Phosphorylation

Gel-based analysis of sarcoplasmic protein phosphorylation was shown in Fig. 3.1 and Table 3.1. As expected, the overall phosphorylation of sarcoplasmic proteins in the minced meat with phosphatase inhibitor was significantly higher ($P < 0.05$) than in control and kinase inhibitor added samples. In addition, the global phosphorylation of sarcoplasmic proteins in the kinase inhibitor group became even lower ($P < 0.05$) than in both control samples.

No difference in protein phosphorylation was detected between the two control samples, which were intermediate among the four treatments. This result confirmed that the overall phosphorylation of sarcoplasmic proteins was altered by inhibitors.

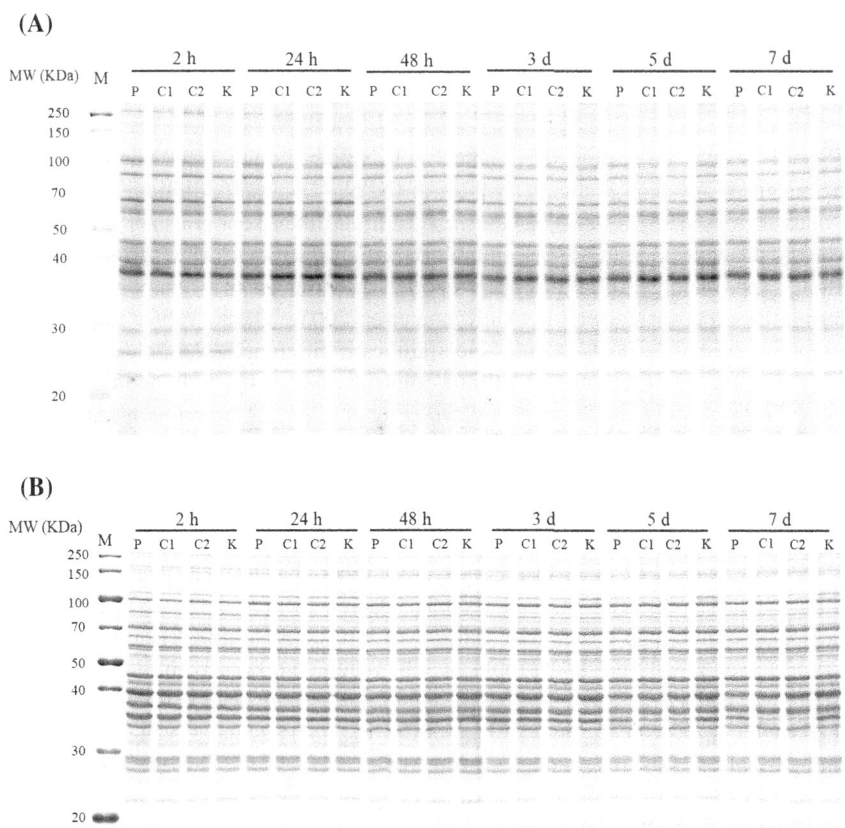

Fig. 3.1 Gel-based analysis of sarcoplasmic protein phosphorylation. (**a**) Image of gels stained with Pro-Q Diamond. (**b**) Image of gels stained with SYPRO Ruby. P, C1, C2, and K represent phosphatase inhibition group, saline control group, DMSO control group, and kinase inhibition group, respectively (Li et al. 2017)

Table 3.1 Effects of protein kinase and phosphatase inhibitors on the global phosphorylation of sarcoplasmic proteins (Li et al. 2017)

Group				
Display time (hours)	Phosphatase inhibition	Saline control	DMSO control	Kinase inhibition
2	0.97 ± 0.002^{aA}	0.80 ± 0.013^{bA}	0.80 ± 0.014^{bA}	0.71 ± 0.015^{bA}
24	1.00 ± 0.009^{aA}	0.84 ± 0.008^{bA}	0.81 ± 0.006^{bA}	0.65 ± 0.011^{cB}
48	1.11 ± 0.011^{aB}	0.83 ± 0.030^{bA}	0.78 ± 0.004^{bA}	0.54 ± 0.008^{bC}
72	1.18 ± 0.002^{aBC}	0.79 ± 0.019^{bA}	0.83 ± 0.016^{abA}	0.54 ± 0.004^{bC}
120	1.17 ± 0.002^{aBC}	0.87 ± 0.011^{bA}	0.84 ± 0.009^{bA}	0.62 ± 0.008^{bB}
168	1.21 ± 0.001^{aC}	0.91 ± 0.006^{aA}	0.87 ± 0.001^{abA}	0.67 ± 0.015^{bAB}

Data are expressed as mean ± standard deviation ($n = 5$). Means with different lower case letters in the same row are significantly different ($P < 0.05$). Means with different upper case letters in the same column are significantly different ($P < 0.05$)

3.2.2 Color of Meat with Different Protein Phosphorylation Level

L^* and b^* values are presented in Table 3.2. L^* values did not differ among the four treatments at any time points. L^* values of the phosphatase inhibition, C1 and C2 groups increased ($P < 0.05$) for the first 24 h and then kept constant afterwards, whereas L^* values of the kinase inhibition group did not change throughout the whole experiment period. It has been reported that changes in L^* value were subtle during the experimental period (McKenna et al. 2005). The b^* values for all meat samples were similar to L^* values. No significant differences were observed in b^* values among the four groups at all time points except for 168 h. In addition, the b^* values in all samples increased ($P < 0.05$) for the first 24 h and then kept stable throughout the rest of the experimental period.

According to Olivera et al. (2013), redness was the most important color parameter for fresh meat, which decreased with the increase of storage time. In consistence, a^* values determined in the present study (Table 3.2) kept decreasing during the whole experimental period. Meat redness was higher ($P < 0.05$) in kinase inhibition group than in control C1 and C2 groups, which was greater ($P < 0.05$) than in phosphatase inhibition group at all times except for 2 h. No difference in a^* values was observed among the four groups at 2 h. When compared with the other three groups, a^* values in the phosphatase inhibition group declined fastest, which displayed a rapid decrease in a^* values for the first 48 h and the a^* values decreased gradually afterwards. The changes in a^* values were similar for the two control groups. Besides, a^* values of the kinase inhibitor added meat declined slowest ($P < 0.05$) among the four groups during the 7 days of setting. This implied that a^* values of meat were obviously affected by protein phosphorylation.

Table 3.2 Effects of inhibitors and storage time (hour) on L^*values, a^*values, b^*values (Li et al. 2017)

		Display time (hours)					
Attribute	Group	2	24	48	72	120	168
L^*value (lightness)	Phosphatase inhibition	42.03 ± 3.14^{aX}	44.65 ± 3.80^{abX}	48.53 ± 3.64^{bX}	47.70 ± 3.22^{bX}	48.25 ± 3.68^{bX}	48.49 ± 3.48^{bX}
	Saline control	43.86 ± 2.61^{aX}	46.33 ± 3.67^{abX}	47.73 ± 2.63^{abX}	48.15 ± 2.82^{bX}	48.81 ± 2.61^{bX}	49.16 ± 2.81^{bX}
	DMSO control	41.10 ± 2.71^{aX}	45.25 ± 2.68^{bX}	46.45 ± 1.84^{bX}	46.12 ± 2.61^{bX}	46.47 ± 2.21^{bX}	47.45 ± 2.27^{bX}
	Kinase inhibition	42.61 ± 2.44^{aX}	45.37 ± 2.77^{aX}	44.78 ± 2.33^{aX}	44.17 ± 2.88^{aX}	44.51 ± 3.78^{aX}	45.16 ± 2.71^{aX}
a^*value (redness)	Phosphatase inhibition	11.50 ± 1.10^{aX}	8.42 ± 1.21^{aX}	6.95 ± 1.05^{bX}	6.04 ± 0.68^{bX}	5.24 ± 0.68^{cX}	4.96 ± 0.16^{cX}
	Blank control	12.43 ± 0.56^{aX}	11.28 ± 0.55^{bY}	9.79 ± 0.49^{cY}	9.47 ± 0.57^{cdY}	8.98 ± 0.42^{deY}	8.65 ± 0.49^{eY}
	DMSO control	12.04 ± 0.81^{aX}	10.60 ± 0.87^{bY}	9.63 ± 0.44^{cY}	9.39 ± 0.45^{cY}	8.94 ± 0.43^{cdY}	8.43 ± 0.71^{dY}
	Kinase inhibition	12.33 ± 0.45^{aX}	12.64 ± 1.10^{bZ}	11.38 ± 11.58^{cZ}	11.16 ± 0.12^{cdZ}	9.93 ± 0.21^{dZ}	9.33 ± 0.47^{dZ}
b^*value (yellowness)	Phosphatase inhibition	9.64 ± 1.49^{aX}	11.14 ± 1.00^{abX}	12.48 ± 2.08^{bcX}	14.00 ± 1.57^{cX}	13.72 ± 2.04^{cX}	13.48 ± 0.93^{cX}
	Saline control	10.02 ± 0.69^{aX}	12.38 ± 0.88^{bX}	11.93 ± 1.54^{bX}	12.30 ± 1.41^{bX}	12.25 ± 1.73^{bX}	12.27 ± 1.22^{bXY}
	DMSO control	9.24 ± 0.85^{aX}	11.20 ± 0.96^{bX}	11.36 ± 1.19^{bX}	12.08 ± 1.58^{bX}	11.75 ± 1.68^{bX}	11.39 ± 1.55^{bY}
	Kinase inhibition	10.06 ± 0.98^{aX}	12.58 ± 1.31^{bX}	12.54 ± 1.14^{bX}	13.44 ± 1.39^{bX}	12.49 ± 0.99^{bX}	12.68 ± 0.92^{bXY}

Data are presented as mean ± standard deviation ($n = 5$). Data with different lower case letters in a row are significantly different ($P < 0.05$). Data with different capital letters in the same column are significantly different ($P < 0.05$)

3.2.3 Role of pH and Lactic Acid Content on Color of Meat with Different Protein Phosphorylation Level

The pH values were shown in Fig. 3.2. There was no difference in pH values between the phosphatase inhibition group and the two control groups. However, the pH of the kinase inhibition group was significantly higher than those of the other three groups ($P < 0.05$) throughout the entire postmortem period except for 2 h postmortem, at which no difference in pH values was found among all the four groups. The pH values of phosphatase inhibition group, C1 and C2 control groups declined to the ultimate values of 5.67, 5.67, and 5.65, respectively, within 24 h, but pH of kinase inhibition group did not declined to the ultimate value of 6.02 till 78 h. This meant the pH of the kinase inhibition group declined much slower during the early postmortem stage ($P < 0.05$) as compared to the other three groups. The difference in meat pH values indicated that the rate and extent of pH decline in postmortem muscle were significantly affected by protein phosphorylation.

Meat quality was largely affected by the rate and extent of pH decline. pH values in postmortem muscle, which was controlled by the accumulation of lactate and H^+ through glycogenolysis and glycolysis (Scheffler and Gerrard 2007), were probably one of the most important factors affecting meat color. The rate and extent of pH

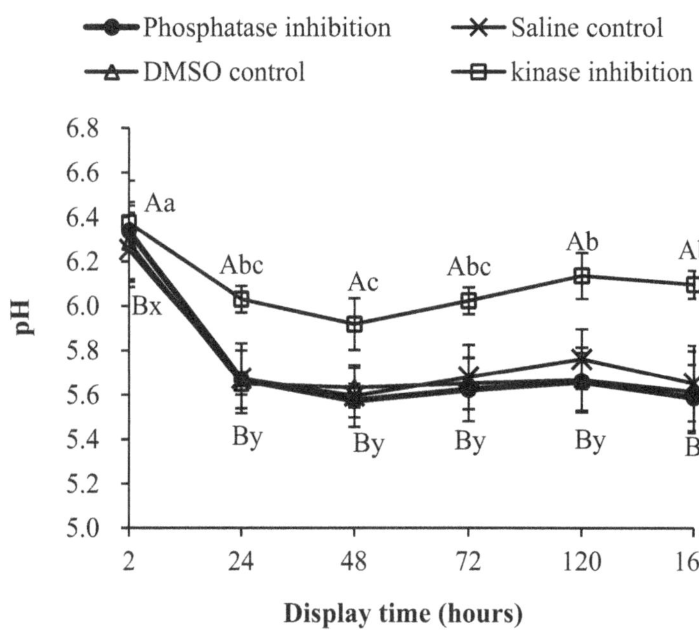

Fig. 3.2 pH values of muscles added with or without enzyme inhibitors. Data with different lower case letters within treatments are significantly different ($P < 0.05$). Data with different capital letters at the same time points are significantly different ($P < 0.05$). Data are presented as mean ± standard deviation, $n = 5$ (Li et al. 2017)

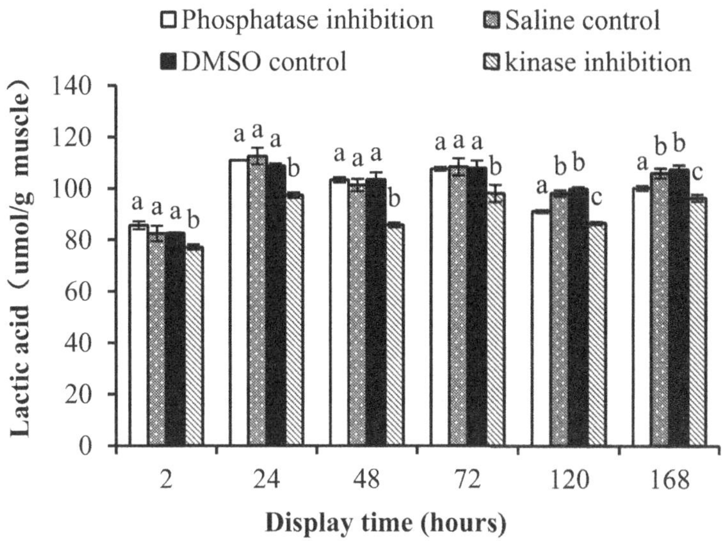

Fig. 3.3 Lactate content in muscles added with or without enzyme inhibitors. Data with different letters at the same time point are significantly different ($P < 0.05$). Data are presented as mean ± standard deviation, $n = 5$ (Li et al. 2017)

decline significantly influenced protein characteristics in postmortem muscle, for example, myoglobin was easier to be oxidized under lower pH conditions and this reaction was very pH dependent. Gutzke and Trout (2002) reported that pH had a consistent effect on the myoglobin autoxidation rate for the different species and temperatures. The effects of pH on the oxidation of myoglobin have been studied from pH 4.8 to pH 12.6, and the result turned out that this autoxidation reaction depended on hydrogen ion concentration directly (Shikama and Sugawara 1978). It is logical to conclude that the lower rate and extent of pH decline were one of the reasons for the higher color stability in kinase inhibition groups comparing to the other three groups. Although both of the phosphorylation level and color stability differ significantly, no difference was found in pH value between phosphatase inhibition group and control groups. This implied that there may be some other reasons for the influence of protein phosphorylation on meat color stability except pH value though the rationale remains unclear.

Lactic acid concentrations in muscle corresponded well with the pH values. No difference was observed in lactic acid content (Fig. 3.3) between C1 and C2 control groups. In addition, no difference in lactic acid was detected between phosphatase inhibition group and the two control groups at most time points, but the kinase inhibition group had lower lactic acid content than the other three groups ($P < 0.05$) throughout the entire experimental period. This result further supported that muscle with lower protein phosphorylation had lower ($P < 0.05$) glycolytic rate when compared with samples with higher protein phosphorylation. Consequently, it was proposed that protein phosphorylation had an influence on the rate and extent of pH decline through regulating glycolysis in postmortem muscle.

Most of the glycometabolic enzymes can be phosphorylated, such as glycogen phosphorylase (GP), pyruvate kinase (PK), and phosphofructokinase (PFK), which were glycometabolic rate-limiting enzymes. Protein phosphorylation was the most common post-translational modifications, which was a key modulator of protein structure, function, and activity. Several previous studies have revealed that protein phosphorylation regulates the activity of those glycometabolic rate-limiting enzymes. For example, after phosphorylation, PK was transformed to a more acid-stable isoform, and maintained high activity in PSE meat (Schwägele et al. 1996). GP can be phosphorylated on serine 14, leading to changes in its structure and activation of its activity (Johnson 1992; Sprang et al. 1988). Phosphorylation of PFK regulated the compartmentalization of the enzyme to provide energy to the cellular component by forming a complex with actin (Cai et al. 1997; Kuo et al. 1986). Protein phosphorylation regulated postmortem glycolysis by regulating the activity or stability of glycolytic enzymes (Huang et al. 2011). Shen and Du (2005) reported that the glycolysis and pH decline were indirectly affected by the phosphorylation status of AMP-activated protein kinase (AMPK) in postmortem muscle. In brief, it can be concluded that protein phosphorylation regulated the rate and extent of pH decline in postmortem muscle by controlling the glycolysis reactions. Since protein phosphorylation altered the structure, activity, and stability of protein, it was likely that phosphorylation of myoglobin influenced meat color stability by regulating its redox stability. However, the phosphorylation of myoglobin was not measured in the present study. Therefore, whether the phosphorylation of myoglobin affected color stability was uncertain, and further research was needed.

3.3 Pattern of Sarcoplasmic Protein Phosphorylation in Meat with Different Color Stability

Protein phosphorylation might have impact on the regulation of meat color stability probably by regulating glycolysis and the redox stability of myoglobin. The phosphorylation of sarcoplasmic proteins in postmortem muscles was investigated in relationship to color stability in the present study. The *longissimus thoracis et lumborum* muscles were used in the study. Each side of the carcasses was fabricated into six steaks, individually wrapped in an oxygen-permeable polyvinylchloride (PVC) film, and stored at 4 °C for 8 days to simulate retail display. The first steak from the left side was used for pH measurement at 45 min and 24 h postmortem. The second steak was used for color determination. The rest of the steaks were used for muscle sample collection at 45 min, 6 h, 24 h, 2 days, 3 days, 4 days, 5 days, 6 days, 7 days, and 8 days postmortem. The meat color parameters were studied to check the color stability in the meat. Liquid chromatography—tandem mass spectrometry (LC-MS/MS) was used for identification of phosphoproteins. The gel electrophoresis was used for the separation of protein and protein phosphorylation level determination. To better understand the biochemistry of meat color stability, protein phosphorylation was comparatively profiled in muscles with different color stability.

3.3.1 Global Sarcoplasmic Protein Phosphorylation Level of Meat with Different Color Stability

Phosphoproteins were detected by SDS-PAGE and fluorescence staining. As shown in Fig. 3.4, a total of 24 individual proteins bands were detected on the SDS-PAGE gels and the intensity of phosphorylated protein to total protein ratios (P/T ratios) of these protein bands was calculated (Fig. 3.4). No significant difference in global phosphorylation level of sarcoplasmic proteins was detected among the three groups during the whole display period.

However, the total phosphorylation level of sarcoplasmic proteins varied with time in all three groups, which decreased ($P < 0.05$) within the first 24 h and then did not change afterwards.

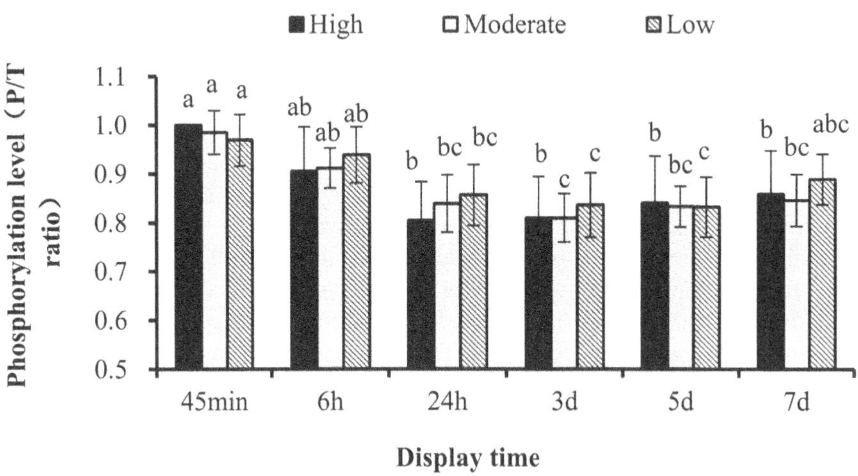

Fig. 3.4 Calculated global phosphorylation levels of sarcoplasmic proteins. High, moderate, and low represent high color stability group, moderate color stability group, and low color stability group, respectively. Data are expressed as mean ± standard deviation ($n = 5$). Values with different letters are significantly different within the same group at different display time ($P < 0.05$) (Li et al. 2018a)

3.3.2 Correlation of Meat Color Parameters and Sarcoplasmic Protein Phosphorylation Level

All the 24 protein bands were subjected to association analysis (Table 3.3). Pearson correlation coefficients between meat color attributes and the total phosphorylation level of sarcoplasmic proteins or the phosphorylation levels of the 24 individual protein bands were shown in Table 3.3. Statistical analysis showed that the total phosphorylation level of sarcoplasmic proteins negatively correlated ($P < 0.01$) with meat L^* (r = −0.897), a^* (r = −0.738), b^* (r = −0.848) values and the relative content of oxymyoglobin (r = − 0.816), but positively correlated ($P < 0.01$) with the relative content of DeoxyMb (r = 0.885).

This suggested that meat color stability was correlated with the total phosphorylation level of sarcoplasmic proteins, which was in accordance with our previous study (Li et al. 2017). The results of Pearson's correlation analysis between the phosphorylation levels of individual protein bands and L^*, a^*, b^*, R630/580 values and the relative content of deoxymyoglobin, oxymyoglobin, and metmyoglobin were shown in Table 3.3. A total of 11 protein bands were found to correlate with instrumental color values, which were band 5, 6, 8, 9, 11, 12, 13, 14, 15, 20, and 23. The P/T ratios of these bands were presented in Fig. 3.5.

In several previous studies, the differential proteome was investigated in relationship to meat color stability. Canto et al. (2015) reported that the expression/abundance of some sarcoplasmic proteins might influence meat color stability. They reported that phosphoglucomutase-1, glyceraldehyde-3-phosphate dehydrogenase, pyruvate kinase M2, and creatine kinase M-type were over-abundant in color-stable steaks, while adenylate kinase isoenzyme 1 and myoglobin were overabundant in color-labile samples. A study by Wu et al. (2015) showed that fructose-bisphosphate aldolase A isoform, pyruvate kinase isozyme M1/M2 isoform, and malate dehydrogenase were negatively correlated, while L-lactate dehydrogenase A chain isoform was positively correlated with beef $a*$ values and metmyoglobin reductase activity (MRA). Joseph et al. (2012) reported that creatine kinase and β-enolase had a positive correlation with meat a^* value. In a study by Gao et al. (2016), it was reported that creatine kinase M-type, malate dehydrogenase, and β-enolase were related to meat color development. Based on all these studies, it can be concluded that proteome played a role in meat color development and discoloration, though the underlying mechanism was not completely understood. The color-related proteins discussed above were also identified in the present study. Most of these proteins are glycolytic enzymes that can be reversibly phosphorylated, such as phosphoglucomutase 1 (PGM1), PK, fructose-bisphosphate aldolase A (ALDOA), glyceraldehyde-3-phosphate dehydrogenase (GAPDH), lactate dehydrogenase (LDHA), and β-enolase (Huang et al. 2011; Kachel et al. 2015). Phosphorylation regulated enzyme activity, stability, and interaction with other proteins (Sale et al. 1987). Several studies have revealed that protein phosphorylation played key roles in regulating the activities of most glycolytic enzymes, like pyruvate kinase, which could be transformed to a more active and acid-stable isoform after phosphorylation

Table 3.3 Pearson correlation coefficients between meat color attributes and the phosphorylation level of total sarcoplasmic proteins or individual protein bands (Li et al. 2018a)

	L^* value	a^* value	b^* value	R630/580	DeoxyMb	OxyMb	MetMb
Total phosphorylation level	−0.897**	−0.738 **	−0.848**	−0.046	0.885**	−0.816**	−0.110
Band1	−0.396	−0.337	−0.462	0.471*	0.534*	−0.373	−0.700**
Band2	−0.252	−0.457	−0.445	−0.138	0.341	−0.450	−0.065
Band3	0.321	−0.180	0.109	−0.758**	−0.346	−0.055	0.689**
Band4	−0.489*	−0.414	−0.497*	0.028	0.577 *	−0.433	−0.205
Band5	−0.664**	−0.387	−0.624**	0.515*	0.720**	−0.509*	−0.676**
Band6	−0.850**	−0.570*	−0.798**	0.357	0.885**	−0.695**	−0.538**
Band7	−0.305	−0.002	−0.122	0.450	0.253	−0.082	−0.408
Band8	−0.787**	−0.505*	−0.717**	0.323	0.854**	−0.604**	−0.462
Band9	0.743**	0.580*	0.777**	−0.441	−0.817**	0.678**	0.718**
Band10	−0.295	−0.35	−0.351	0.051	0.395	−0.356	−0.190
Band11	−0.854**	−0.656**	−0.821**	0.267	0.872**	−0.739**	−0.484*
Band12	−0.543*	−0.078	−0.348	0.447	0.564**	−0.205	−0.387
Band13	0.722**	0.467	0.708**	−0.518*	−0.775**	0.571*	0.776**
Band14	−0.324	−0.097	−0.261	0.509*	0.402	−0.166	−0.571*
Band15	−0.250	0.191	−0.082	0.530*	0.224	0.067	−0.390
Band16	−0.520*	−0.410	−0.512*	0.060	0.500*	−0.498*	−0.182
Band17	−0.094	−0.113	−0.090	−0.041	0.051	−0.140	0.021
Band18	−0.585*	−0.338	−0.525*	0.495*	0.654**	−0.421	−0.623**
Band19	−0.093	0.170	−0.024	0.560*	0.162	0.121	−0.578*
Band20	0.391	0.496*	0.383	0.564*	−0.356	0.463	−0.543*
Band21	−0.489 *	−0.541*	−0.556*	−0.029	0.460	−0.567*	−0.179
Band22	−0.443	−0.340	−0.384	−0.212	0.328	−0.397	0.190

(continued)

Table 3.3 (continued)

	L^* value	a^* value	b^* value	R630/580	DeoxyMb	OxyMb	MetMb
Band23	0.393	0.340	0.490*	−0.358	−0.566*	0.385	0.595**
Band24	0.313	0.184	0.314	−0.441	−0.334	0.240	0.593**

Significant levels: *, $0.01 < P < 0.05$; **, $P < 0.01$
R630/580 was calculated by the ratio of reflectance at 630 to 580 nm as an indicator of meat color stability, DeoxyMb means deoxymyoglobin, OxyMb means oxymyoglobin, and MetMb means metmyoglobin

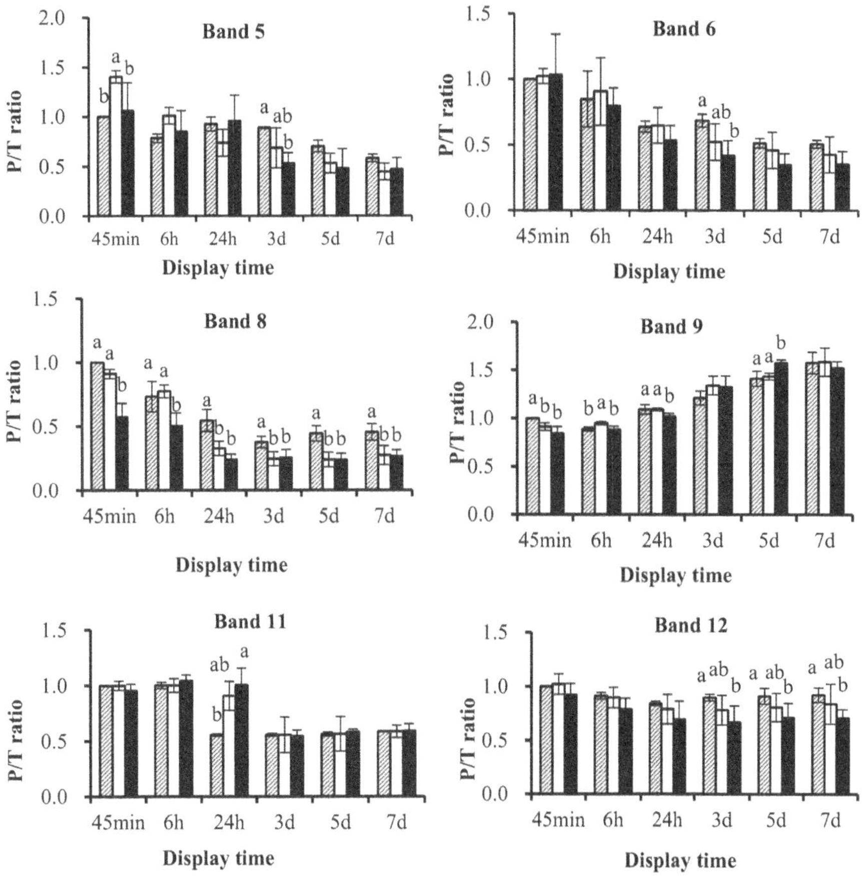

Fig. 3.5 The phosphorylation levels of 11 color stability related protein bands. High, moderate, and low represent high color stability group, moderate color stability group, and low color stability group, respectively. Data are presented as mean ± standard deviation ($n = 5$). Different letters at the same time mean significant difference among groups ($P < 0.05$) (Li et al. 2018a)

(Schwägele et al. 1996). As phosphorylation regulated the activity and stability of glycolytic enzymes, the reversible phosphorylation of these color-related proteins may play a role in the meat color development by regulating the glycolysis. The main protein identified in band 23 (Fig. 3.1a, b) was myoglobin in the present study, which was the primarily pigment responsible for meat color. The phosphorylation of myoglobin has been confirmed in our lab (unpublished) and the phosphorylation sites of myoglobin of some species are available on the PhosphoSitePlus Web site. According to the results of one-way analysis of variance (Fig. 3.5, band 23), the phosphorylation level of myoglobin in the high color stability muscle was significantly lower ($P < 0.05$) than that in the low color stability group within the first 24 h postmortem. Although no difference in the phosphorylation level of myoglobin

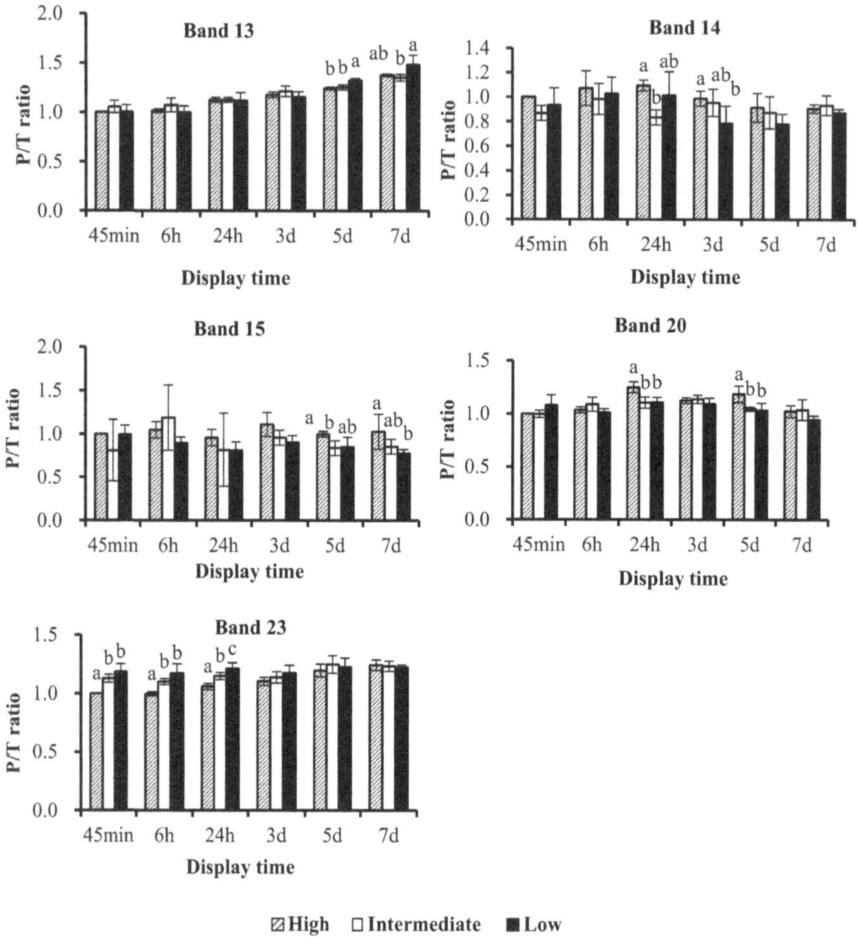

Fig. 3.5 (continued)

existed between the moderate and low color stability groups at 45 min and 6 h postmortem (Fig. 3.5, band 23), higher ($P < 0.05$) phosphorylation level was detected in low color stability muscle at 24 h. This result indicated that the phosphorylation level of myoglobin was inversely related to meat color stability. In addition, correlation analysis (Table 3.3) showed that the phosphorylation level of myoglobin (band 23) was positively correlated with ($r = 0.595$) the content of metmyoglobin, suggesting that phosphorylated myoglobin might be more susceptible to oxidation. As phosphorylation changed the structure and stability of proteins, it was likely that phosphorylation changed the structure of myoglobin and its susceptibility to oxidation as observed in the present study (Díaz-Moreno et al. 2009; Zhang and Liu 2017). A recent study has reported that human neuroglobin underwent hypoxia-dependent phosphorylation (Ascenzi et al. 2013), it was possible that the difference in myoglobin phosphorylation between treatments was contributed by the difference in oxygen availability. In summary, the phosphorylation of myoglobin may play an important role in keeping meat color stable. This was consistent with a previous study (Canto et al. 2015) which observed an acidic shift of isoelectric point (pI) of myoglobin on 2-DE gels in greater color lability muscle. As a pI acidic shift was a symbol for phosphorylation of proteins with pI >6.4 (Zhu et al. 2005). Thus, phosphorylation modification could lead to this kind of pI shift in myoglobin, and phosphorylation might play a negative role in the regulation of meat color stability.

3.3.3 Identification of Phosphorylated Sarcoplasmic Proteins Related with Meat Color

The eleven interested protein bands were excised from gels and proteins were identified by LC-MS/MS (Table 3.4). More than one protein was identified in every protein band. Proteins that had an inappropriate molecular weight or a very low percentage of coverage of the entire amino acid sequence or a very low score were excluded. Some proteins were also identified to be presented in more than one band, which may be caused by proteolytic cleavage of the proteins or protein polymorphisms. The functions of all identified proteins were annotated using Protein Knowledgebase (UniProtKB in www.uniprot.org). Most of the identified proteins were glycolytic enzymes, which included glucose-6-phosphate isomerase (GPI), fructose-bisphosphate aldolase (ALDOA), phosphoglycerate kinase 1, L-lactate dehydrogenase A chain (LDHA), alpha-enolase, beta-enolase, phosphoglucomutase-1 (PGM1), pyruvate kinase (PK), and glyceraldehyde-3 phosphate dehydrogenase (GAPDH). Besides, myoglobin, adenylate kinase isoenzyme 1, malate dehydrogenase, creatine kinase M-type, alpha-1, 4 glucan phosphorylase, sarcoplasmic/ endoplasmic reticulum calcium ATPase 1, calcium-transporting ATPase, and several other proteins were also identified.

Creatine kinase M-type was identified in both band 12 and band 13 (Table 3.4). Although no difference was observed in the phosphorylation level of band 12 among the three groups within the first 24 h PM (Fig. 3.5, band 12), it was lower ($P < 0.05$) in the low color stability group than that in the high color stability group on day 3, 5, and 7 (Fig. 3.5, band 12). Adenylate kinase isoenzyme 1 was the main protein identified in band 20 (Table 3.4). The phosphorylation level of band 20 was higher ($P < 0.05$) in the high color stability group than in the low color stability group at 24 h and 5 days PM (Fig. 3.5, band 20). Besides, the phosphorylation level of band 20 was positively ($P < 0.05$) correlated with a^* and R630/580 values, and negatively ($P < 0.05$) correlated with the content of metmyoglobin (Table 3.3). The main proteins identified in band 5 were uncharacterized protein (GN = AKAP4), sarcoplasmic/endoplasmic reticulum calcium ATPase 1, and calcium-transporting ATPase (Table 3.4). The main proteins identified in band 6 were uncharacterized

Table 3.4 Identified proteins in the 11 color stability related protein bands by LC-MS/MS (Li et al. 2018a)

Protein name[a]	Band no[b]	Accession no[a]	Score[c]	MW (Da)[d]	MP[e]	Seq.cov (%)[f]	Organism[g]	Function
Uncharacterized protein (GN = AKAP4)	5	F1MYH5	7300	95,481	4	13	Bos	Protein kinase A binding
Sarcoplasmic/endoplasmic reticulum calcium ATPase 1	5	Q0VCY0	4894	110,532	36	35.1	Bos	Ca^{2+} signaling
Calcium-transporting ATPase	5	F1MPR3	1890	113,512	22	27.6	Bos	Calcium transport
Alpha-1,4 glucan phosphorylase	6	F1MJ28	11,923	97,674	46	54.4	Bos	Glycometabolism
Uncharacterized protein(GN = AKAP4)	6	F1MYH5	5115	95,481	5	16.5	Bos	Protein kinase A binding
Phosphoglucomutase-1	8	Q08DP0	16,360	61,836	35	55.3	Bos	Glycolytic enzyme
Very long-chain specific acyl-CoA dehydrogenase	8	P48818	577	71,003	9	28.4	Bos	Lipid metabolism Oxidoreductase
Pyruvate kinase (Fragment)	8	Q3ZC87	421	62,016	12	34.7	Bos	Glycolytic enzyme
Pyruvate kinase (Fragment)	9	Q3ZC87	13,687	62,016	32	54.7	Bos	Glycolytic enzyme
Pyruvate kinase (Fragment)	9	B3IVN4	5448	16,688	14	85.2	Bos	Glycolytic enzyme
Glucose-6-phosphate isomerase	9	Q3ZBD7	804	63,043	12	25.1	Bos	Glycolytic enzyme
Phosphoglucomutase-1	9	Q08DP0	797	61,836	11	33.5	Bos	Glycolytic enzyme
Beta-enolase	11	Q3ZC09	34,925	47,409	28	58.5	Bos	Glycolytic enzyme
Alpha-enolase	11	F1MB08	18,978	47,596	7	28.1	Bos	Glycolytic enzyme
Phosphoglycerate kinase 1	11	Q3T0P6	1242	44,908	13	49.9	Bos	Glycolytic enzyme
Creatine kinase M-type	12	Q9XSC6	39,535	43,190	31	80.6	Bos	ATP regeneration
Beta-enolase	12	Q3ZC09	19,316	47,409	27	54.6	Bos	Glycolytic enzyme
Fructose-bisphosphate aldolase	13	A6QLL8	27,484	39,925	28	73.4	Bos	Glycolytic enzyme
Creatine kinase M-type	13	Q9XSC6	22,291	43,190	18	63.3	Bos	ATP regeneration
Glyceraldehyde-3-phosphate dehydrogenase	14	P10096	8470	36,073	15	44.1	Bos	Glycolytic enzyme
Malate dehydrogenase	14	Q3T145	990	36,700	10	32.6	Bos	Glycometabolism
L-lactate dehydrogenase A chain	15	P19858	3556	36,916	10	34.6	Bos	Glycolytic enzyme

Adenylate kinase isoenzyme 1	20	P00570	2027	21,764	8	34.5	Bos	Creatine energy metabolism
Myoglobin	23	P02192	5897	17,067	8	57.1	Bos	Oxygen transport

[a]Protein names and accession numbers were derived from the UniProt database
[b]The number of band containing the identified proteins
[c]For the proteins identified in more than one band, the highest score was presented
[d]Theoretical molecular weight (recorded in UniProt database)
[e]Number of matched peptides
[f]Percentage of coverage of the entire amino acid sequence
[g]Bos, *Ovis aries*
[i]GN, gene name

protein (GN = AKAP4) and alpha-1, 4 glucan phosphorylase (Table 3.4). The phosphorylation of proteins in band 5 and band 6 exhibited a negative ($P < 0.05$) correlation with L^* value, b^* value, the relative content of oxymyoglobin, and the relative content of metmyoglobin (Table 3.3). Besides, phosphorylation of protein in band 5 was positively ($P < 0.05$) correlated with R630/580 value (Table 3.3), while that of band 6 was negatively ($P < 0.05$) correlated with a^* value (Table 3.3). Malate dehydrogenase was one of the main proteins identified in band 14 (Table 3.4). All those identified proteins may contribute to differences in meat color, though the underlying mechanism through which the phosphorylation of those protein influences meat color stability was still unclear.

3.4 Quantitative Phosphoproteome of Meat with Different Color

Phosphorylation of some glycolytic enzymes and myoglobin at specific serine residues may play critical roles in the regulation of meat color stability. A quantitative analysis of protein phosphorylation in ovine LTL muscle with different color stability was performed using TMT labeling in combination with TiO_2 phosphopeptide enrichment. The pH and instrumental color measurement and phosphoproteomic analysis were carried out. Proteins involved in carbohydrate metabolism, especially glycolytic enzymes, were the largest cluster of protein determined to be color-related. The phosphorylation of myoglobin at Ser133 played a negative role in the regulation of meat color stability. Thus, the phosphorylation of some glycolytic enzymes and myoglobin at specific serine residues may play critical roles in the regulation of meat color stability. To further understand the mechanism by which protein phosphorylation regulated meat color stability, in the present study, a quantitative MS-based phosphoproteomic analysis of ovine muscle with different color stability at 24 h postmortem was performed.

3.4.1 Differentially Expressed Phosphoproteins Between Meat with Different Color

A total of 3412 phosphopeptides were identified, including 1070 phosphoproteins. Among these, 243 phosphoproteins were significantly different in abundance between low and high color stability groups, of which 98 proteins were up-regulated and 145 were down-regulated compared to high color stability group. One hundred and thirty five phosphoproteins were determined to be different in abundance between medium and high color stability group, of which 63 proteins were up-regulated and 72 were down-regulated in medium color stability group. One hundred and two phosphoproteins were differentially expressed in low and medium color stability muscles, of which 47 proteins were up-regulated and the other 55 were down-regulated compared to medium color stability group.

In total, 432 unique phosphopeptides were determined to be different in abundance between the three groups of muscle, which were assigned to 273 phosphoproteins. To show the phosphorylation pattern of the 432 phosphopeptides, a hierarchical clustering analysis (HCA) was applied. As shown in Fig. 3.6, the 432 phosphopeptides were marked from red to green in abundance from high to low. A redder color meant a higher phosphorylation level and a greener color meant a lower phosphorylation level of the phosphopeptide. Similar color layouts were observed within the three biological replicates of each group and different color was

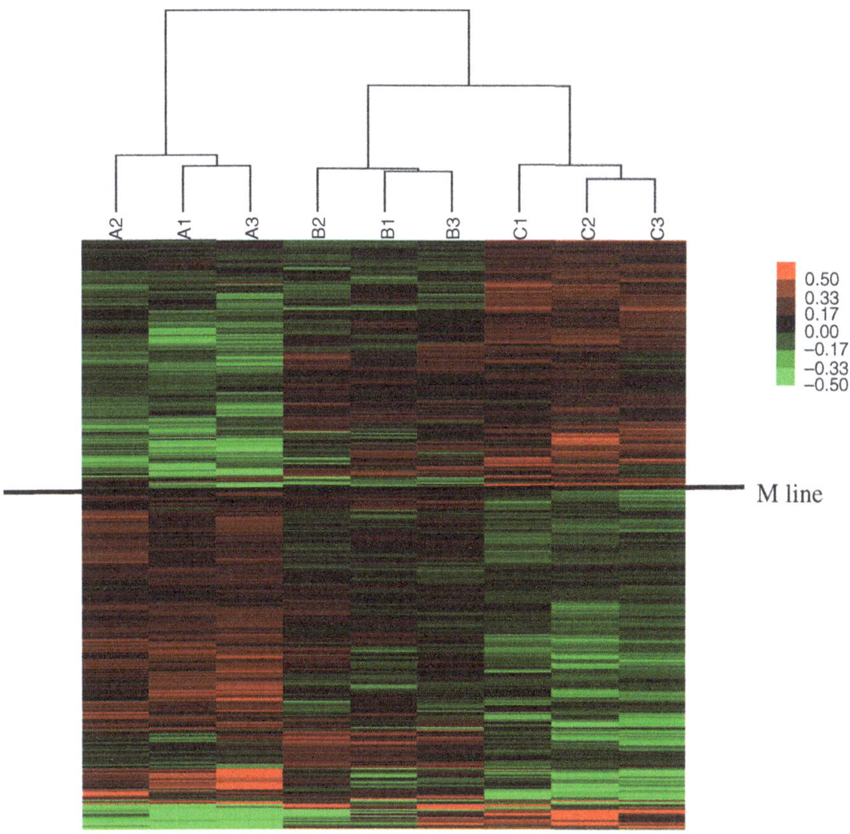

Fig. 3.6 Clustering of differently expressed phosphopeptides in ovine muscles of different color stability. Note: Muscle samples are displayed in columns and classified by phosphoproteomic subtypes as indicated by different colors. A1, A2, and A3 are the three samples in high color stability group. B1, B2, and B3 are the three samples from moderate color stability group. C1, C2, and C3 represent the three samples in low color stability group. Deeper red represents higher phosphorylation level and deeper green represents lower phosphorylation level. Protein names are listed in Supplementary material 1. (For interpretation of the references to color in this figure legend, the reader is referred to the web version of this article.) (Li et al. 2018b)

displayed among different groups, revealing a characteristic pattern in protein phosphorylation in muscles of different color stability. Specifically, protein phosphorylation displayed a "reverse" pattern between the H and L groups. Highly phosphorylated sites in the H group were lowly phosphorylated in the L group, vice versa. In order to improve our understanding of the underlying mechanism by which protein phosphorylation regulated meat color stability, the differentially expressed phosphoproteins between the H and L groups were selected for bioinformatic analysis in the study. Twenty four putative phosphorylation motifs were identified by using Motif-X software which included 16 serine motifs, seven threonine motifs, and one tyrosine motif (Supplementary material 2). This indicated that phosphoproteins with the same phosphorylation motifs might affect meat color development.

3.4.2 Functional Enrichment Analysis of Differentially Expressed Phosphoproteins

Functional enrichment analysis of differentially expressed phosphoproteins was performed to obtain some information about the mechanism regulating meat color stability (Fig. 3.7). Gene Ontology (GO) terms enrichment revealed that most differentially expressed phosphoproteins were involved in cellular carbohydrate metabolic process, glucan metabolic process, polysaccharide metabolic process, cellular glucan metabolic process, cellular polysaccharide metabolic process, single-organism carbohydrate metabolic process, carbohydrate metabolic process, and glyceraldehyde-3-phosphate metabolic process. Kyoto encyclopedia of genes and genomes (KEGG) pathway enrichment showed that the primary eight pathways that these phosphoproteins participated were starch and sucrose metabolism, pentose phosphate pathway, galactose metabolism, glycolysis/gluconeogenesis, fructose and mannose metabolism, amino sugar and nucleotide sugar metabolism, purine metabolism, and biosynthesis of amino acids.

In summary, the functional enrichment analysis revealed that the carbohydrate metabolism-related proteins were the largest cluster of proteins that were color-related, showing that meat color stability was probably primarily influenced by carbohydrate metabolism in postmortem muscle. All differentially expressed phosphoproteins of these color-related metabolic processes were selected as key color-related phosphoproteins. A total of 27 key color-related phosphoproteins were selected in this study.

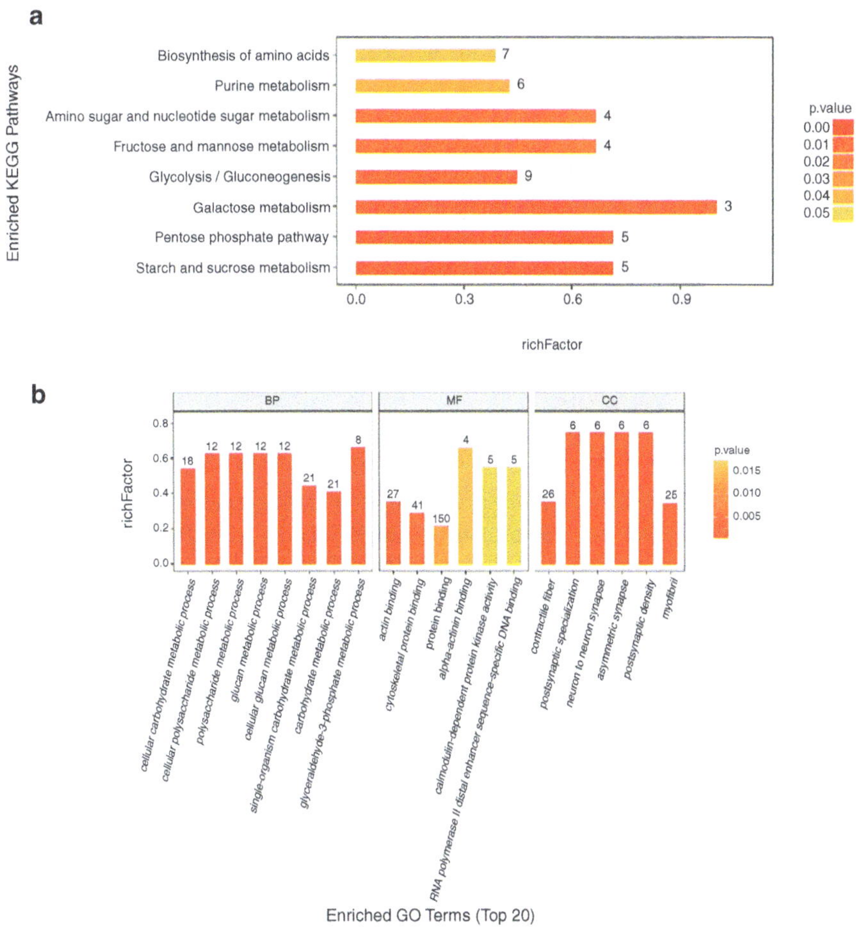

Fig. 3.7 The results of GO terms enrichment and KEGG pathway enrichment. (**a**)The enriched KEGG pathways of differently expressed phosphoprotein between high and low color stability group. (**b**)The enriched GO terms of differently expressed phosphoprotein between high and low color stability group. (Li et al. 2018b)

3.4.3 Functional Annotation Analysis of Key Color-Related Phosphoproteins

Gene ontology annotation of the key color-related phosphoproteins was shown in Table 3.3. The metabolic process grouping of the 27 key color-related phosphoproteins was shown in Fig. 3.8. As in Fig. 3.8, 19 out of the 27 key color-related phosphoproteins were carbohydrate metabolic enzymes, including nine glycolytic enzymes, six galactose, glycogen, polysaccharide and glucan metabolism enzymes,

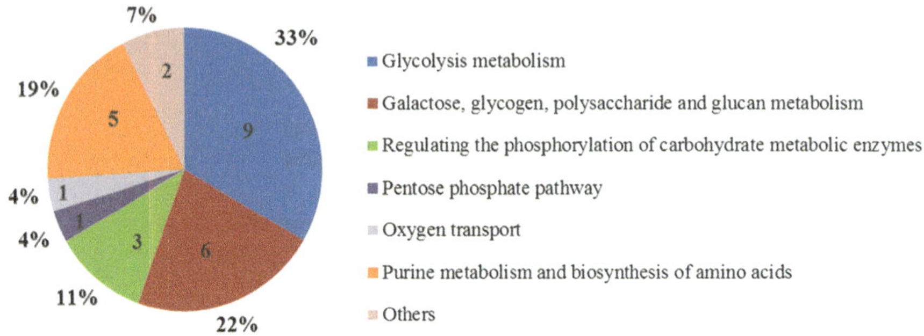

Fig. 3.8 Metabolic pathway classification of the 27 key color-related phosphoproteins (Li et al. 2018b)

three phosphorylation regulation related enzymes, and one enzymes of tricarboxylic acid cycle.

Myoglobin, five purine metabolism and amino acid biosynthesis related enzymes, and two other proteins were also identified as key color-related phosphoproteins. More than one phosphopeptides and phosphorylation sites from one key color-related phosphoprotein were identified in this study. However, not all the phosphopeptides and phosphorylation sites from one key color-related phosphoprotein were differentially expressed in muscles with different color stability.

3.5 Conclusions

Adding additional inhibitors could regulate sarcoplasmic protein phosphorylation level. The phosphatase inhibition group had a significantly higher phosphorylation level than control group, and that of kinase inhibition group was significantly lower than control group. Compared with high phosphorylation level group, the group with low phosphorylation level had low glycolysis rate, slow pH value decline rate, high limit pH value, and high redness value. Therefore, protein phosphorylation negatively regulated meat color stability, and it might be caused by regulating glycolysis metabolism. There was no difference in the whole phosphorylation level of sarcoplasmic protein from *longissimus thoracis et lumborum* muscle between the high, middle, and low meat color-stable groups, and all reached the lowest level at 24 h after slaughter, and then basically kept the same level. However, there was significant difference in the phosphorylation level of some sarcoplasmic protein bands among the three groups. The results of mass spectrometry showed that most of the meat-color-related protein bands were glycolytic enzymes. Moreover, the Mb phosphorylation level was low in the high meat color-stable group, and high in the low meat color-stable group. There were two possible ways of protein phosphorylation controlling meat color stability. First, it affected meat color stability by regulating glycolysis process. Second, Mb phosphorylation might affect meat color stability.

Three thousand four hundred and twelve phosphorylated peptide segments and 1070 phosphorylated proteins were identified. The results of Gene Ontology and KEGG Pathway Analysis showed that the phosphorylated differential proteins related to meat color were mainly involved in the glycometabolism pathway. The key phosphorylated proteins that regulated meat color stability included pyruvate kinase, pyruvate dehydrogenase, glucose phosphomutase, phosphopropyl isomerase, fructose diphosphate aldolase, enolase, 6-phosphoglucose isomerase, adenylate kinase isoenzyme 1 and Mb. Most of the phosphorylation of meat color occurred on specific serine residues. The phosphorylation of these key meat-color-related phosphorylation proteins at specific sites might be an important factor for phosphorylation to control meat color stability. In addition, phosphorylation negatively regulated meat color stability happened on the 133 serine of myoglobin.

Acknowledgments Parts of this chapter are reprinted from Food Chemistry, 240, Li, M., et al., Comparative profiling of sarcoplasmic phosphoproteins in ovine muscle with different color stability, 104-111; Food Chemistry, 249, Li, Z., et al., Quantitative phosphoproteomic analysis among muscles of different color stability using tandem mass tag labeling, 8-15; Food Chemistry, 219, Li, M., et al., Effects of protein phosphorylation on color stability of ground meat, 304–310.

References

Ascenzi, P., Marino, M., Polticelli, F., Coletta, M., Gioia, M., Marini, S., et al. (2013). Non-covalent and covalent modifications modulate the reactivity of monomeric mammalian globins. *Biochimica et Biophysica Acta, 1834*(9), 1750–1756.

Cai, G. Z., Callaci, T. P., Luther, M. A., & Lee, J. C. (1997). Regulation of rabbit muscle phosphofructokinase by phosphorylation. *Biophysical Chemistry, 64*(1–3), 199–209.

Canto, A. C., Suman, S. P., Nair, M. N., Li, S., Rentfrow, G., Beach, C. M., et al. (2015). Differential abundance of sarcoplasmic proteome explains animal effect on beef Longissimus lumborum color stability. *Meat Science, 102*, 90–98.

Clydesdale, F. M. (1978). Colorimetry – Methodology and applications. *Critical Review in Food Science and Nutrition, 10*(3), 243–301.

Cohen, P. (2002). Protein kinases—The major drug targets of the twenty-first century? *Nature Reviews Drug Discovery, 1*(4), 309–315.

Díaz-Moreno, I., Hollingworth, D., Frenkiel, T. A., Kelly, G., Martin, S., Howell, S., et al. (2009). Phosphorylation-mediated unfolding of a KH domain regulates KSRP localization via 14-3-3 binding. *Nature Structural & Molecular Biology, 16*(3), 238–246.

Fischer, E. H. (1993). Protein phosphorylation and cellular regulation II (Nobel lecture). *Angewandte Chemie International Edition, 32*(8), 1130–1137.

Gao, X., Wu, W., Ma, C., Li, X., & Dai, R. (2016). Postmortem changes in sarcoplasmic proteins associated with color stability in lamb muscle analyzed by proteomics. *European Food Research and Technology, 242*(4), 527–535.

Gutzke, D., & Trout, G. R. (2002). Temperature and pH dependence of the autoxidation rate of bovine, ovine, porcine, and cervine oxymyoglobin isolated from three different muscles longissimus dorsi, gluteus medius, and biceps femoris. *Journal of Agricultural and Food Chemistry, 50*(9), 2673–2678.

Huang, H. G., Larsen, M. R., Karlsson, A. H., Pomponio, L., Costa, L. N., & Lametsch, R. (2011). Gel-based phosphoproteomics analysis of sarcoplasmic proteins in postmortem porcine muscle with pH decline rate and time differences. *Proteomics, 11*(20), 4063–4076.

Jeyamkondan, S., Jayas, D. S., & Holley, R. A. (2000). Review of centralized packaging systems for distribution of retail-ready meat. *Journal of Food Protection, 63*(6), 796–806.

Johnson, L. N. (1992). Glycogen phosphorylase: Control by phosphorylation and allosteric effectors. *FASEB Journal, 6*(6), 2274–2282.

Joseph, P., Suman, S. P., Rentfrow, G., Li, S. T., & Beach, C. M. (2012). Proteomics of muscle-specific beef color stability. *Journal of Agricultural and Food Chemistry, 60*(12), 3196–3203.

Kachel, P., Trojanowicz, B., Sekulla, C., Prenzel, H., Dralle, H., & Hoang-Vu, C. (2015). Phosphorylation of pyruvate kinase M2 and lactate dehydrogenase A by fibroblast growth factor receptor 1 in benign and malignant thyroid tissue. *BMC Cancer, 15*(1), 140–153.

Krebs, E. G. (1993). Protein phosphorylation and cellular regulation I (Nobel lecture). *Angewandte Chemie International Edition, 32*(8), 1122–1129.

Kuo, H. J., Malencik, D. A., Liou, R. S., & Anderson, S. R. (1986). Factors affecting the activation of rabbit muscle phosphofructokinase by actin. *Biochemistry, 25*(6), 1278–1286.

Li, M., Li, X., Xin, J. Z., Li, Z., Li, G. X., Zhang, Y., et al. (2017). Effects of protein phosphorylation on color stability of ground meat. *Food Chemistry, 219*, 304–310.

Li, M., Li, Z., Li, X., Xin, J. Z., Wang, Y., Li, G. X., et al. (2018a). Comparative profiling of sarcoplasmic phosphoproteins in ovine muscle with different color stability. *Food Chemistry, 240*, 104–111.

Li, Z., Li, M., Li, X., Xin, J. Z., Wang, Y., Shen, Q. W., et al. (2018b). Quantitative phosphoproteomic analysis among muscles of different color stability using tandem mass tag labeling. *Food Chemistry, 249*, 8–15.

Mancini, R. A., & Hunt, M. C. (2005). Current research in meat color. *Meat Science, 71*(1), 100–121.

Mckenna, D. R., Mies, P. D., Baird, B. E., Pfeiffer, K. D., Ellebracht, J. W., & Savell, J. W. (2005). Biochemical and physical factors affecting discoloration characteristics of 19 bovine muscles. *Meat Science, 70*(4), 665–682.

Olivera, D. F., Bambicha, R., Laporte, G., Cárdenas, F. C., & Mestorino, N. (2013). Kinetics of colour and texture changes of beef during storage. *Journal of Food Science and Technology, 50* (4), 821–825.

Sale, E. M., White, M. F., & Kahn, C. R. (1987). Phosphorylation of glycolytic and gluconeogenic enzymes by the insulin receptor kinase. *Journal of Cellular Biochemistry, 33*(1), 15–26.

Scheffler, T. L., & Gerrard, D. E. (2007). Mechanisms controlling pork quality development: The biochemistry controlling postmortem energy metabolism. *Meat Science, 77*(1), 7–16.

Schwägele, F., Haschke, C., Honikel, K. O., & Krauss, G. (1996). Enzymological investigations on the causes for the PSE-syndrome, I. Comparative studies on pyruvate kinase from PSE- and normal pig muscles. *Meat Science, 44*(1–2), 27–40.

Shen, Q. W., & Du, M. (2005). Role of AMP-activated protein kinase in the glycolysis of postmortem muscle. *Journal of the Science of Food and Agriculture, 85*(14), 2401–2406.

Shikama, K., & Sugawara, Y. (1978). Autoxidation of native oxymyoglobin. Kinetic analysis of the pH profile. *European Journal of Biochemistry, 91*(2), 407–413.

Sprang, S. R., Acharya, K. R., Goldsmith, E. J., Stuart, D. I., Varvill, K., Fletterick, R. J., et al. (1988). Structural changes in glycogen phosphorylase induced by phosphorylation. *Nature, 336* (6196), 215–221.

Wu, W., Gao, X. G., Dai, Y., Fu, Y., Li, X. M., & Dai, R. T. (2015). Post-mortem changes in sarcoplasmic proteome and its relationship to meat color traits in M. semitendinosus of Chinese Luxi yellow cattle. *Food Research International, 72*, 98–105.

Yin, H., Zhang, Z., Lan, X., Zhao, X., Wang, Y., & Zhu, Q. (2011). Association of MyF5, MyF6 and MyOG gene polymorphisms with carcass traits in Chinese meat type quality chicken populations. *Journal of Animal and Veternary Advances, 10*(10), 704–708.

Zhang, B., & Liu, J. Y. (2017). Serine phosphorylation of the cotton cytosolic pyruvate kinase GhPK6 decreases its stability and activity. *FEBS Open Bio, 7*(3), 358–366.

Zhu, K., Zhao, J., Lubman, D. M., Miller, F. R., & Barder, T. J. (2005). Protein pI shifts due to posttranslational modifications in the separation and characterization of proteins. *Analytical Chemistry, 77*(9), 2745–2755.

Chapter 4
Protein Phosphorylation Affects Meat Tenderness

Abstract Tenderness is one of the most important quality attributes especially for beef and lamb. As protein phosphorylation and dephosphorylation regulate glycolysis, muscle contraction, and turnover of proteins within living cells, it may contribute to the conversion of muscle to meat. The protein phosphorylation level (P/T ratio) of the tender group increased from 0.5 to 12 h postmortem and then decreased. The P/T ratio of tough group increased during 24 h postmortem, with higher increase rate from 0.5 to 4 h postmortem than from 4 to 24 h postmortem. The global phosphorylation level of tough meat was higher than tender meat at 4, 12, and 24 h postmortem. Protein identification revealed that most of the phosphoproteins were proteins with sarcomeric function; the others were involved in glycometabolism, stress response, etc. The phosphorylation levels of myofibrillar proteins, e.g. myosin light chain 2 and actin, were significantly different among groups of different tenderness and at different postmortem time points. Protein phosphorylation may influence meat rigor mortis through contractile machinery and glycolysis, which in turn affected meat tenderness.

Keywords Ovine muscle · Tenderness · Aging time · Sarcoplasmic proteins · Myofibrillar protein · Protein phosphorylation

4.1 Introduction

Tenderness is one of the most important quality indicators that affect consumers' choices when purchase meat and meat products. Usually, the most unacceptable thing to consumers is meat with low tenderness (Jeremiah, 1982). Myofibrillar proteins are the largest protein class in skeletal muscle, and account for 55–60% of total muscle protein by mass. Myofibrillar proteins are responsible for contractile, functional, and culinary properties of muscle and meat (Lee, Joo, & Ryu, 2010). Myofibrillar proteins that regulate muscle contraction could be involved in rigor mortis development and affect meat tenderness (Hopkins, Toohey, Lamb, Kerr, & Refshauge, 2011; Huang, Huang, Xu, & Zhou, 2011; Li et al., 2013). Post-translational modifications are key modulators of protein structure, function,

D. Zhang et al., *Protein Phosphorylation and Meat Quality*,
https://doi.org/10.1007/978-981-15-9441-0_4

signaling, and regulation. Protein phosphorylation, a common post-translational modification, plays a regulatory role in the contraction of skeletal muscle and myofibrillar protein degradation. Phosphorylated skeletal muscle proteins, e.g. myomesin and myosin regulatory light chain 2 (MYLC2), were reported to affect skeletal muscle contraction and protein interaction (Obermann, Gautel, Weber, & F Rst, 1997; Ryder, Lau, Kamm, & Stull, 2007; Zhi et al., 2005). Phosphorylation of some myofibrillar proteins has been proved to reduce the degradation of these proteins by calpains (Di Lisa et al., 1995; Zhang, Lawrence, & Stracher, 1988). Phosphorylation of calpastatin (endogenous calpain inhibitor) rendered it a more effective inhibitor of m-calpain than unphosphorylated calpastatin (Salamino et al., 1994). However, whether there are differences in protein phosphorylation levels and the change in protein phosphorylation levels with postmortem time between muscles with different tenderness levels is unknown.

Study of the dynamic change of protein phosphorylation postmortem in muscles with different tenderness levels can identify candidate regulatory proteins and help to understand the underlying mechanisms of meat quality development. In this study, ovine muscle samples were divided into two groups according to tenderness. A combination of sodium dodecyl Diamond-SYPRO Ruby staining and tandem mass spectrometric strategy was used to detect the phosphoproteins. Phosphorylation of myofibrillar proteins was comparatively studied between tender and tough muscles and among muscles at different postmortem time points.

Meat tenderness relies mainly on changes postmortem that affect the contractile system of the muscle or myofibrils. Warner–Bratzler shear force (WBSF) was one of the first instrumental methods developed for tenderness measurement, and has remained the most commonly used one globally (Silva et al., 2015). Another most extensively studied event taking place in muscles after slaughter is proteolysis, which leads to myofibrillar degradation (Taylor, Geesink, Thompson, Koohmaraie, & Goll, 1995). As the main aim of this study was to investigate the myofibrillar proteins postmortem, the myofibrillar fragmentation index (MFI) was measured as another indicator to determine tenderness.

The objective of this chapter was to investigate the relation between protein phosphorylation and meat tenderness. *Longissimus dorsi* muscles of 40 sheep (fat-tailed sheep × tail sheep) were collected at 0.5, 1, 4, 12, and 24 h postmortem, and were divided into two groups (high-level-of-tenderness and low-level-of-tenderness) based on the shear force measurement and MFI at 24 h postmortem. Sodium dodecyl sulfate-polyacrylamide gel electrophoresis (SDS-PAGE) was combined with fluorescence staining in order to analyze the level of phosphorylation of sarcoplasmic proteins and myofibrillar proteins.

4.2 Changes of Proteins Phosphorylation Levels in Postmortem Muscle with Different Tenderness

There were four samples in the tough group and tender group, respectively. Tender meat was characterized by significantly lower WBSF (7.57 vs. 13.06 kg, $P < 0.01$), higher MFI (99.55 vs. 73.40, $P < 0.01$), and longer sarcomere length (1.15 vs. 0.94 μm, $P < 0.01$) (Fig. 4.1) at 24 h postmortem when compared with tough meat. Muscle pH values at 4, 12, and 24 h postmortem were also recorded (Table 4.1).

4.2.1 Changes of Sarcoplasmic Proteins Phosphorylation Levels in Postmortem Muscle with Different Tenderness

4.2.1.1 SDS-PAGE of Phosphorylated and Total Sarcoplasmic Proteins of Muscle

The SDS-PAGE electrophoresis patterns of sarcoplasmic proteins in the tough and tender groups at 0.5 h, 1 h, 4 h, 12 h, and 24 h postmortem were shown in Figs. 4.2 and 4.3. Figure 4.2 was phosphorylated proteins stained with Pro-Q Diamond, and Fig. 4.3 was SYPRO Ruby-stained sarcoplasmic total proteins. It can be seen from the images that the bands were straight and clear, and the separation effect was good. Eighteen clearer bands were selected from the images and their relative optical densities were analyzed.

Phosphoproteins and the total proteins were detected using Pro-Q Diamond dye and SYPRO Ruby dye respectively, and the ratio of the fluorescence intensity of Pro-Q Diamond staining to the fluorescence intensity of SYPRO Ruby staining in each band was used as the protein phosphorylation level by semi-quantitative method. Because the protein phosphorylation level (P/T) ratios of the bands 1, 4,

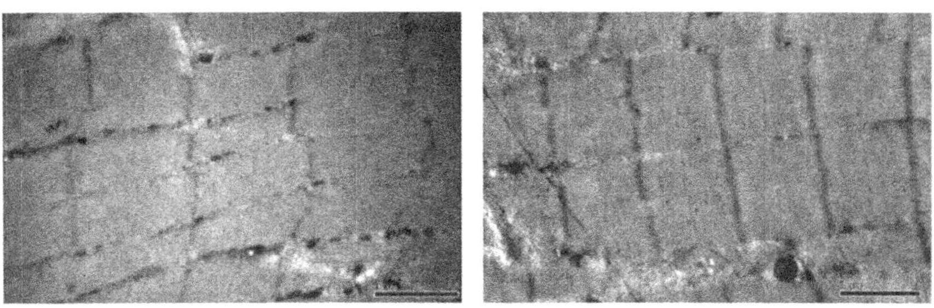

(A) (B)

Fig. 4.1 The ultrastructure of myofibril of longissimus muscles in groups of different tenderness (30000×): (**a**) ultrastructure of myofibril of tender group; (**b**) Ultrastructure of myofibril of tough group (Chen et al., 2016)

Table 4.1 Meat tenderness indicators (means ± SD) (Chen, Li, Ni, et al., 2016)

Group	WBSF (kg)	MFI (24 h)	Sarcomere length (μm)	pH (4 h)	pH (12 h)	pH (24 h)
Tender	7.57 ± 7.39^{b}	99.55 ± 4.31^{a}	1.15 ± 0.06^{a}	6.49 ± 0.18	5.91 ± 0.09	5.67 ± 0.06
Tough	13.06 ± 2.42^{a}	73.40 ± 6.20^{b}	0.94 ± 0.08^{b}	6.69 ± 0.22	6.03 ± 0.14	5.71 ± 0.05

Different letters (a, b) within a column indicate that mean values differ ($P < 0.01$)

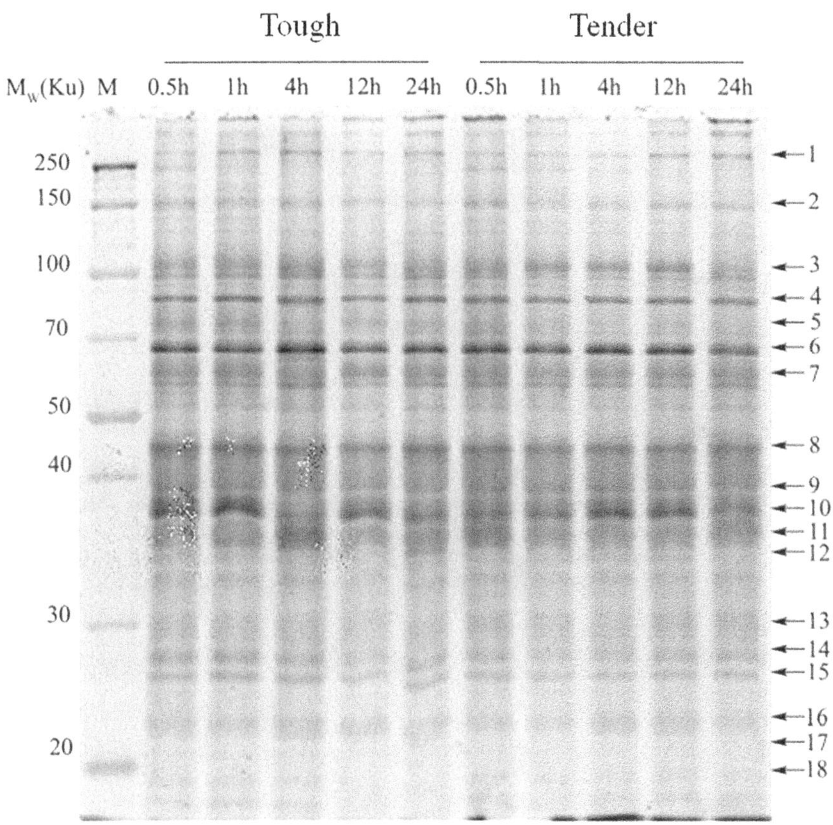

Fig. 4.2 SDS-PAGE image displaying the phosphorylated sarcoplasmic proteins present in mutton. Translated from Chen et al. (2015)

6, 7, 8, 10, 15, and 16 were higher than 1.0, which showed the proteins in these bands had higher phosphorylation levels. The bands 4, 6, and 10 that were intensively stained in the Pro-Q Diamond images were lighter when they were stained in SYPRO Ruby, while more intensive-stained bands 3, 7, and 9 in the SYPRO Ruby images were lighter in Pro-Q Diamond staining. These differences indicated that the Pro-Q Diamond staining was specific for phosphorylated proteins.

4.2.1.2 Comparison of Global Sarcoplasmic Proteins Phosphorylation Levels in Postmortem Muscle with Different Tenderness

It can be seen from Fig. 4.4 that the phosphorylation levels of the tough group and tender group all increased and then decreased. The highest protein phosphorylation level in tough group appeared at 4 h postmortem, and the highest protein phosphorylation level in tender group appeared at 12 h postmortem. The global

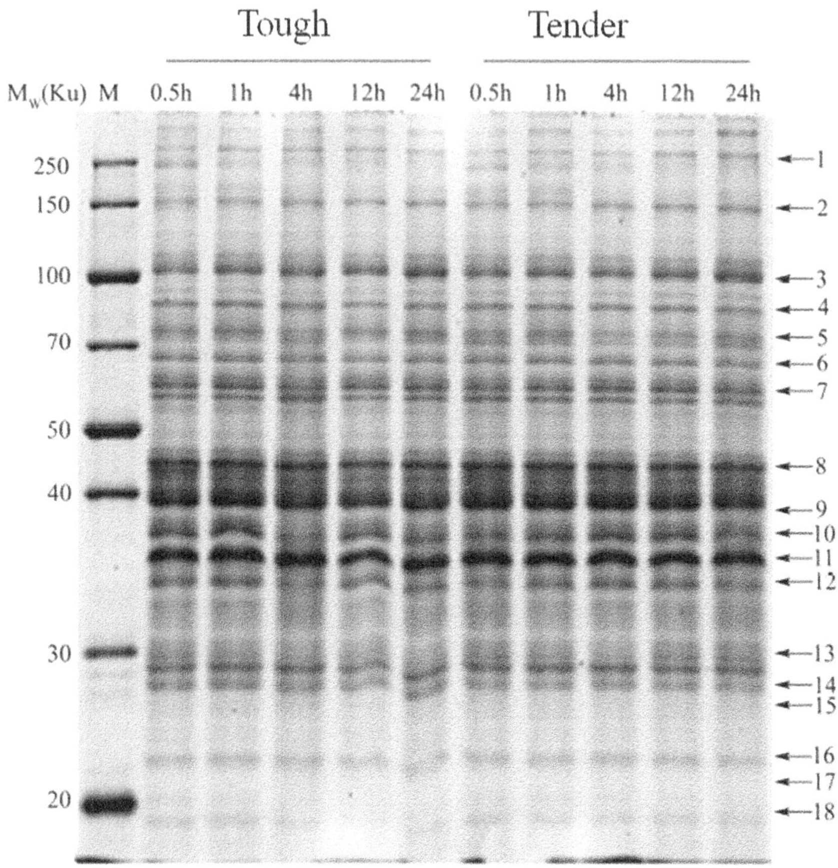

Fig. 4.3 SDS-PAGE image displaying the total sarcoplasmic protein content present in mutton. Translated from Chen et al. (2015)

phosphorylation level of tough group was significantly higher than that of tender group at 0.5 h, 1 h, and 4 h postmortem ($P < 0.05$). The global phosphorylation level of sarcoplasmic protein was significantly affected by the tenderness, postmortem maturity time, and the interaction between tenderness and postmortem maturity ($P < 0.05$).

The reason why the phosphorylation level of sarcoplasmic protein in tough group was higher than that in tender group can be analyzed from the following aspects: First, according to D'Alessandro and Zolla (2013) and Li et al. (2012), due to their perspective of "lower phosphorylation-induced enzyme activity," the higher the phosphorylation level, the lower the activity of most glycolytic enzymes. The results showed that the tough group had higher phosphorylation level, and tender group had lower phosphorylation level. Therefore, under the influence of phosphorylation, the glycolytic enzyme activity of the tender group was higher, the glycolysis rate was

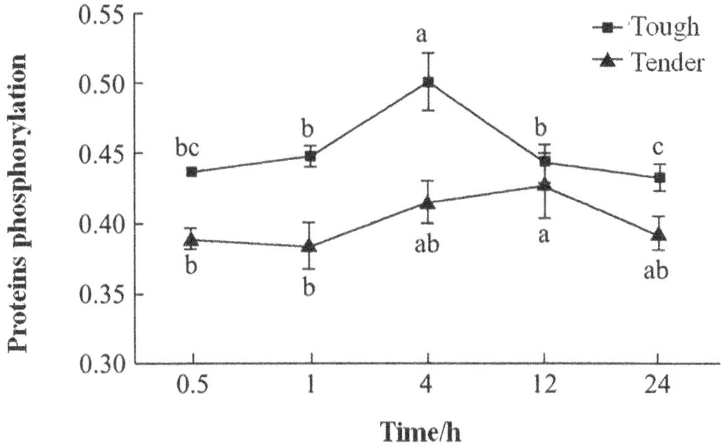

Fig. 4.4 Comparison of global sarcoplasmic proteins phosphorylation levels between two mutton groups with different degrees of tenderness. Letters a, b in the same row indicate a significant difference ($P < 0.05$). Translated from Chen et al. (2015)

faster, and the pH was lower, which was more conducive to the release and activation of cathepsins, thus promoting protein degradation, and was beneficial to meat tenderization (Xia, Li, Chen, Chen, & Zhang, 2014). Second, Doumit and Bates (2000) pointed out that calpastatin would phosphorylate, and calpastatin phosphorylation could increase the inhibition of calpain and reduce the degradation of myofibrillar protein by calpain which was conducive to the tenderization of muscles. Third, phosphorylation of heat shock protein 27 was a key step in protecting actin from disruption (Loktionova & Kabakov, 1998). In this experiment, the phosphorylation level of heat shock protein 27 was higher in tough group. The more heat shock protein 27 was phosphorylated, the more actin was protected and in a complete state, which was not conducive to muscle tenderization. Fourth, muscle contraction increased the phosphorylation level of phosphofructokinase (Luther & Lee, 1986), and the muscle contraction of tough group was severe, so the phosphorylation of phosphofructokinase was high.

4.2.1.3 Sarcoplasmic Protein Phosphorylation Related to Meat Tenderness and Postmortem Maturation

There were 18 bands detected and selected for image analysis of the total sarcoplasmic protein content (Table 4.2). The phosphorylation levels of bands 3, 6, and 18 were significantly different between different tenderness groups and postmortem maturity groups ($P < 0.05$). The phosphorylation levels of bands 1, 2, 5, 9, 14, and 15 were significantly influenced by the interactive effects of tenderness and postmortem time ($P < 0.05$), but the phosphorylation levels of bands 8, 10, and 16 were

Table 4.2 Comparison of the sarcoplasmic protein phosphorylation between two mutton groups with different degrees of tenderness. Translated from Chen et al. (2015)

	Tender					Tough					Effect level (*P* value)		
Band	0.5 h	1 h	4 h	12 h	24 h	0.5 h	1 h	4 h	12 h	24 h	Tenderness	Time	Tenderness × Time
1	0.86 ± 0.05	0.86 ± 0.12	1.00 ± 0.07	1.29 ± 0.15	1.02 ± 0.11	0.60 ± 0.06	1.22 ± 0.09	1.38 ± 0.02	0.97 ± 0.04	0.89 ± 0.05	0.951	0.008	0.014
2	0.37 ± 0.02	0.42 ± 0.03	0.39 ± 0.05	0.36 ± 0.04	0.32 ± 0.02	0.54 ± 0.04	1.10 ± 0.07	0.60 ± 0.06	0.28 ± 0.03	0.20 ± 0.01	0.720	<0.0001	0.001
3	0.40 ± 0.03	0.70 ± 0.08	0.51 ± 0.05	0.38 ± 0.03	0.36 ± 0.04	0.46 ± 0.02	0.60 ± 0.05	0.70 ± 0.05	0.39 ± 0.02	0.55 ± 0.03	0.001	<0.0001	0.003
4	0.92 ± 0.08	1.39 ± 0.14	1.45 ± 0.21	1.18 ± 0.04	1.09 ± 0.09	0.74 ± 0.09	0.77 ± 0.12	0.87 ± 0.07	0.84 ± 0.05	0.87 ± 0.04	0.069	0.109	0.234
5	0.40 ± 0.05	0.37 ± 0.02	0.27 ± 0.03	0.19 ± 0.02	0.21 ± 0.03	0.39 ± 0.02	0.35 ± 0.04	0.43 ± 0.03	0.32 ± 0.01	0.26 ± 0.01	0.630	0.001	0.002
6	2.09 ± 0.17	2.40 ± 0.07	3.08 ± 0.24	3.08 ± 0.22	2.19 ± 0.19	1.65 ± 0.08	2.46 ± 0.20	2.46 ± 0.18	2.14 ± 0.13	2.14 ± 0.05	0.011	0.017	0.020
7	0.53 ± 0.06	0.55 ± 0.03	0.58 ± 0.06	0.58 ± 0.05	0.52 ± 0.04	0.36 ± 0.04	0.39 ± 0.02	0.42 ± 0.03	0.39 ± 0.04	0.41 ± 0.03	0.001	0.291	0.235
8	0.56 ± 0.04	0.57 ± 0.06	0.60 ± 0.11	0.77 ± 0.05	0.63 ± 0.02	0.51 ± 0.07	0.66 ± 0.13	0.59 ± 0.04	0.76 ± 0.03	0.62 ± 0.07	0.891	0.010	0.775
9	0.26 ± 0.01	0.36 ± 0.02	0.52 ± 0.006	0.38 ± 0.06	0.27 ± 0.01	0.40 ± 0.05	0.32 ± 0.04	0.29 ± 0.02	0.35 ± 0.04	0.33 ± 0.01	0.562	0.349	0.020
10	0.94 ± 0.07	1.04 ± 0.09	1.06 ± 0.05	1.19 ± 0.15	1.13 ± 0.05	0.91 ± 0.05	1.02 ± 0.05	0.98 ± 0.06	1.16 ± 0.09	1.07 ± 0.07	0.152	0.002	0.687
11	0.38 ± 0.01	0.41 ± 0.03	0.42 ± 0.06	0.42 ± 0.05	0.34 ± 0.01	0.36 ± 0.02	0.40 ± 0.06	0.41 ± 0.05	0.38 ± 0.02	0.38 ± 0.04	0.643	0.096	0.531
12	0.25 ± 0.03	0.26 ± 0.02	0.25 ± 0.04	0.21 ± 0.03	0.12 ± 0.01	0.11 ± 0.02	0.10 ± 0.01	0.13 ± 0.02	0.16 ± 0.03	0.11 ± 0.01	0.799	0.058	0.060
13	0.24 ± 0.02	0.42 ± 0.04	0.50 ± 0.06	0.36 ± 0.03	0.20 ± 0.04	0.16 ± 0.01	0.59 ± 0.07	0.56 ± 0.03	0.34 ± 0.04	0.22 ± 0.01	0.774	0.118	0.937
14	0.43 ± 0.03	0.49 ± 0.05	0.70 ± 0.07	0.58 ± 0.03	0.44 ± 0.06	0.20 ± 0.02	0.29 ± 0.01	0.34 ± 0.04	0.28 ± 0.01	0.58 ± 0.06	0.703	0.002	0.002
15	3.65 ± 0.24	3.39 ± 0.31	3.30 ± 0.05	3.76 ± 0.07	5.33 ± 0.13	2.12 ± 0.11	1.95 ± 0.06	2.24 ± 0.12	2.59 ± 0.17	2.67 ± 0.13	0.471	<0.0001	0.001
16	0.59 ± 0.07	0.57 ± 0.05	0.57 ± 0.04	0.67 ± 0.03	0.77 ± 0.09	0.74 ± 0.04	0.70 ± 0.08	0.72 ± 0.06	0.85 ± 0.07	0.87 ± 0.03	0.725	0.018	0.493
17	0.50 ± 0.06	0.51 ± 0.07	0.76 ± 0.05	0.80 ± 0.09	0.66 ± 0.08	0.17 ± 0.01	0.29 ± 0.04	0.25 ± 0.01	0.42 ± 0.02	0.35 ± 0.02	0.677	0.272	0.754
18	0.24 ± 0.03	0.12 ± 0.01	0.19 ± 0.02	0.20 ± 0.03	0.14 ± 0.02	0.28 ± 0.03	0.21 ± 0.01	0.37 ± 0.05	0.38 ± 0.02	0.33 ± 0.04	0.001	0.002	0.032

Note: Mean ± SD

only affected by postmortem time ($P < 0.05$), and the phosphorylation levels of bands 7 were only affected by tenderness ($P < 0.05$).

The time of the highest phosphorylation was different in different protein bands. The phosphorylation level of band 6 and 14 was highest at 4 h postmortem, and the phosphorylation level of band 8, 10, and 17 was highest at 12 h postmortem. In addition, the protein phosphorylation level of the band 16 and 18 in the tender group was higher than that in the tough group, and the protein phosphorylation level of the band 4, 6, 7, 12, 14, 15, and 17 in the tough group was higher than that in the tender group. Thus, the global protein phosphorylation level of the tough group was higher, but not all proteins in the tough group had higher phosphorylation level than the tender group.

It has been proved that the main protein identified in band 3 (102 ku) might be glycogen phosphorylase, of which the molecular weight was about 97.4 ku (Huang, Larsen, Karlsson, Pomponio, & Lametsch, 2011). The molecular weight in this experiment was large, and the reason might be that the glycogen phosphorylase in the mutton was highly phosphorylated, causing a change in molecular weight (Huang, Larsen, Karlsson, Pomponio, & Lametsch, 2011). The same phenomenon occurred in the study of Huang, Martin, and Lametsch (2012). Glycogen degradation was carried out by the kinetics process of phosphorylation and dephosphorylation. Glycogen phosphorylase usually existed in the inactive dephosphorylation state b. Under the action of phosphorylase kinase and Ca^{2+}, the serine-14 site of glycogen phosphorylase was phosphorylated, which was converted from inactive b to active a, and catalyzed the degradation of glycogen to 1-phosphate glucose, and then to 6-phosphate glucose. Band 4 (83 ku) might be phosphofructokinase, which acted on fructose-6-phosphate and was one of the rate-limiting enzymes for glycolysis. Therefore, the regulation of this enzyme was a key step affecting glycolysis. On the one hand, the activation of AMP-dependent protein kinase (AMPK) could cause the phosphorylation of phosphofructokinase 2, resulting in the activity of phosphofructokinase 2 and the decreasing production of fructose 2,6-diphosphate, thereby inhibiting glycolysis and slowing down the drop rate of muscle pH; on the other hand, phosphorylated phosphofructokinase could alter the affinity for F-actin, which increased phosphorylation of phosphofructokinase and binding to F-actin (Luther & Lee, 1986). Band 7 (58 ku) might be pyruvate kinase, which was also one of the rate-limiting enzymes in the glycolytic pathway. Pyruvate kinase catalyzed the conversion of phosphoenolpyruvate to pyruvate, which can be converted to lactic acid under the catalysis of lactate dehydrogenase, while nicotinamide adenine dinucleotide (NADH) being oxidized to NAD+. However, studies have shown that AMPK inhibits the activity of L-pyruvate kinase, thereby inhibiting glycolysis. Band 8 (48 ku) might be β-enolase, the rate-limiting enzyme in the glycolysis process, which catalyzed the conversion of 2-phosphoglycerate to phospho-enolpyruvate during glycolysis. It can also catalyze the reverse reaction in the process of glycogen synthesis, that is, as a phosphopyruvate hydratase, the phospho-enolpyruvate was converted to 2-phosphoglycerate (Bolten et al., 2008). The band 14 (27 ku) might be heat shock protein 27, which was increased in expression and phosphorylated to make it active, and functioned to protect actin and the like. Most of the above enzymes were related to glycolysis, and most of them in the muscle existed in the sarcoplasm, and the glycolytic enzymes accounted for two-thirds of the sarcoplasmic proteins. The rate of glycolysis was determined by the activity of the glycolytic enzyme. Studies have shown that the increase in glycolytic enzyme activity during the early postmortem period had an important effect on the muscle maturation process (Shen, Underwood, Means, Mccormick, & Du, 2007). Whether in vivo or in vitro, most glycolytic enzymes could be phosphorylated by different kinases, and phosphorylation affected enzyme activity or stability. By regulating the activity of enzymes, protein phosphorylation could regulate the postmortem muscle glycolysis reaction, which in turn affected the postmortem rigor process of the muscle and played a decisive role of the meat tenderness.

4.2.2 Changes of Myofibrillar Proteins Phosphorylation Levels in Postmortem Muscle with Different Tenderness

4.2.2.1 Comparison of Global Myofibrillar Proteins Phosphorylation Levels in Postmortem Muscle with Different Tenderness

The SDS-PAGE electrophoresis patterns of myofibrillar proteins in the tough and the tender groups at 0.5, 1, 4, 12, and 24 h postmortem were shown in Fig. 4.5. It could be seen from the figures that the strips were straight and clear, and the separation effect was good. Since the protein phosphorylation level (P/T) of band 5, 7, 10, 12, 13, 14, 15, 16, and 21 were higher than 1.0, the proteins in these bands were considered to be highly phosphorylated. The global protein phosphorylation level in tender group increased slowly from 0.5 h to 12 h postmortem and then gradually decreased from 12 h to 24 h. However, the global protein phosphorylation level in tough group increased rapidly from 0.5 h to 4 h and then increased slowly from 4 h to 24 h (Fig. 4.3). Among them, the phosphorylation level of protein in tough group was significantly higher than that in the tender group at 4 h, 12 h, and 24 h postmortem ($P < 0.05$) (Figs. 4.5 and 4.6).

4.2.2.2 Identification of Phosphorylated Myofibrillar Proteins Related to Meat Tenderness

There were 21 clearer protein bands excised from gels, then analyzed their relative optical density and examined them with mass spectrometry. The functional classifications of the proteins were assigned based on the information from the protein database (UniProtKB), and 34 proteins were finally identified, most of which were related to sarcomeric and structural function (Table 4.3). There were several sarcoplasmic proteins in the final experimental results, which may be caused by two reasons: on the one hand, some sarcoplasmic proteins were contained in the precipitated protein pallet during centrifugation due to solubility. On the other hand, some

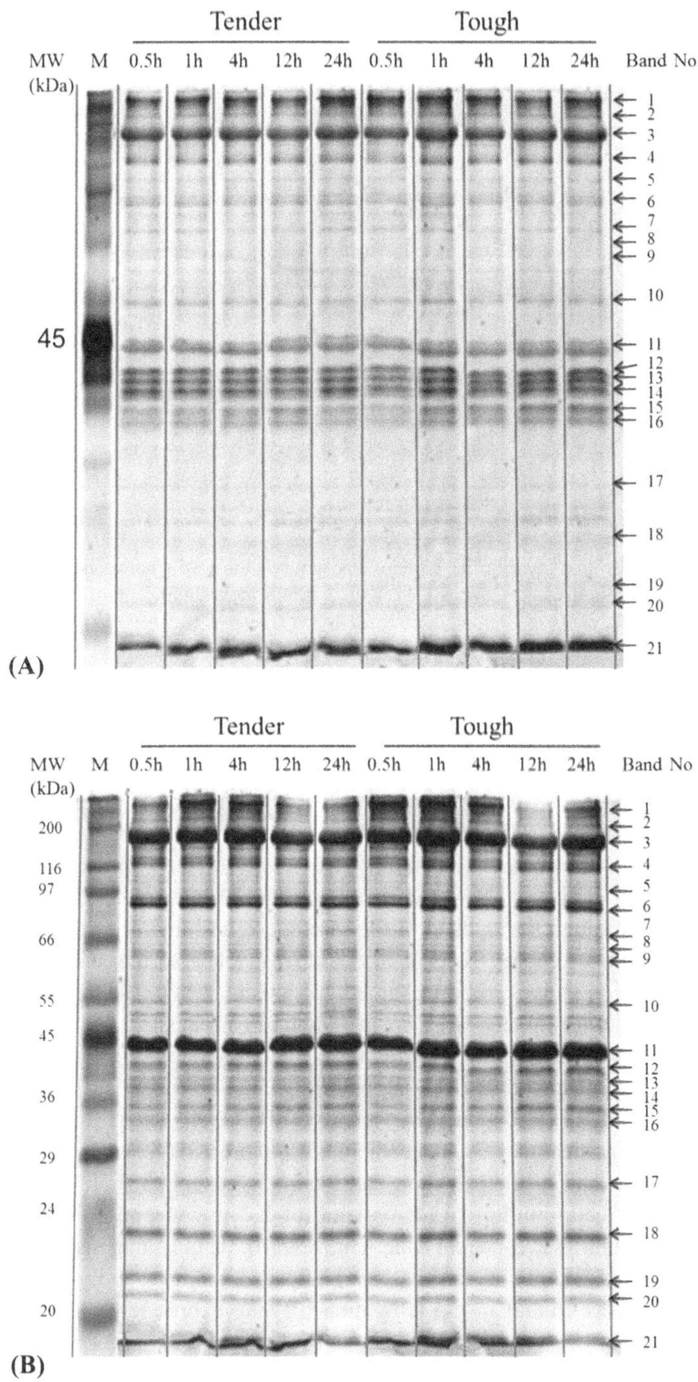

Fig. 4.5 Gel images of phosphoproteins and total proteins separated by 1-DE. (**a**) Image of phosphoproteins stained with Pro-Q Diamond. (**b**) Image of total proteins stained with SYPRO Ruby (Chen, Li, Ni, et al., 2016)

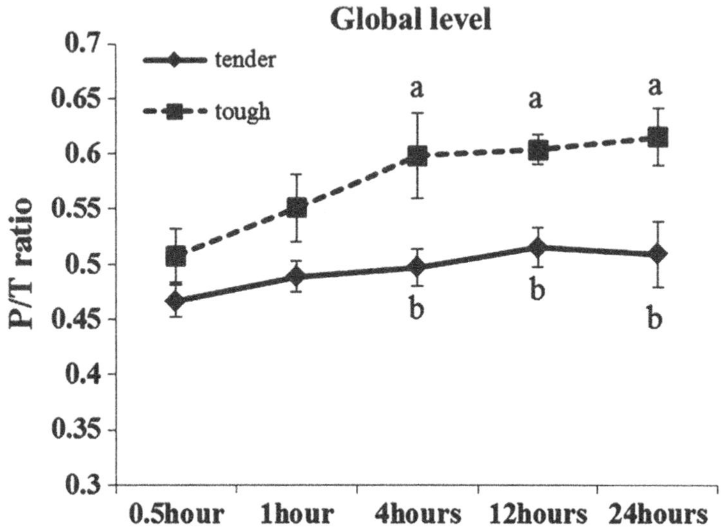

Fig. 4.6 Comparison of myofibrillar proteins phosphorylation levels between tender and tough groups. Letters a, b in the same row at 4, 12, and 24 h indicate a significant difference ($P < 0.05$) (Chen, Li, Ni, et al., 2016)

denatured sarcoplasmic proteins were combined with myofibrillar proteins to form complexes. Some glycolytic enzymes, especially aldolase, glyceraldehyde-3-phosphate dehydrogenase, and phosphofructokinase, had significant binding ability with myofibrillar, such as actin, troponin, and tropomyosin (Arnold, Henning, & Pette, 2005; Clarke & Masters, 1975; Dagher & Hultin, 1975; Melnick & Hultin, 1973).

The main proteins in band 1 were uncharacterized protein (gene name: TTN) and titin-like protein (TTN). TTN was derived from the M-line and extended along the myosin fibers, through the A-band of the sarcomere, and finally to the Z-line. TTN was highly elastic, keeping myosin fibers at the center of the sarcomere during muscle contraction and relaxation (Ahmed & Lindsey, 2009). Because of the direct and indirect relationship between TTN and signaling molecules, and multiple phosphorylation sites in the Z-line, M-band, and I bands segments, TTN was considered to be a major regulator of myocyte signaling (Linke, 2008). The proteins in band 2 were uncharacterized protein (gene name: NEB), uncharacterized protein (gene name: FLNC), and histone H3. The uncharacterized protein (gene name: NEB) was a protein related to the Z-line function, and the uncharacterized protein (gene name: FLNC) was an actin-binding protein (UniProtKB in www.uniprot.org). Certain residues of histone H3, especially the Thr3, Ser10, Thr11, and Ser28 sites, were highly phosphorylated in the interphase and mitosis of animals, and phosphorylation of histone H3 was crucial in the regulation of gene expression and chromosome condensation or isolation (Cerutti & Casas-Mollano, 2009). The proteins in band 3 and 21 were uncharacterized protein (gene name: MYH1), myosin heavy chain (MHC), myosin light chain 2 (MYLC2). Studies showed that MYLC2

Table 4.3 Unique proteins identified from 1-DE gel and effect analysis (Chen, Li, Ni, et al., 2016)

Protein name[a]	Band no.[b]	Accession no.	Organism[c]	MW (Da)[d]	Score[e]	MP[f]	Seq. cov.[g]	Function	Effect level[h]		
									Time	Tenderness	Time × Tenderness
Uncharacterized protein (GNi: TTN)	1	W5Q754	OS	3,937,785	129,795	4532	63%	Sarcomeric function	< 0.001	0.370	0.003
Titin-like protein (fragment) (TTN)	1	P79395	OS	7620	243	10	61%	Sarcomeric function	< 0.001	0.370	0.003
Uncharacterized protein (GN: NEB)	2	W5PFV9	OS	804,157	9886	404	49%	Sarcomeric function	< 0.001	< 0.001	< 0.001
Uncharacterized protein (fragment) (GN: FLNC)	2	W5NZK9	OS	278,503	8908	293	66%	Sarcomeric function	< 0.001	< 0.001	< 0.001
Histone H3 (fragment)	2	B0LRN3	OS	14,760	166	4	43%	DNA binding	< 0.001	< 0.0001	< 0.001
Uncharacterized protein (GN: MYH1)	3	W5PPG6	OS	223,600	79,339	2184	75%	Sarcomeric function	0.018	0.085	0.001
Myosin heavy chain 2 × (fragment) (MYHC2x)	3	Q9BE43	OS	11,397	2272	94	55%	Sarcomeric function	0.018	0.085	0.001
Myosin heavy chain 2a (fragment) (MYHC2a)	3	Q9BE44	OS	11,439	1921	90	55%	Sarcomeric function	0.018	0.085	0.001
Myosin heavy chain slow (fragment) (MYHCs)	3	Q9BE42	OS	11,544	316	9	43%	Sarcomeric function	0.018	0.085	0.001
Uncharacterized protein (GN: MYBPC2)	4	W5PX04	OS	128,132	6791	253	73%	Sarcomeric function	< 0.001	0.437	0.001
Glycogen phosphorylase, muscle form (GP)	5	O18751	OS	97,702	1169	42	75%	Glycogen metabolism	< 0.001	< 0.001	< 0.001

(continued)

Table 4.3 (continued)

Protein name[a]	Band no.[b]	Accession no.	Organism[c]	MW (Da)[d]	Score[e]	MP[f]	Seq. cov.[g]	Function	Effect level[h]		
									Time	Tenderness	Time × Tenderness
Glycogen phosphorylase, brain form (GP)	5	Q5MIB6	OS	96,881	237	8	14%	Glycogen metabolism	< 0.001	< 0.001	< 0.001
6-Phosphofructokinase (PFK)	6	W5QDD4	OS	94,291	556	25	36%	Glycolysis	0.003	0.025	0.007
Heat shock protein α (HSPα)	7	A8DR93	OS	85,077	240	8	14%	Stress response	< 0.001	0.907	0.117
Heat shock protein 70 (HSP70)	8	H2DGR2	OS	70,536	225	8	7%	Stress response blood	< 0.001	0.008	< 0.001
Serum albumin (ALB)	8	P14639	OS	71,139	2131	79	51%	Pressure regulation	< 0.001	0.008	< 0.001
Uncharacterized protein (fragment) (GN: PGM1)	9	W5PJB6	OS	65,544	2307	97	78%	Glucose metabolism	< 0.001	0.070	0.001
Glutamate dehydrogenase (fragment) (GDH)	9	Q8MIP8	OS	16,893	124	2	8%	Glycogen metabolism	< 0.001	0.070	0.001
Enolase (fragment) (ENO)	10	Q9N113	OS	11,952	490	19	79%	Glycolysis	0.001	< 0.001	0.001
Beta-actin (fragment) (ACTB)	11	Q9MZW1	OS	25,476	4538	159	85%	Sarcomeric function	0.019	0.318	0.007
Actin (fragment) (ACT)	11	C7F7J7	OS	38,516	1907	111	48%	Sarcomeric function	0.019	0.318	0.007
β-Actin variant 2 (ACTB2)	12	D7RIF5	OS	42,052	592	23	17%	Sarcomeric function	0.024	0.005	0.003
Phosphoglycerate kinase (PGK)	13	B7TJ13	OS	44,936	1461	46	74%	Glycolysis	< 0.001	0.939	< 0.001
Lactate dehydrogenase A (fragment) (LDHA)	13	Q9N109	OS	13,391	871	29	80%	Glycolysis	< 0.001	0.939	< 0.001

Aldolase A (fragment) (ALDOA)	14	Q9N116	OS	15,432	1960	66	67%	Glycolysis	0.341	0.521	0.001
Glyceraldehyde-3-phosphate dehydrogenase (GAPDH)	15	D7R7V6	OS	36,110	490	17	54%	Glycolysis	0.128	0.150	0.306
TPM1	16	B2LU28	OS	32,732	1438	45	45%	Sarcomeric function	0.071	0.654	0.224
SLC25A4	17	B2MVW8	OS	33,198	952	44	41%	Transmembrane transport	0.473	0.733	0.353
SLC25A5	17	B2MVW9	OS	33,162	761	29	28%	Transmembrane transport	0.473	0.733	0.353
SLC25A6	17	B2MVX0	OS	33,065	725	28	24%	Transmembrane transport	0.473	0.733	0.353
Adenylate kinase isoenzyme 1 (AK1)	18	C5IJA8	OS	21,750	331	13	47%	Glycogen metabolism	< 0.001	0.076	0.001
Putative tropomyosin (TPM)	19	Q3BJY7	OS	32,894	43	2	3%	Sarcomeric function	0.001	0.004	0.002
Troponin C (TNC)	20	A8WEG2	OS	18,141	1064	27	73%	Sarcomeric function	0.047	0.150	0.040
Fast skeletal myosin light chain 2 (MYLC2)	21	B9VGZ8	OS	19,204	2634	78	74%	Sarcomeric function	< 0.001	0.007	0.003

[a] Protein names and accession numbers were derived from the UniProt database
[b] The number of bands containing the identified proteins
[c] OS, *Ovis aries*
[d] Theoretical molecular weight (recorded in UniProt database)
[e] For the proteins identified in more than one band, the highest score was presented
[f] Number of matched peptides
[g] Percentage of coverage of the entire amino acid sequence
[h] Significant levels of different effects on the protein phosphorylation level of individual bands
[i] GN, gene name

phosphorylated during rigor formation and the ratio of phosphorylated MYLC2 to unphosphorylated MYLC2 increased at 24 h postmortem, suggesting that MYLC2 phosphorylation might be involved in rigor mortis processes (Huang et al., 2012; Muroya et al., 2007; Sawdy, Kaiser, St-Pierre, & Wick, 2004). The protein in band 4 was the uncharacterized protein (gene name: MYBPC2).

The phosphorylation of binding protein C disrupted its interaction with myosin S2, and the distance between myosin S1 and actin was shortened. This shortening of the distance might increase the possibility of cross-bridge formation, and accelerate contraction force formation (Taylor et al., 1995). The main proteins in band 5, 9, and 18 were uncharacterized protein (gene name: PGM1), glycogen phosphorylase (GP, muscle type and brain type), glutamate dehydrogenase (GDH), and adenylate kinase isoenzyme 1 (AK1) respectively, which were all involved in energy metabolism. For example, GPb could be phosphorylated by phosphorylase kinase at the serine 14 site, and its structure was changed and converted to an active state after phosphorylation (Johnson, 1992; Sprang et al., 1988). The proteins in band 6, 10, 13, 14, and 15 were 6-phosphate fructokinase (PFK), enolase (ENO), phosphoglycerate kinase (PGK), lactate dehydrogenase A (LDHA), aldolase A (ALDOA), and glyceraldehyde-3-phosphate dehydrogenase (GAPDH), respectively. These enzymes were all glycolytic enzymes, of which PFK was the rate-limiting enzyme of glycolysis, which could form a complex with actin after phosphorylated. And it was important to provide energy to cellular components (Cai, Callaci, Luther, & Lee, 1997; Kuo, Malencik, Liou, & Anderson, 1986; Luther & Lee, 1986). The proteins in band 7 and 8 were heat shock protein alpha (HSPα) and heat shock protein 70 (HSP70). They were important proteins that protected against the stress conditions, and heat shock proteins 70 in certain tissues were prone to phosphorylation, such as liver, extensor digitorum longus, soleus muscles, and myocardium (González & Manso, 2004; Melling, Thorp, & Noble, 2009). The proteins in band 11 and 12 were β-actin (ACTB), actin and β-actin variant 2 (ACTB2). Studies have found significant differences in actin phosphorylation levels in muscles of young rabbits and aging rabbits, suggesting that aging-induced changes in phosphorylated proteins might affect muscle contraction (Gannon, Staunton, O'Connell, Doran, & Ohlendieck, 2008). The proteins in band 16 and 19 were the tropomyosin α-1 chain (TPM1) and putative tropomyosin (TPM). TPM was one of the most phosphorylated proteins that comprised the thin filaments of muscle. The interactions of adjacent TPM molecules were more strengthened when phosphorylation occurred, so phosphorylation was required for long-range collaborative activation of the thin filaments of muscle (Rao, Marongelli, & Guilford, 2009). The proteins in band 17 were the solute carrier family 25 protein (SLC25), including the solute carrier family 25A4 protein (SLC25A4), the solute carrier family 25 A5 protein (SLC25A5), and the solute carrier family 25 A6 protein (SLC25A6). They were transporters on the mitochondrial membrane and composed of proteins that can transport a variety of molecular functions (Pebay-Peyroula et al., 2003). The proteins in band 20 were troponin C (TNC), which can prevent phosphorylase b kinase from catalyzing the phosphorylation of troponin I (Cole & Perry, 1975; Perry & Cole, 1974).

According to the functions of identified proteins, these proteins can be divided into three groups: (1) proteins related to sarcomere function, for example TTN, MYHC, β-ATBC, TPM1, YNC, MYLC2. (2) Proteins related to glycolysis or glycogen metabolism, e.g. GP, ENO, PGK, LDHA, ALDOA. (3) Proteins with other functions, for example, stress-related HSPα and HSP70, and SLC25 proteins associated with transmembrane transport.

4.2.2.3 Protein Phosphorylation Related to Meat Tenderness and Postmortem Maturation

The bands in Fig. 4.7 had significant differences in phosphorylation levels which were affected by tenderness and postmortem time. The P/T ratios of band 5, 7, 10, 12, 13, 14, 15, 16, and 21 were greater than 1, so the proteins in these bands were highly phosphorylated. The main proteins identified in these bands were GP, ENO, β-ATBC2, GPK, LDHA, ALDOA, GAPDH, TPM1, and MYLC2, respectively.

The P/T ratios of band 2, 5, 6, 8, 10, 12, 14, 19, and 21 were affected by different tenderness and postmortem time ($P < 0.05$, Table 4.3). The phosphorylated proteins in these bands may have a great impact on the tenderness of the meat. The P/T ratios of the band 5, 6, 10, 19, and 21 showed similar changing trends between the two groups, and the ratios of one group was higher than that of the other group (Fig. 4.4). The protein phosphorylation level of band 2, 8, 12, and 14 initially showed different patterns of variation between the two groups and eventually tended to be the same (Fig. 4.7). The P/T ratios of band 1, 3, 4, 11, 13, and 20 were influenced by the interaction of tenderness and postmortem time ($P < 0.05$, Table 4.3), while the ratios of band 7, 9, and 18 were only affected by postmortem time ($P < 0.05$, Table 4.3). The trend of protein phosphorylation level over time in the above 18 bands was shown in Fig. 4.7. The protein phosphorylation level of these bands had significant differences with postmortem time, but most of these bands had similar changing trends between the two groups with different tenderness. Thus, postmortem time should be the main factor affecting the protein phosphorylation level.

The protein phosphorylation levels of band 2, 5, 8, 10, 12, 16, and 21 were significantly different between the tender and tough groups ($P < 0.05$). The following aspects could explain the relationship between proteins phosphorylation and muscle tenderness:

First, many studies have demonstrated that MYLC2 phosphorylation was associated with muscle contraction. Because myosin light chain could be phosphorylated under the action of Ca^{2+}/calmodulin-dependent myosin light chain kinase, the repeated low-frequency stimulation can cause increasing contractility of fast-twitch muscle (Ryder et al., 2007). The degree of stimulation was positively correlated with the phosphorylation level of MYLC2 (Klug, Botterman, & Stull, 1982). On the other hand, MYLC2 phosphorylation was a biochemical memory factor that enhanced muscle performance (Sweeney, Bowman, & Stull, 1993). Therefore, the higher phosphorylation level of MYLC2 in the tough group may be one of the most important reasons for the poor tenderness group of muscles than the other group.

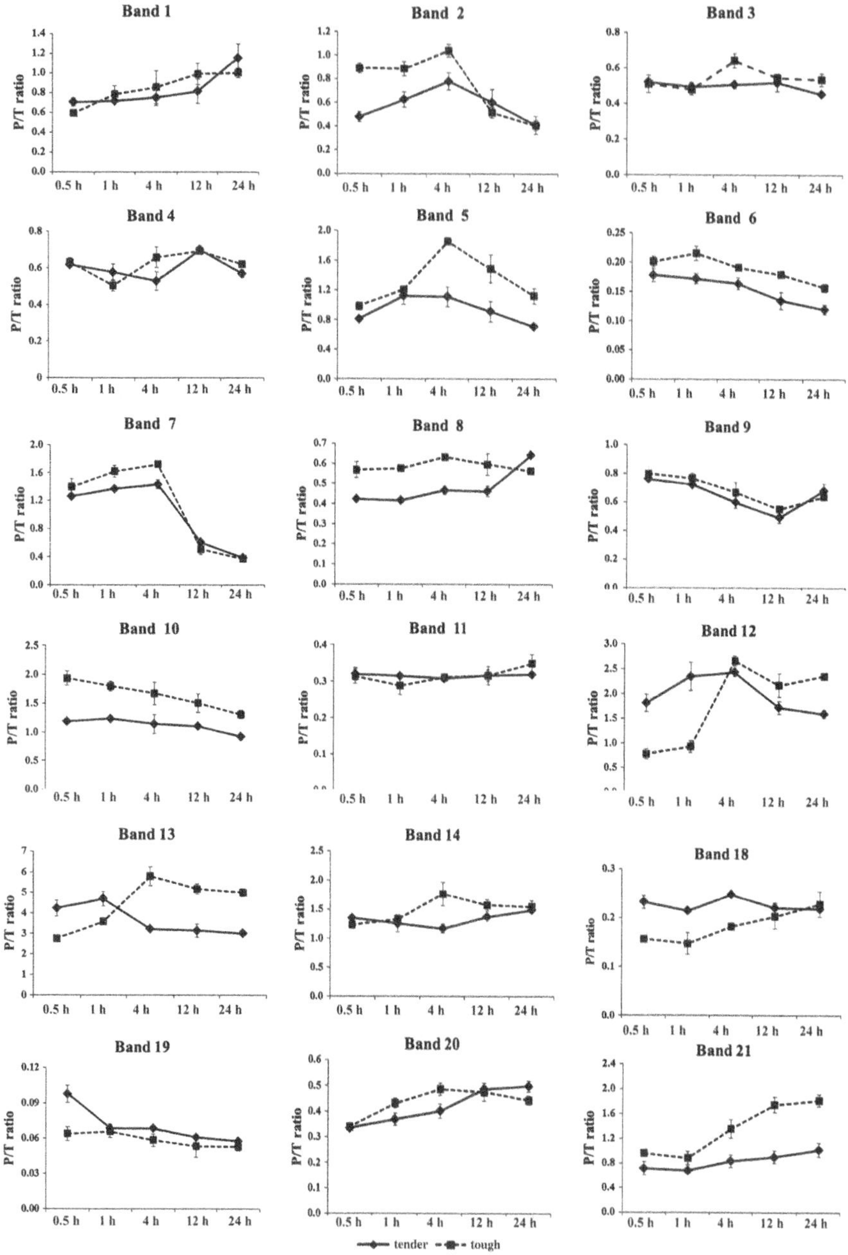

Fig. 4.7 Change patterns of protein phosphorylation level in 18 individual bands (Chen, Li, Ni, et al., 2016)

Second, protein phosphorylation played an important regulatory role in glycolysis in postmortem muscles. The phosphorylation levels of GP, PFK, ENO, ALDOA, PGK, and LDHA in the tough group were significantly higher than those in the tender group ($P < 0.05$, Table 4.3). The high phosphorylation levels of glycolytic enzymes caused their decreased enzyme activities (D'Alessandro et al., 2012; Li et al., 2012). Thus, the glycolytic enzyme activity of the tender group was higher than that of tough group, which further affected the rate of glycolysis and the tenderization of muscle (Hopkins, Toohey, Lamb, & Refshauge, 2009).

Third, the P/T ratios of PFK in the tough group were higher than that in the tender group (Table 4.3). Studies have shown that phosphorylated PFK had a stronger affinity PFK than the unphosphorylated form. Once a complex was formed, the enzyme activity was increased as well as the energy production (Luther & Lee, 1986). Stimulation of muscle contraction increased the phosphorylation level of PFK, which promoted its ability to bind to F-actin (Clarke, Shaw, & Morton, 1980), which may be the reason that the PFK P/T ratios in the tough group were higher than the tender group.

Fourth, the interaction between adjacent TPM molecules was enhanced after the phosphorylation of TPM. Phosphorylation was a requirement for remote collaborative activation of the thin filament (Rao et al., 2009). The TPM phosphorylation level in the tough group was higher than that in the tender group, so the long-range cooperative activation of the muscle filaments in tough group was much stronger. The higher levels of TPM phosphorylation in the tough group may be an important factor of this group with poorer tenderness than the other group.

4.2.3 Functional Verification of Protein Phosphorylation Regulation of Myofibrillar Protein

4.2.3.1 Effects of Different Activators on Protein Kinase Activity

After hatching for 0–24 h, the activity of the kinase in the control group decreased, while the 2-methoxy-1-methylethylacetate (PMA) group and Forskolin group increased at the beginning and then decreased (Fig. 4.8). After incubation for 1 h, 2 h, and 4 h, the protein kinase activity of PMA group was significantly higher than the control group ($P < 0.05$). After 1 h and 4 h, the protein kinase activity of Forskolin group was significantly higher than control group ($P < 0.05$). The highest protein kinase activity of the PMA group and Forskolin group appeared at 1 h, indicating that protein kinase C activator PMA and Forskolin were involved with activating kinase activity, leading to the increase of kinase activity. However, the enzyme activity decreased gradually at 2–24 h, which may be due to the fact that with the prolongation of postmortem time, the activity of kinase itself decreased more than the activation of kinase activator, and the activator was consumed gradually, and the activation ability decreased gradually.

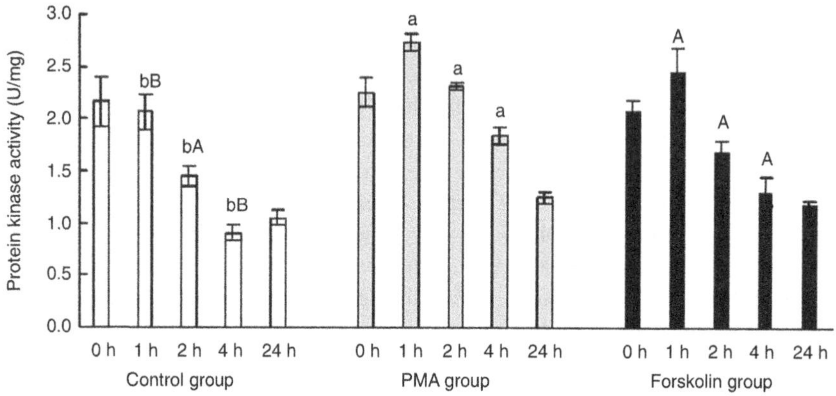

Fig. 4.8 The protein kinases' activities of different activators' groups. Letters a, b represent the significant difference between PMA group and control group ($P < 0.05$), Letters A, B represent the significant difference between Forskolin group and control group ($P < 0.05$). The same as below. Translated from Chen et al. (2015)

4.2.3.2 Effect of Different Activators on the Phosphorylation Level of Myofibrillar Protein

The gray value of the electrophoretic graph in Fig. 4.9 was analyzed, and the overall phosphorylation level of different activator treatment groups was shown in Fig. 4.10. The whole phosphorylation level of PMA group (protein kinase C (PKC) activated group) was significantly higher than the control group at 2 h and 4 h ($P < 0.05$), Forskolin group (protein kinase A (PKA) activated group) was significantly higher than control at 2 h and 24 h ($P < 0.05$). It is suggested that PMA played an important role after incubation for 2 h and 4 h, while Forskolin played a role at 4 h and 24 h, the time of action of these two activators was different. The overall phosphorylation level of the control group slowly increased within 2 h and decreased gradually at 2–24 h, the PMA group increased 0–2 h and then decreased gradually. In Forskolin group, it increased at 0–1 h and then gradually decreased, but the rate of decline was lower than the PMA group.

PKA and PKC can promote the phosphorylation level of some proteins (Fig. 4.11). It was found that PMA promoted the phosphorylation level of band 1 at 2 h, but there was no significant difference between the treatment group and the control group at other time points ($P > 0.05$). Forskolin promoted the phosphorylation of band 1 at 1 h, 2 h, and 4 h ($P < 0.05$). PMA promoted phosphorylation of band 3 at 2 h and 4 h, and of band 5 at 2 h and 24 h, and Forskolin promoted the band 5 at 4 h and 24 h. PMA significantly increased the phosphorylation level of band 7 ($P < 0.05$). PMA increased the phosphorylation level of band 8 at 2 h, and Forskolin increased the phosphorylation level of band 8 at 1, 4 h, and 24 h. A new

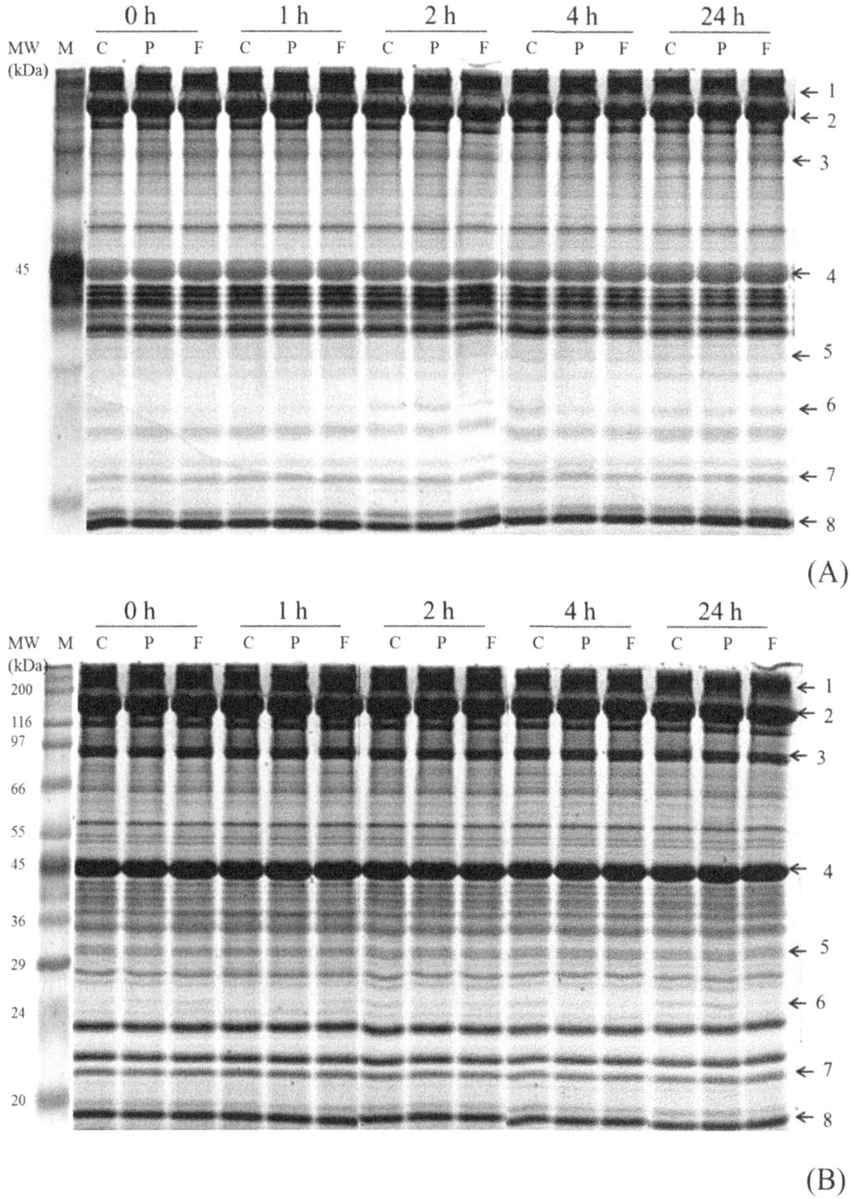

Fig. 4.9 Gel images of phosphoproteins and total proteins of different activators group separated by SDS-PAGE. A: Image of phosphoproteins stained with Pro-Q Diamond, B: Image of total proteins stained with SYPRO Ruby, C: Control group, P: PMA group, F: Forskolin group. Translated from Chen et al. (2016)

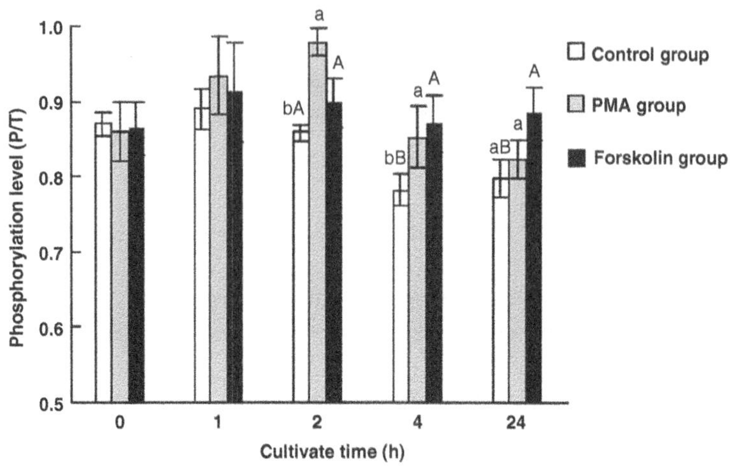

Fig. 4.10 The global protein phosphorylation level of different activators' groups. Translated from Chen, Li, Li, et al. (2016)

protein band appeared at band 6 (25 kDa, Fig. 4.9a) in three different treatment groups after incubation for 2 h. It was speculated that it may be a small protein produced by degradation, and Forskolin significantly increased the protein phosphorylation level at 4 h and 24 h incubation, and PMA significantly increased the phosphorylation level of the protein at 24 h (Fig. 4.11).

4.2.3.3 Effect of Different Activators on Protein Degradation

From Fig. 4.9b, it can be observed that a new band appears at the molecular weight of 25 kDa in the control group at 2 h, as well as the control and PMA group at 24 h.

The MFI of three different treatment groups did not significantly change within 0 h–24 h incubation, and there was no significant difference between the three treatment groups (Fig. 4.12, $P > 0.05$).

4.2.3.4 Effects of Different Activators on Muscle Contraction

The results of transmission electron microscopy (TEM) were shown in Fig. 4.13. The results of the sarcomere length were shown in Table 4.4. With the prolongation of incubation time, the sarcomere length of each treatment group showed a trend of slight increase firstly and then sharply decreased, there was no significant difference in the length of sarcomere between 0–4 h in the control group, and there were significant differences at 24 h, 2 h, and 4 h ($P < 0.05$). In the PMA group, there was no significant difference in the length of sarcomere between 0–4 h, and a significant difference at 24 h and 2 h ($P < 0.05$). There was no significant difference in the

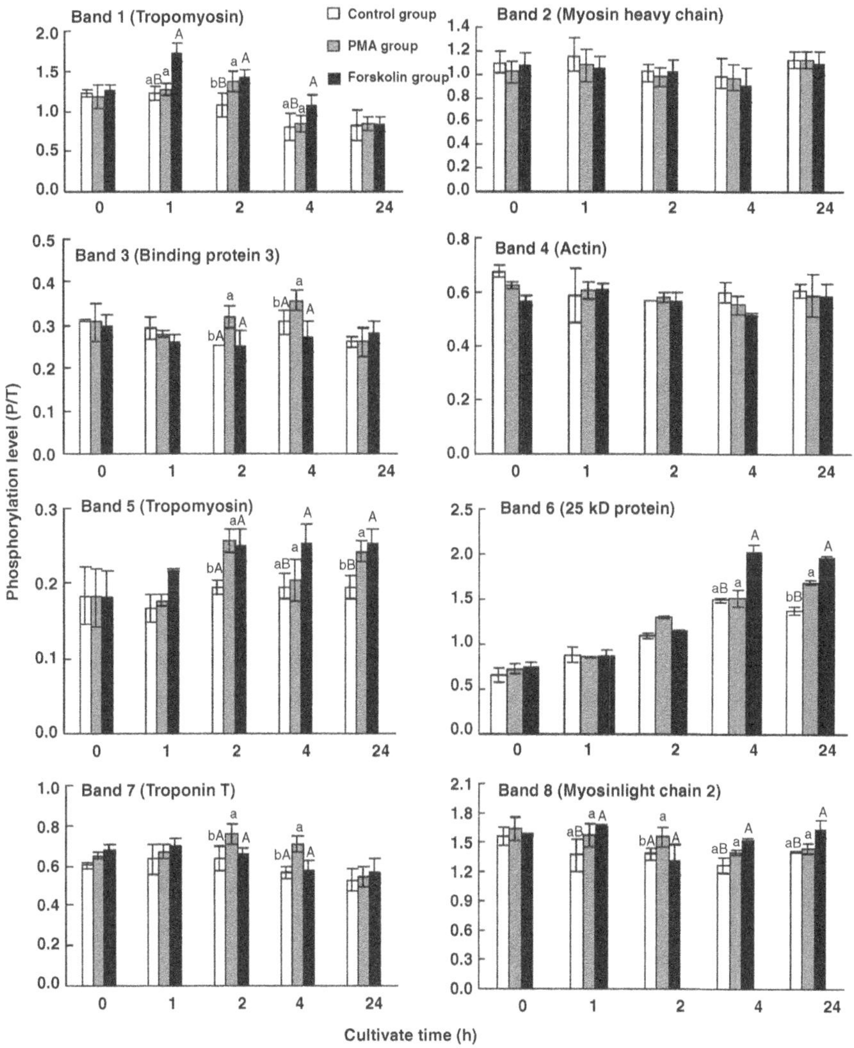

Fig. 4.11 The protein phosphorylation level of some important myofibrillar proteins in different activators' groups. Translated from Chen, Li, Li, et al. (2016)

length of the sarcomere in Forskolin group within 0–2 h, but a significant difference between the length of sarcomere at 4 h, 24 h, and 0–2 h ($P < 0.05$). According to the comparison of different treatment groups, there was no significant difference in the length of sarcomere between different treatment groups within 0–2 h of incubation. At 4 h of incubation, the length of the sarcomere in the activator treatment group was significantly shorter, and lasted until 24 h, indicating that the activator could result in the shortening of sarcomere length.

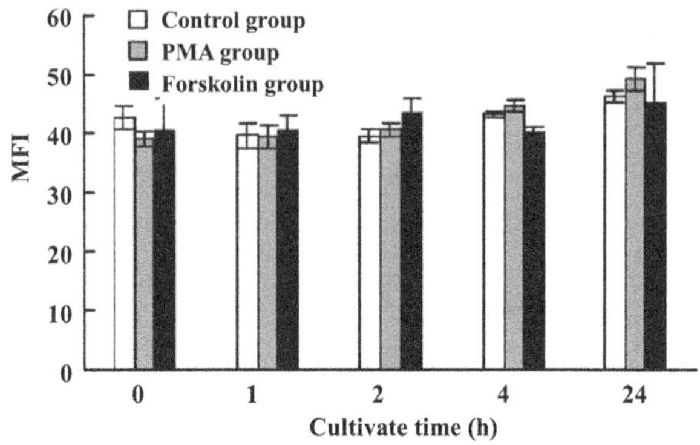

Fig. 4.12 The myofibrillar fragmentation index of different activators' groups. Translated from Chen, Li, Li, et al. (2016)

4.3 Conclusions

The level of protein phosphorylation was significantly affected by the varying degrees of tenderness and the different postmortem times. The protein phosphorylation level (P/T ratio) of the tender group increased from 0.5 to 12 h postmortem and then decreased. The P/T ratio of tough group increased during 24 h postmortem, increasing faster from 0.5 to 4 h postmortem than from 4 to 24 h postmortem. The global phosphorylation level of tough meat was significantly higher than tender meat at 4, 12, and 24 h postmortem. The highest level of sarcoplasmic protein phosphorylation in the tough meat and tender meat was observed at 4 and 12 h postmortem, respectively. Moreover, protein identification revealed that most of the phosphoproteins were proteins with sarcomeric function. Protein phosphorylation may influence meat rigor mortis through contractile machinery and glycolysis, which in turn affected meat tenderness. PKA and PKC phosphorylated TTN, MYBPC, TPM, MYLC2 and a protein of 25 kDa. Phosphorylation of some myofibrillar proteins resulted in reduced degradation of these proteins by the calpains, shortening of sarcomere length, and muscle contraction.

Acknowledgments Parts of this chapter are reprinted from Journal of the Science of Food and Agriculture, 96(5), Chen, L., et al., Phosphorylation of myofibrillar proteins in postmortem ovine muscle with different tenderness, 1474–1483. Copyright (2020), with permission from Elsevier. Parts of this chapter are translated from Scientia Agricultura Sinica, 49, CHEN, L. J., et al., Protein phosphorylation on the function of myofibrillar proteins in mutton muscle (Chinese), 1360–1370; Modern Food Science and Technology, 31, CHEN, L. J., et al., Analyzing the changes in sarcoplasmic protein phosphorylation with respect to postmortem ageing times in mutton with different levels of tenderness (Chinese), 95–101. Copyright (2020), with permission from two journals.

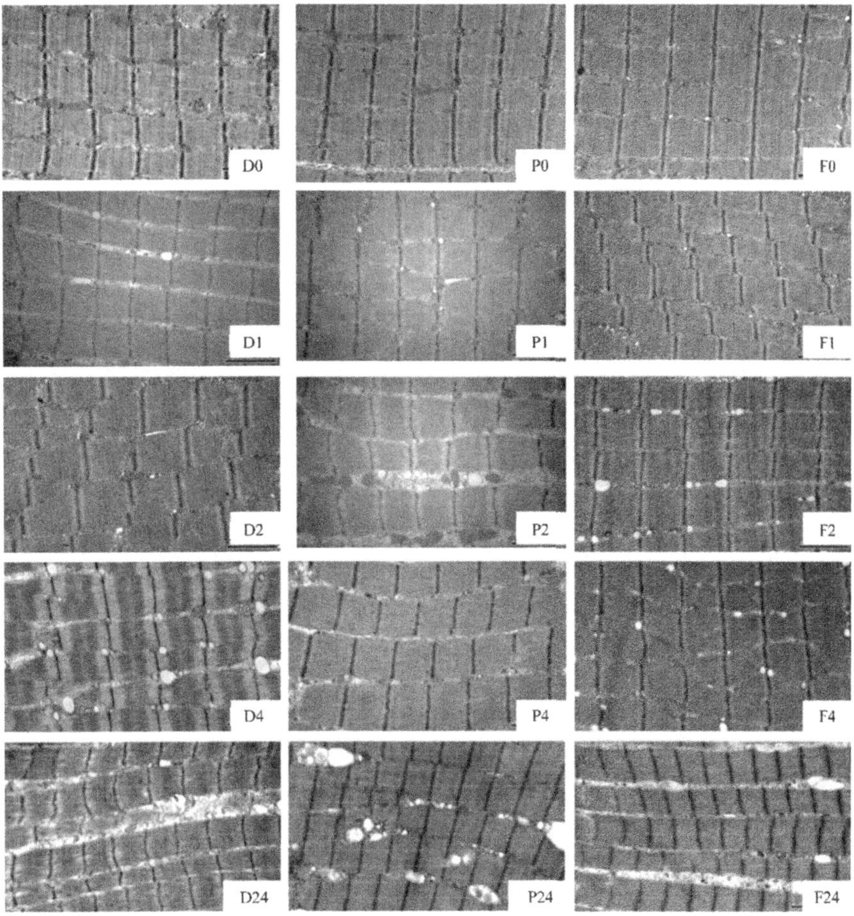

Fig. 4.13 Transmission electron microscopy photos of different activators' groups in different time. D0: Control group 0 h; P0: PMA group 0 h; F0: Forskolin group 0 h; D1: Control group 1 h; P0: PMA group 1 h; F1: Forskolin group 1 h; D2: Control group 2 h; P2: PMA group 2 h; F2: Forskolin group 2 h; D4: Control group 4 h; P4: PMA group 4 h; F4: Forskolin group 4 h; D24: Control group 24 h; P24: PMA group 24 h; F24: Forskolin group 24 h. Translated from Chen, Li, Li, et al. (2016)

Table 4.4 Analysis of the sarcomere length of different activators' groups. Translated from Chen, Li, Li, et al. (2016)

	Time (h)				
Treatment	0	1	2	4	24
Control group	1.71 ± 0.06^{xy}	1.62 ± 0.08^{xy}	1.95 ± 0.10^{x}	1.92 ± 0.04^{a}	1.46 ± 0.26^{aAy}
PMA group	1.71 ± 0.06^{xy}	1.67 ± 0.07^{xy}	1.89 ± 0.29^{x}	$1.61 \pm 0.10b^{xy}$	1.14 ± 0.06^{by}
Forskolin group	1.71 ± 0.06^{x}	1.75 ± 0.04^{x}	1.74 ± 0.06^{x}	1.56 ± 0.07^{By}	0.96 ± 0.04^{Bz}

The letters a, b represent the significant difference between PMA group and control group ($P < 0.05$), the letters A, B represent the significant difference between Forskolin group and control group ($P < 0.05$), the letters x, y, z represent the significant difference of the same group in different time points ($P < 0.05$).

References

Ahmed, S. H., & Lindsey, M. L. (2009). Titin phosphorylation: Myocardial passive stiffness regulated by the intracellular giant. *Circulation Research, 105*, 611–613.

Arnold, H., Henning, R., & Pette, D. (2005). Quantitative comparison of the binding of various glycolytic enzymes to F-actin and the interaction of aldolase with G-actin. *FEBS Journal, 22*, 121–126.

Bolten, K. E., Marsh, A. E., Reed, S. M., Dubey, J. P., Toribio, R. E., & Saville, W. J. A. (2008). Sarcocystis neurona: Molecular characterization of enolase domain I region and a comparison to other protozoa. *Experimental Parasitology, 120*, 108–112.

Cai, G. Z., Callaci, T. P., Luther, M. A., & Lee, J. C. (1997). Regulation of rabbit muscle phosphofructokinase by phosphorylation. *Biophysical Chemistry, 64*, 199–209.

Cerutti, H., & Casas-Mollano, J. A. (2009). Histone H3 phosphorylation: Universal code or lineage specific dialects? *Epigenetics, 4*, 71–75.

Chen, L., Li, X., Ni, N., Liu, Y., Chen, L., Wang, Z., et al. (2016). Phosphorylation of myofibrillar proteins in post-mortem ovine muscle with different tenderness. *Journal of the Science of Food and Agriculture, 96*, 1474–1483.

Chen, L.-J., Li, X., Li, Z., Li, P.-D., Li, Z.-W., & Zhang, D.-Q. (2016). Protein phosphorylation on the function of myofibrillar proteins in mutton muscle. *Scientia Agricultura Sinica, 49*, 1360–1370.

Chen, L. J., Li, X., Yang, Y., Chen, L., Ni, N., & Zhang, D. Q. (2015). Analyzing the changes in sarcoplasmic protein phosphorylation with respect to postmortem ageing times in mutton with different levels of tenderness. *Modern Food Science & Technology, 31*, 95–101.

Clarke, F. M., & Masters, C. J. (1975). On the association of glycolytic enzymes with structural proteins of skeletal muscle. *Biochimica et Biophysica Acta, 381*, 37–46.

Clarke, F. M., Shaw, F. D., & Morton, D. J. (1980). Effect of electrical stimulation post mortem of bovine muscle on the binding of glycolytic enzymes. Functional and structural implications. *Biochemical Journal, 186*, 105–109.

Cole, H. A., & Perry, S. V. (1975). The phosphorylation of troponin I from cardiac muscle. *Biochemical Journal, 149*, 525–533.

Dagher, S. M., & Hultin, H. O. (1975). Association of glyceraldehyde-3-phosphate dehydrogenase with the particulate fraction of chicken skeletal muscle. *European Journal of Biochemistry, 55*, 185–192.

D'Alessandro, A., Rinalducci, S., Marrocco, C., Zolla, V., Napolitano, F., & Zolla, L. (2012). Love me tender: An omics window on the bovine meat tenderness network. *Journal of Proteomics, 75*, 4360–4380.
D'Alessandro, A., & Zolla, L. (2013). Meat science: From proteomics to integrated omics towards system biology. *Journal of Proteomics, 78*, 558–577.
Di Lisa, F., De Tullio, R., Salamino, F., Barbato, R., Melloni, E., Siliprandi, N., et al. (1995). Specific degradation of troponin T and I by Mu-Calpain and its modulation by substrate phosphorylation. *Biochemical Journal, 308*, 57–61.
Doumit, M., & Bates, R. (2000). Regulation of pork water holding capacity, color, and tenderness by protein phosphorylation. *Pork Quality, 63*(1), 17–22.
Gannon, J., Staunton, L., O'Connell, K., Doran, P., & Ohlendieck, K. (2008). Phosphoproteomic analysis of aged skeletal muscle. *International Journal of Molecular Medicine, 22*, 33–42.
González, B., & Manso, R. (2004). Induction, modification and accumulation of Hsp70s in the rat liver after acute exercise: Early and late responses. *Journal of Physiology, 556*, 369–385.
Hopkins, D. L., Toohey, E. S., Lamb, T. A., Kerr, M. J., & Refshauge, G. (2011). Explaining the variation in the shear force of lamb meat using sarcomere length, the rate of rigor onset and Ph. *Meat Science, 88*, 794–796.
Hopkins, D. L., Toohey, E. S., Lamb, T. A., & Refshauge, G. (2009). Alternative methods for determining the temperature at Ph 6. In Proceedings of the 55th International Congress of Meat Science and Technology, 1.
Huang, F., Huang, M., Xu, X., & Zhou, G. (2011). Influence of heat on protein degradation, ultrastructure and eating quality indicators of pork. *Journal of the Science of Food & Agriculture, 91*, 443–448.
Huang, H., Larsen, M. R., Karlsson, A. H., Pomponio, L., & Lametsch, R. (2011). Gel-based phosphoproteomics analysis of sarcoplasmic proteins in postmortem porcine muscle with Ph decline rate and time differences. *Proteomics, 11*, 4063–4076.
Huang, H., Martin, R. L., & Lametsch, R. (2012). Changes in phosphorylation of myofibrillar proteins during postmortem development of porcine muscle. *Food Chemistry, 134*, 1999–2006.
Jeremiah, L. E. (1982). A review of factors influencing consumption, selection and acceptability of meat purchases. *Journal of Consumer Studies & Home Economics, 6*, 137–154.
Johnson, L. N. (1992). Glycogen phosphorylase: Control by phosphorylation and allosteric effectors. *FASEB Journal, 6*, 2274–2282.
Klug, G. A., Botterman, B. R., & Stull, J. T. (1982). The effect of low-frequency stimulation on myosin light chain phosphorylation in skeletal-muscle. *Journal of Biological Chemistry, 257*, 4688–4690.
Kuo, H. J., Malencik, D. A., Liou, R. S., & Anderson, S. R. (1986). Factors affecting the activation of rabbit muscle phosphofructokinase by actin. *Biochemistry, 25*, 1278–1286.
Lee, S. H., Joo, S. T., & Ryu, Y. C. (2010). Skeletal muscle fiber type and myofibrillar proteins in relation to meat quality. *Meat Science, 86*, 166–170.
Li, C., Wang, D., Dong, H., Xu, W., Gao, F., Zhou, G., et al. (2013). Effects of different cooking regimes on the microstructure and tenderness of duck breast muscle. *Journal of the Science of Food & Agriculture, 93*, 1979–1985.
Li, C. B., Li, J., Zhou, G. H., Lametsch, R., Ertbjerg, P., Brüggemann, D. A., et al. (2012). Electrical stimulation affects metabolic enzyme phosphorylation, protease activation, and meat tenderization in beef. *Journal of Animal Science, 90*, 1638–1649.
Linke, W. A. (2008). Sense and stretchability: The role of titin and titin-associated proteins in myocardial stress-sensing and mechanical dysfunction†. *Cardiovascular Research, 77*, 637.
Loktionova, S. A., & Kabakov, A. E. (1998). Protein phosphatase inhibitors and heat preconditioning prevent Hsp27 dephosphorylation, F-actin disruption and deterioration of morphology in Atp-depleted endothelial cells. *FEBS Letters, 433*, 294–300.
Luther, M. A., & Lee, J. C. (1986). The role of phosphorylation in the interaction of rabbit muscle phosphofructokinase with F-actin. *Journal of Biological Chemistry, 261*, 1753–1759.

Melling, C. W. J., Thorp, D. B., & Noble, M. E. G. (2009). Myocardial Hsp70 phosphorylation and Pkc-mediated cardioprotection following exercise. *Cell Stress & Chaperones, 14*, 141–150.

Melnick, R. L., & Hultin, H. O. (1973). Studies on the nature of the subcellular localization of lactate dehydrogenase and glyceraldehyde-3-phosphate dehydrogenase in chicken skeletal muscle. *Journal of Cellular Physiology, 81*, 139–147.

Muroya, S., Ohnishi-Kameyama, M., Oe, M., Nakajima, I., Shibata, M., & Chikuni, K. (2007). Double phosphorylation of the myosin regulatory light chain during rigor mortis of bovine longissimus muscle. *Journal of Agricultural & Food Chemistry, 55*, 3998–4004.

Obermann, W. M. J., Gautel, M., Weber, K., & F Rst, D. O. (1997). Molecular structure of the sarcomeric M band: Mapping of titin and myosin binding domains in myomesin and the identification of a potential regulatory phosphorylation site in myomesin. *EMBO Journal, 16*, 211–220.

Pebay-Peyroula, E., Dahout-Gonzalez, C., Kahn, R., Tr, Z., Guet, V., Lauquin, G. J.-M., et al. (2003). Structure of mitochondrial Adp/Atp carrier in complex with carboxyatractyloside. *Nature, 426*, 39–44.

Perry, S. V., & Cole, H. A. (1974). Phosphorylation of troponin and the effects of interactions between the components of the complex. *Biochemical Journal, 141*, 733–743.

Rao, V. S., Marongelli, E. N., & Guilford, W. H. (2009). Phosphorylation of Tropomyosin extends cooperative binding of myosin beyond a single regulatory unit. *Cell Motility & the Cytoskeleton, 66*, 10–23.

Ryder, J. W., Lau, K. S., Kamm, K. E., & Stull, J. T. (2007). Enhanced skeletal muscle contraction with myosin light chain phosphorylation by a calmodulin-sensing kinase. *Journal of Biological Chemistry, 282*, 20447–20454.

Salamino, F., Detullio, R., Michetti, M., Mengotti, P., Melloni, E., & Pontremoli, S. (1994). Modulation of calpastatin specificity in rat tissues by reversible phosphorylation and dephosphorylation. *Biochemical and Biophysical Research Communications, 199*, 1326–1332.

Sawdy, J. C., Kaiser, S., St-Pierre, N. R., & Wick, M. P. (2004). Myofibrillar 1-D fingerprints and myosin heavy chain Ms analyses of beef loin at 36 H postmortem correlate with tenderness at 7 days. *Meat Science, 67*, 421–426.

Shen, Q. W., Underwood, K. R., Means, W. J., Mccormick, R. J., & Du, M. (2007). The halothane gene, energy metabolism, adenosine monophosphate-activated protein kinase, and glycolysis in postmortem pig Longissimus dorsi muscle. *Journal of Animal Science, 85*, 1054–1061.

Silva, D. R. G., Torres Filho, R. A., Cazedey, H. P., Fontes, P. R., Ramos, A. L. S., & Ramos, E. M. (2015). Comparison of Warner–Bratzler shear force values between round and square cross-section cores from cooked beef and pork longissimus muscle. *Meat Science, 103*, 1–6.

Sprang, S. R., Acharya, K. R., Goldsmith, E. J., Stuart, D. I., Varvill, K., Fletterick, R. J., et al. (1988). Structural changes in glycogen phosphorylase induced by phosphorylation. *Nature, 336*, 215–221.

Sweeney, H. L., Bowman, B. F., & Stull, J. T. (1993). Myosin light chain phosphorylation in vertebrate striated muscle: Regulation and function. *American Journal of Physiology, 264*, 1085–1095.

Taylor, R. G., Geesink, G. H., Thompson, V. F., Koohmaraie, M., & Goll, D. E. (1995). Is Z-disk degradation responsible for postmortem tenderization? *Journal of Animal Science, 73*, 1351–1367.

Xia, A. Q., Li, X., Chen, L., Chen, L. J., & Zhang, D. Q. (2014). Effect of pre-slaughter transport time on lamb quality. *Modern Food Science & Technology, 30*, 230–235.

Zhang, Z., Lawrence, J., & Stracher, A. (1988). Phosphorylation of platelet actin binding protein protects against proteolysis by calcium dependent sulfhydryl protease. *Biochemical & Biophysical Research Communications, 151*, 355–360.

Zhi, G., Ryder, J. W., Huang, J., Ding, P., Chen, Y., Zhao, Y., et al. (2005). Myosin light chain kinase and myosin phosphorylation effect frequency-dependent potentiation of skeletal muscle contraction. *Proceedings of the National Academy of Sciences of the United States of America, 102*, 17519–17524.

Chapter 5
Protein Phosphorylation Affects Meat Water Holding Capacity

Abstract Water holding capacity and its corollary, drip loss, are important meat quality indicators. The high drip loss muscle had lower pH, higher brightness, higher shear force value, and shorter sarcomere length than the low drip loss muscle. The protein phosphorylation level of different drip loss muscles was detected via the sodium dodecyl sulfate-polyacrylamide gel electrophoresis (SDS-PAGE) in this study. The results showed that the phosphorylation level of sarcoplasmic and myofibrillar in the high drip loss group was higher than that in the low drip loss group. Combined with the gray value analysis, the bands with higher phosphorylation levels of sarcoplasmic proteins were mainly enzymes involved in glycolysis which could affect the pH; while the bands with higher phosphorylation levels of myofibrillar proteins were mainly cytoskeletal proteins and contractile proteins. It can be concluded that the low pH and increased muscle contraction had a major influence on muscle drip loss.

Keywords Water holding capacity · Protein phosphorylation · Drip loss · Meat quality · Sarcoplasmic protein · Myofibrillar protein1 · Introduction

5.1 Introduction

Water holding capacity (WHC) and its corollary, drip loss, were important meat quality indicators. Low drip loss often indicated a high WHC of meat, and the tissue is juicy, fresh, and tender with a dry surface. The genetics (the halothane and rendement napole genes), feed composition, pre-slaughter management, and post-mortem handling could influence the WHC (Apple and Yancey 2013; Bond and Warner 2007; Huff-Lonergan and Lonergan 2005). At a molecular level, the WHC was related to the denaturation and oxidation of myofibrillar protein, and actomyosin interaction (Huynh et al. 2013; Liu et al. 2010). Although studies have found out that desmin, actomyosin, μ-calpain, integrin, and dystrophin were crucial to the WHC of meat (Bond and Warner 2007; Huff-Lonergan and Lonergan 2005), it was not enough to fully reveal the mechanism of drip loss.

D. Zhang et al., *Protein Phosphorylation and Meat Quality*,
https://doi.org/10.1007/978-981-15-9441-0_5

Protein phosphorylation plays an important role in muscles after slaughter. However, whether there are differences in protein phosphorylation level between muscles with different WHC levels is unknown.

In this chapter, the *longissimus dorsi* muscles of three breeds of Boer goat, Huang-Huai goat, and Boer × Laoshan white goat were collected. After measuring the drip loss, the two samples with the highest and lowest drip loss were selected and divided into high drip loss group and low drip loss group. The physicochemical properties and muscle fiber structure of the two mutton groups were compared, and the protein phosphorylation level was studied by sodium dodecyl sulfate-polyacrylamide gel electrophoresis (SDS-PAGE).

5.2 Meat Quality Traits with Different Drip Loss Values

5.2.1 Comparison of Physical and Chemical Indicators Between Different Drip Loss Meat

The $pH_{45\ min}$ value, the pH of muscle at 45 min after slaughter, was related to drip loss (Otto et al. 2004). The $pH_{45\ min}$ value of the high drip loss group was significantly lower than that of the low drip loss group ($P < 0.05$), but there was no significant difference among the three breeds within each group ($P > 0.05$) (Table 5.1). The L* value of the high drip loss group was significantly higher than of the low drip loss group for all breeds ($P < 0.05$), because of more water flowing onto the muscle surface of high drip loss group and increasing light reflection. There was no significant difference of a* and b* values between the high drip loss group and the low drip loss group ($P > 0.05$).

Delta E (ΔE) was defined as the difference between two colors in L*, a*, and b* color space. The ΔE values of the high drip loss group were significantly higher than those in the low drip loss group only for Boer goats ($P < 0.05$). There was no difference of Warner Bratzler shear force between the high drip loss group and the low drip loss group ($P > 0.05$), perhaps because water loss could lead to a tougher muscle.

The sarcomere length of the high drip loss group was significantly shorter than that of the low drip loss group ($P < 0.05$). With an increase in drip loss, the sarcomere length decreased, due to greater muscle contraction.

5.2.2 Comparison of Muscle Fiber Ultrastructure Between Different Drip Loss Meat

In Fig. 5.1a–c were obtained by using a transmission electron microscope to amplify 40,000 × to observe the mutton with different drip loss. The differences between the

Table 5.1 Meat quality traits of goat *longissimus dorsi* muscles with different drip loss values. (Wang et al. 2016)

		High drip loss	Low drip loss
Drip loss/%	Huang-Huai goat	2.27 ± 0.31a	0.33 ± 0.02b
	Boer goat	2.20 ± 0.16a	0.82 ± 0.16b
	Boer × Laoshan white goat	4.68 ± 0.71a	0.36 ± 0.09b
$pH_{45\ min}$	Huang-Huai goat	6.10 ± 0.07b	6.35 ± 0.04a
	Boer goat	6.19 ± 0.05b	6.34 ± 0.08a
	Boer × Laoshan white goat	6.11 ± 0.04b	6.22 ± 0.06a
L*	Huang-Huai goat	55.89 ± 3.83a	52.47 ± 2.76b
	Boer goat	54.07 ± 1.95a	45.33 ± 0.65b
	Boer × Laoshan white goat	49.82 ± 1.98a	40.86 ± 1.15b
a*	Huang-Huai goat	10.35 ± 0.20	13.29 ± 1.52
	Boer goat	10.90 ± 0.35	11.20 ± 0.40
	Boer × Laoshan white goat	9.47 ± 0.67	12.55 ± 1.41
b*	Huang-Huai goat	11.84 ± 0.30	13.35 ± 2.07
	Boer goat	11.64 ± 0.35	13.15 ± 0.47
	Boer × Laoshan white goat	10.91 ± 0.70	11.16 ± 1.26
ΔE	Huang-Huai goat	54.20 ± 1.29	53.13 ± 1.22
	Boer goat	53.01 ± 1.56a	44.04 ± 0.67b
	Boer × Laoshan white goat	50.52 ± 1.77	47.68 ± 2.04
Shear force/kg	Huang-Huai goat	7.18 ± 0.60	6.66 ± 1.49
	Boer goat	7.47 ± 0.50	7.06 ± 0.67
	Boer × Laoshan white goat	7.43 ± 1.07	7.01 ± 1.26
Sarcomere length/μm	Huang-Huai goat	1.27 ± 0.33	1.38 ± 0.31
	Boer goat	1.26 ± 0.15	1.36 ± 0.11
	Boer × Laoshan white goat	1.37 ± 0.26b	1.97 ± 0.30a

Data is reported as mean ± SD ($n = 3$). Different superscripts in the same row indicate significant differences ($P < 0.05$)

high and low drip loss mutton tissue structure can be visualized. In the low drip loss group, the sarcomere was relatively complete and clear, the muscle cells were arranged closely, and there was almost no gap. While the high drip loss group was partially distorted, deformed, arranged disorderly, z-line partially deviated or broken, and other proteins near the z-line might be degraded and then attached to the z-line, then more voids were formed between the muscle cells, and certain metabolites were accumulated. In addition, it can be seen from Table 5.1 that the length of the sarcomere in the high drip loss group was lower than that in the low drip loss group. As the drip loss increased, the length of the sarcomere decreased, especially in Boer × Laoshan white goats with the highest drip loss. The length of sarcomere was significantly lower than that in meat samples with low drip loss ($P < 0.05$), indicating that the sarcomere became shorter and the muscle contracted in the sample with high drip loss.

Figure 5.1d–f showed that the arrangement of muscle fibers in mutton with different drip loss was different. Observed by scanning electron microscopy for

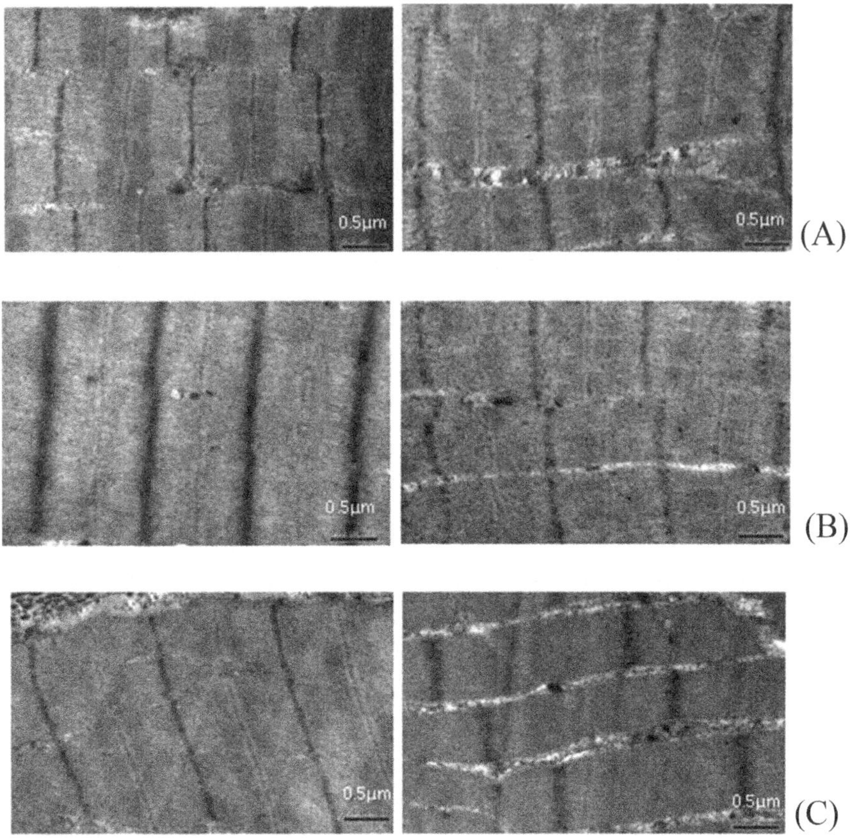

Fig. 5.1 Comparison of muscle fiber ultrastructure between drip loss. (**a**) TEM of Boer Goats with low drip loss and high drip loss; (**b**) TEM of Huang-Huai Goats with low drip loss and high drip loss; (**c**) TEM of Boer × Laoshan Goats with low drip loss and high drip loss; (**d**) SEM of Boer Goats with low drip loss and high drip loss; (**e**) SEM of Huang-Huai Goats with low drip loss and high drip loss; (**f**) SEM of Boer × Laoshan goats with low drip loss and high drip loss. Translated from (He 2014)

500 ×, it was found that the muscle fibers in the low drip loss group were densely arranged, and there was almost no gap between the muscle fibers. However, in the high drip loss group, the muscle fibers were disordered, loose in structure, irregular in distribution, and the diameter became smaller, the gap between muscle fibers gradually increased, the homogeneity decreased, and the perimysium also partially fractured and disintegrated. This indicated that with the increase of drip loss, the water held in muscle decreased, and the muscle fibers contracted, making the gap between the muscle fibers become larger, and the integrity of tissue structure was destroyed.

The connective tissue membrane in the muscle was elastic, which could affect the maintenance of muscle integrity and damage prevention of muscle fibers. It can be

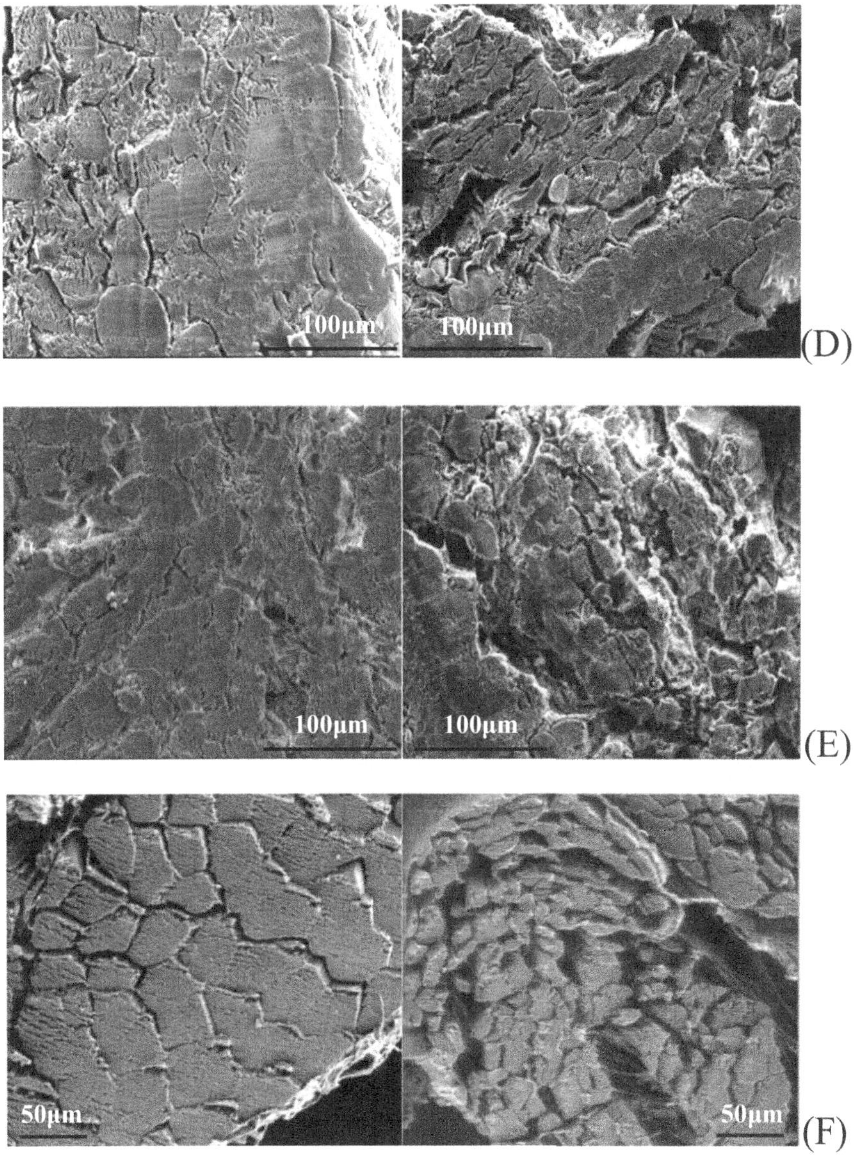

Fig. 5.1 (continued)

seen from the figure that the connective tissue of the high drip loss group was ruptured, indicating that the muscle tissue integrity was lost and the dense structure was destroyed. The muscle fiber boundary was blurred, the arrangement was disordered, part of the protein was broken, segregated, and disconnected.

5.3 Changes of Proteins Phosphorylation Levels Meat with Different Drip Loss

5.3.1 Comparison of Sarcoplasmic Proteins Phosphorylation Levels Between Different Drip Loss Meat

Figures 5.2 and 5.3 showed SDS-PAGE of sarcoplasmic protein in lamb with different drip loss, the left one was the phosphorylated protein dyed by Pro-Q Diamond. Figure 5.4 was the global protein dyed by SYPRO Ruby, the bands were straight and 14 clear bands can be seen in the figure.

As presented in Fig. 5.2 and Table 5.2, the P/T values of the band 3, 5, 6, 7, 8, 9, 10, 11, and 13 were greater than 0.5, which meant that the proteins in these bands may be phosphorylated to a higher level. The Pro-Q Diamond dye was specific for phosphorylated proteins. Compared to data in Fig. 5.2, the darker bands in Pro-Q Diamond dye were lighter in SYPRO Ruby, while the darker bands in SYPRO Ruby dye were lighter in Pro-Q Diamond.

Compare to the global phosphorylation levels of sarcoplasmic proteins in different drip loss samples from the three breeds, Table 5.2 shows that the phosphorylation level of the high drip loss group was generally higher than that of the low drip loss group. Among them, the band 7, 10, and 13 in the high drip loss group were significantly higher than the low drip loss group ($P < 0.05$), indicating that the phosphorylation level of individual sarcoplasmic protein in the high drip loss group was higher.

According to the results of the P/T value of each band in Table 5.2 and other studies, band 7 (58 kDa) may be pyruvate kinase (PK), which was also one of the three rate-limiting enzymes in the glycolytic pathway (Huang et al. 2011; Li et al. 2012). PK catalyzed the conversion of phosphoenol pyruvate (PEP) to pyruvate,

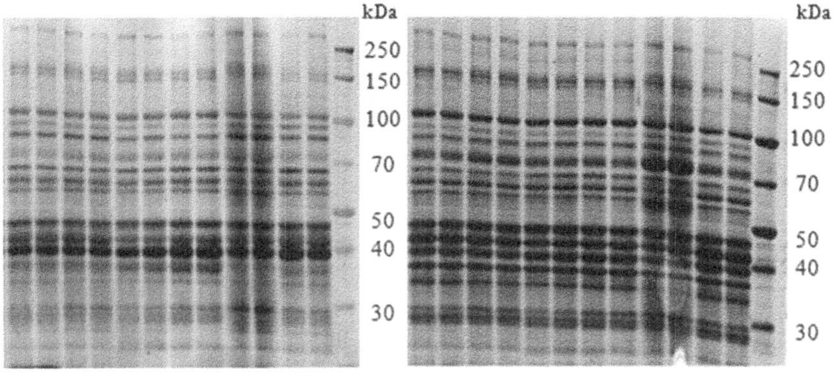

Fig. 5.2 SDS-PAGE of phosphorylated protein (left) and total protein in sarcoplasmic protein. 1–2 is low drip loss group, 3–4 is high drip loss group, H is Huang-Huai goat, B is Boer goat, L is Lushan goat. Translated from (He 2014)

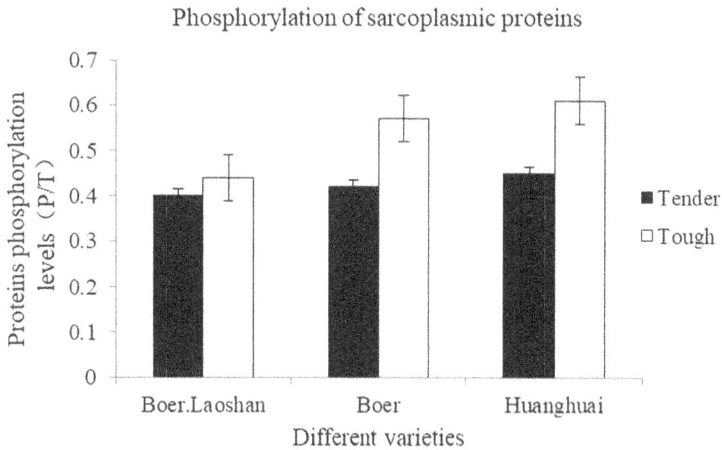

Fig. 5.3 Phosphorylation of sarcoplasmic proteins between high, low drip loss group in different varieties. Translated from (He 2014)

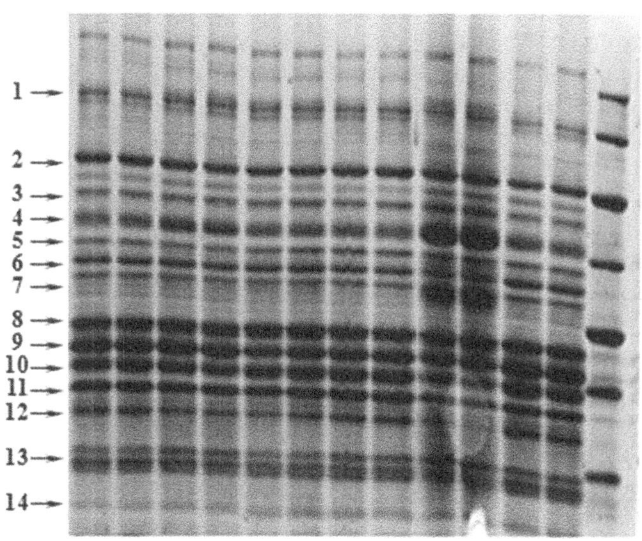

Fig. 5.4 SDS-PAGE of total protein in sarcoplasmic protein. Translated from (He 2014)

which can be converted to lactic acid under the catalysis of lactic dehydrogenase (LDH), and nicotinamide adenine dinucleotide (NADH) was oxidized to NAD+.

Band 10 (40 kDa) was presumed to be creatine kinase (CK), which as an important phosphokinase, might directly involve in intracellular energy transfer, muscle contraction, and adenosine triphosphate (ATP) regeneration. With the participation of ATP, it can catalyze the phosphorylation of creatine to generate

Table 5.2 P/T value of each band in the sarcoplasmic protein. Translated from (He 2014)

Band	MW/kDa	High drip loss	Low drip loss
1	173.571	0.280	0.346
2	107.306	0.428	0.404
3	89.270	0.902	0.895
4	77.150	0.546	0.164
5	68.143	0.817	1.356
6	62.854	0.851	0.862
7	58.769	1.863	1.570
8	46.768	2.283	2.892
9	43.250	0.862	1.744
10	40.473	7.312	0.840
11	37.500	1.214	1.869
12	34.410	0.302	0.620
13	27.250	3.849	1.012
14	24.760	0.526	0.080

phosphocreatine and adenosine diphosphate (ADP). It can also reversibly catalyze the transfer of high-energy phosphate bond of phosphocreatine to ADP to produce ATP and creatine. When ATP was rapidly consumed, ADP combined with phosphate to form ATP. When ATP was sufficient, creatine combined with phosphoric acid to form phosphocreatine. Phosphocreatine was a high-energy phosphate substance that could rapidly recover ATP, a form of energy storage. Studies have shown that CK and actin, alpha β-crystallin were the substrates of proteolysis during pork aging. And CK, actin and -crystal protein were the substrates of proteolysis during pork aging (Lametsch et al. 2002).

Band 11 was glyceraldehyde-3-phosphate dehydrogenase (GAPDH), a key enzyme involved in glycolysis, catalyzing the oxidative phosphorylation of glyceraldehyde 3-phosphate with the participation of NAD+ and inorganic phosphate. Its glycolytic activity was regulated by F-actin and plasma membrane. Choudhary et al. (2000) demonstrated that the GAPDH phosphorylation level increased and the pH decreased in vivo. GAPDH has also been reported to regulate the structure of the cytoskeleton by forming actin fiber grids.

Most of the enzymes in the muscle could be found in sarcoplasm, and glycolytic enzymes accounted for two-thirds of the sarcoplasmic proteins (Scopes and Stoters 1982). Glycolytic enzyme activity was the main factor affecting the decline rate of muscle pH after slaughter, which further affected the process of postmortem rigor and played a decisive role in the WHC of meat. The activity of glycolytic enzymes was regulated by phosphorylation levels. Therefore, studying the phosphorylation level of postmortem sarcoplasmic proteins was a crucial step in regulating meat quality.

5.3.2 *Comparison of Myofibrillar Proteins Phosphorylation Levels Between Different Drip Loss Meat*

Figures 5.5 and 5.6 showed SDS-PAGE of myofibrillar proteins in lambs with different drip loss, the left figure was the phosphorylated protein dyed by Pro-Q Diamond. Figure 5.7 was the global protein dyed by SYPRO Ruby, the bands were straight and 17 clear bands can be seen in the figure.

In the high drip loss group, the P/T values of the band 2, 3, 5, 7, 9, 10, 11, 15, and 17 were greater than 0.5, while those of band 9, 11, 15, and 17 in low drip loss group were greater than 0.5 (Fig. 5.5 and Table 5.3). And the P/T values of the band 2, 3, 5, 7, 10, and 17 in the high drip loss group were higher than those in the low drip loss group, indicating that the phosphorylation level of those bands was higher. The global phosphorylation level of myofibrillar proteins was consistent with that of sarcoplasmic proteins, and the phosphorylation level of the high drip loss group was generally higher than that of the low drip loss group.

There were seven bands in myofibrillar proteins that differ greatly, and five of them were proteins involved in muscle contraction. Band 2 (140 kDa) should be a C-protein located in the myofibril A-band, interspersed between thin filaments and thick filaments. It was a muscle-regulating protein that maintained the stability of thick filaments (Craig and Offer 1976). Previous study has shown that the phosphorylation level of C-proteins was related to the contraction rate of the heart and regulated the interaction of actomyosin during diastole (Hartzell 1984).

As we can see in Fig. 5.7 and Table 5.3, band 3 (105 kDa) should be α-actin, an important contractile protein in muscle cells, and one of the major components of the cytoskeleton. Band 10 (36.7 kDa) may be Troponin T, and band 17 (22 kDa) should be Troponin I. Troponin was a kind of calcium ion receptor protein, which was composed of three subunits: I, T, and C. Troponin played an important role in muscle

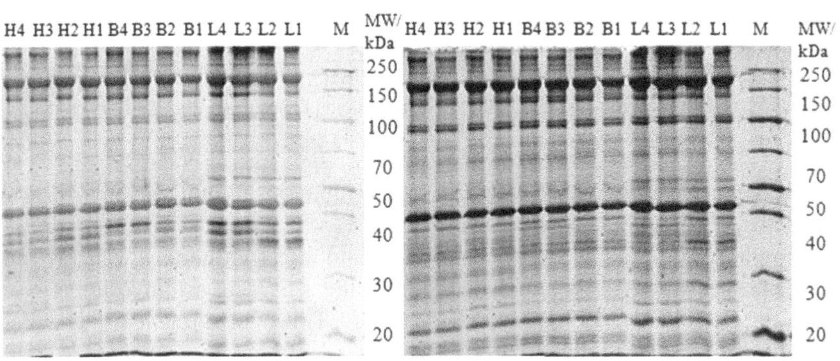

Fig. 5.5 SDS-PAGE of phosphorylated protein (left) and total protein in myofibrillar protein. 1–2 is low drip loss group, 3–4 is high drip loss group, H is Huang-Huai goat, B is Boer goat, L is Lushan goat. Translated from (He 2014)

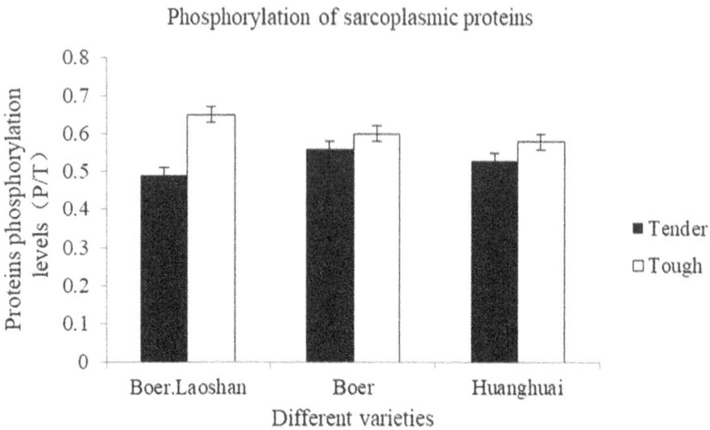

Fig. 5.6 Phosphorylation of myofibrillar protein between high, low drip loss group in different varieties. Translated from (He 2014)

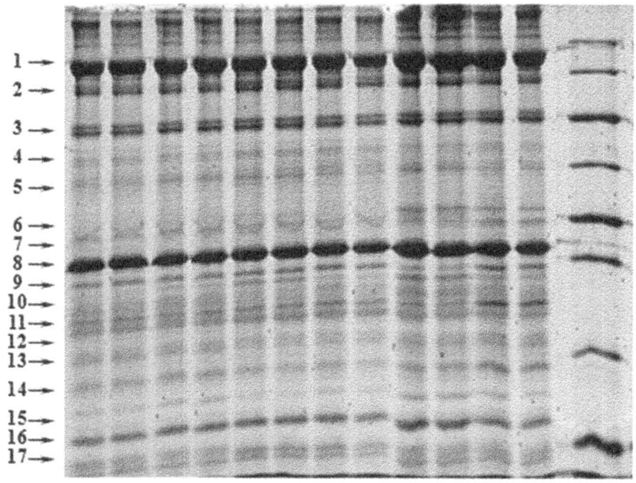

Fig. 5.7 SDS-PAGE of total protein in myofibrillar protein. Translated from (He 2014)

contraction. When combined with Ca^{2+}, its inhibition of actin and myosin was weakened, resulting in muscle contraction.

Phosphorylation of the troponin subunit could affect its gliding with myosin, altered the sensitivity of myofibrils to Ca^{2+} and the activity of actomyosin ATPase. When the activation of ATPase by Ca^{2+} decreased, TnI could re-regulate the binding of actomyosin (Solaro et al. 1976).

Muscle contraction and relaxation were the results of the mutual sliding of myofibril thick filaments and thin filaments. A cross-bridge was formed between the myosin head in the thick myofilament and the actin in the thin filaments. The

Table 5.3 P/T value of each band in the myofibrillar protein. Translated from (He 2014)

Band	MW/kDa	High drip loss	Low drip loss
1	182.964	0.647	0.456
2	140.298	0.834	0.423
3	105.555	1.944	0.220
4	85.224	0.107	0.303
5	70.866	2.964	0.257
6	60.343	0.196	0.162
7	55.201	1.330	0.593
8	42.961	0.515	0.330
9	38.417	2.475	5.358
10	36.711	1.418	0.202
11	34.904	1.145	2.250
12	31.394	0.249	0.389
13	29.105	0.114	0.220
14	28.833	0.183	0.204
15	27.112	0.786	0.799
16	24.270	0.350	0.260
17	22.278	0.822	0.661

ATP required for muscle contraction was derived from the hydrolysis and catalysis of actomyosin ATPase. The activity of the muscle contraction system was regulated by troponin–tropomyosin located on the thin filament. When the nerve stimulation signal was transmitted, the troponin combined with the Ca^{2+} released by the sarcoplasmic reticulum and changed its configuration, and then drove the thick and thin muscle filaments to slide and contract, resulting in the combination of actin and myosin to cause muscle contraction.

According to Fig. 5.7 and Table 5.3, band 5 (70 kDa) may be HSP 70 and band 7 (55 kDa) was presumed to be desmin. HSP 70 was important for muscle function and played an important role in the regulation of structural proteins. In addition, HSP70 had the function of anti-apoptosis. Many studies have shown that it was one of the most important proteins that could affect meat tenderness. Brigitte et al. (2010) found that its expression decreased in the low shear force group. As a major component of the intermediate filaments in the muscle cytoskeleton structure, desmin was closely connected with the z-line, and was a cytoskeletal protein associated with tenderness (Robson et al. 1981). It interacted with other intermediate filament proteins to form a reticular formation within the cytoplasm that maintained the link between the cell's contractile device and other structural elements. When it was degraded, the integrity of the myofibrillar protein could be destroyed. However, if the degradation degree of desmin was low, the intact desmin in the postmortem muscle could transform the contraction of myofibrillar protein into the contraction of the whole muscle cell, resulting in water loss (Huff-Lonergan and Lonergan 2005).

5.4 Conclusion

The phosphorylation level of sarcoplasmic and myofibrillar in high drip loss meat was higher than that in low drip loss meat. The bands with higher phosphorylation levels of sarcoplasmic proteins were mainly enzymes involved in glycolysis which had effects on the pH, while the bands with higher phosphorylation levels of myofibrillar proteins were mainly cytoskeletal proteins and contractile proteins. It can be concluded that the low pH and increased muscle contraction had a major influence on drip loss.

Acknowledgments Parts of this chapter are reprinted from Food Science & Biotechnology, 25, Wang, Z., et al., Proteomic analysis of goat *longissimus dorsi* muscles with different drip loss values related to meat quality traits, 425-431. Copyright (2020), with permission from Springer. Parts of this chapter are translated from Master Thesis, 2014, HE, F., P Mechanism of drip loss in mutton postmortem based on proteomics (Chinese), Yunnan Agricultural University.

References

Apple, J. K., & Yancey, J. W. S. (2013). Water-holding capacity of meat. In *The science of meat quality* (pp. 119–145). Ames, IA: Wiley.

Bond, J. J., & Warner, R. D. (2007). Ion distribution and protein proteolysis affect water holding capacity of Longissimus thoracis et lumborum in meat of lamb subjected to antemortem exercise. *Meat Science, 75*, 406–414.

Brigitte, P., C Cile, B., Louis, L., Caroline, M., Thierry, S., & Claudia, T. (2010). Skeletal muscle proteomics in livestock production. *Briefings in Functional Genomics, 9*, 259–278.

Choudhary, S., De, B. P., & Banerjee, A. K. (2000). Specific phosphorylated forms of glyceraldehyde 3-phosphate dehydrogenase associate with human parainfluenza virus type 3 and inhibit viral transcription in vitro. *Journal of Virology, 74*, 3634–3641.

Craig, R., & Offer, G. (1976). He location of C-protein in rabbit skeletal muscle. *Proceedings of the Royal Society of London. Series B. Biological Sciences, 192*, 451–461.

Hartzell, C. (1984). Phosphorylation of C-protein in intact amphibian cardiac muscle. Correlation between 32P incorporation and twitch relaxation. *The Journal of General Physiology, 83*, 563–588.

He, F. (2014). *Mechanism of drip loss in mutton post-mortem based on proteomics*. Kunming: Yunnan Agricultural University.

Huang, H., Larsen, M. R., Karlsson, A. H., Pomponio, L., & Lametsch, R. (2011). Gel-based phosphoproteomics analysis of sarcoplasmic proteins in postmortem porcine muscle with pH decline rate and time differences. *Proteomics, 11*, 4063–4076.

Huff-Lonergan, E., & Lonergan, S. M. (2005). Mechanism of water-holding capacity of meat: The role of postmortem biochemical and structural changes. *Meat Science, 71*, 194–204.

Huynh, T. P. L., Mur Ni, E., Maak, S., Ponsuksili, S., & Wimmers, K. (2013). UBE3B and ZRANB1 polymorphisms and transcript abundance are associated with water holding capacity of porcine M. longissimus dorsi. *Meat Science, 95*, 166–172.

Lametsch, R., Roepstorff, P., & Bendixen, E. K. (2002). Identification of protein degradation during post-mortem storage of pig meat. *Journal of Agricultural and Food Chemistry, 50*, 5508–5512.

Li, C. B., Li, J., Zhou, G. H., Lametsch, R., Ertbjerg, P., Brüggemann, D. A., et al. (2012). Electrical stimulation affects metabolic enzyme phosphorylation, protease activation, and meat tenderization in beef. *Journal of Animal Science, 90*, 1638–1649.

Liu, Z., Xiong, Y. L., & Chen, J. (2010). Protein oxidation enhances hydration but suppresses water-holding capacity in porcine longissimus muscle. *Journal of Agricultural and Food Chemistry, 58*, 10697–10704.

Otto, G., Roehe, R., Looft, H., Thoelking, L., & Kalm, E. (2004). Comparison of different methods for determination of drip loss and their relationships to meat quality and carcass characteristics. *Meat Science, 68*, 401–409.

Robson, R. M., Yamaguchi, M., Huiatt, T. W., Richardson, F. L., O'shea, J. M., Hartzer, M. K., et al. (1981). Biochemistry and molecular architecture of muscle cell 10-nm filaments and Z-line: Roles of desmin and alpha-actinin. In *Proceeding of 34th Annual Reciprocal Meat Conference* (p. 5). Chicago, IL: National Livestock and Meat Board.

Scopes, R. K., & Stoter, A. (1982). Purification of all glycolytic enzymes from one muscle extract. *Methods in Enzymology, 90*, 479–490.

Solaro, R. J., Moir, A. J., & Perry, S. V. (1976). Phosphorylation of troponin I and the inotropic effect of adrenaline in the perfused rabbit heart. *Nature, 262*, 615–617.

Wang, Z., He, F., Rao, W., Ni, N., Shen, Q., & Zhang, D. (2016). Proteomic analysis of goat Longissimus dorsi muscles with different drip loss values related to meat quality traits. *Food Science and Biotechnology, 25*, 425–431.

Part II
Mechanism of the Effect of Protein Phosphorylation on Meat Quality

Chapter 6
Mechanism of the Effect of Protein Phosphorylation on Postmortem Glycolysis

Abstract Glycolysis, the core metabolic pathway in muscle, is very significant to the ultimate meat quality. The objective of this chapter was to confirm the involvement of protein phosphorylation on postmortem glycolysis and the possible regulatory mechanism from the perspective of enzyme activity. The global phosphorylation level of sarcoplasmic protein increased early postmortem and then decreased afterwards in muscle. Protein bands with significantly different phosphorylation level were identified as glycometabolism related enzymes. By applying quantitative proteomic tools (isobaric Tags for Relative and Absolute Quantitation, iTRAQ) and bioinformatics analysis, 24 phosphoproteins clustered in glycolysis and muscle contraction were identified to be glycolytic rate related proteins, and phosphorylation of pyruvate kinase at Thr 157 was negatively correlated with glycolytic rate. Furthermore, the animal muscle was treated with a kinase inhibitor, dimethyl sulfoxide, or a phosphatase inhibitor. Protein phosphorylation was positively correlated with the activity of glycogen phosphorylase, pyruvate kinase, and phosphofructokinase. In conclusion, protein phosphorylation played a role in postmortem glycolysis through the regulation of enzyme activity in muscle. Protein phosphorylation was related to postmortem glycolysis and glycolytic enzymes, and the phosphorylation level of glycolytic enzymes may influence its activities, further regulating postmortem glycolysis.

Keywords Protein phosphorylation · Glycolysis · Glycolytic enzyme · Glycolytic rate · Phosphoproteomic · Glycogen phosphorylase · Phosphofructokinase · Pyruvate kinase

6.1 Introduction

The biochemical changes related to postmortem metabolism occurring in muscle after slaughter were very important to the conversion of muscle into the meat (Paredi et al. 2012). Generally, they included glycogenolysis converting glycogen into glucose and glycolysis converting glucose into lactate and H^+, with a drop in pH value. Many literatures have reported that pH decline in the early postmortem was

D. Zhang et al., *Protein Phosphorylation and Meat Quality*,
https://doi.org/10.1007/978-981-15-9441-0_6

critical to meat quality development (Huang et al. 2011; Lindahl et al. 2006). Rapid and excessive glycolysis leaded to pale, soft, and exudative (PSE) meat, while insufficient glycolysis resulted in dark, firm, and dry (DFD) meat (Shen et al. 2006a; Immonen and Puolanne 2000). Very minute differences in the ultimate pH and rates of pH decline could cause great differences in meat quality. The underlying mechanisms are still not fully understood, however, pH and carbohydrate metabolism were always of key importance (Pösö and Puolanne 2005).

Posttranslational modifications had the ability to regulate protein structure, function, signalling, and activity. Protein phosphorylation, one of the most frequent posttranslational modification, was a key regulator in glycolysis metabolism (Graves and Krebs 1999). Most glycolytic enzymes were phosphoproteins, including the three rate-limiting enzymes, hexokinase, phosphofructokinase (PFK), and pyruvate kinase. PFK was activated after phosphorylation by AMP-activated protein kinase (AMPK) to upregulate glycolysis and maintain intracellular ATP homeostasis in the ischemic heart (Marsin et al. 2000). Phosphorylation of phosphoglycerate mutase at threonine 466 significantly increased the enzyme activity (Gururaj et al. 2004). Consistently, glyceraldehyde-3-phosphate dehydrogenase purified from the skeletal muscle of hibernating mammals showed lower phosphorylation levels compared to the control (Bell and Storey 2014). However, these studies have not been carried out in postmortem muscle, as differences may exist in the machinery regulating glycolysis antemortem and postmortem. During the postmortem time under anaerobic conditions, Shen et al. (2006b) also found out that AMPK regulated glycolysis in postmortem muscle at least partially through phosphorylation. Lametsch et al. (2011) proved that the pH decline of the rendement napole (RN) genotype could be a consequence of phosphorylation of glycolytic enzymes during the postmortem metabolism. All these studies showed that protein phosphorylation had a wide-range effect on the postmortem conversion of muscle to meat. Thus, it is necessary to have an overview of protein phosphorylation in postmortem muscle to understand its function.

The influence of protein phosphorylation on meat quality, including meat tenderness, color stability, juiciness, and so on, has been intensively studied (Li et al. 2017; Chen et al. 2016). The influence of phosphorylation on glycolysis has been proposed to be one possible mechanism. To further understand the mechanism regulating postmortem changes and meat quality development, especially the biochemistry of postmortem glycolysis, we profiled phosphoproteins in ovine muscle with the different glycolytic rate in the early postmortem to see the role of protein phosphorylation on the glycolytic rate and the effect of related glycolytic enzymes in ovine muscle. Furthermore, we profiled protein phosphorylation in postmortem muscle with kinase/phosphatase inhibitors to verify the effect of protein phosphorylation on glycolysis and elucidate the regulatory mechanisms from the perspective of enzyme activity. This research broadened our knowledge about the regulation of protein phosphorylation in postmortem muscle.

6.2 The Effect of Sarcoplasmic Protein Phosphorylation on Glycolysis in Postmortem Muscle

The phosphorylation level of sarcoplasmic proteins was investigated in relationship to the glycolysis rate in postmortem ovine muscle. The *longissimus thoracis* muscles of 60 sheep were removed immediately after evisceration and kept at 4 °C. The pH values of muscle were determined at 0.5, 2, 6, 12, 24, 48, and 72 h postmortem (PM) with a portable pH meter (Testo205 pH meter, Lenzkirch, Germany). Muscle samples at 0.5, 2, 6, 12, 24, 48, and 72 h PM were collected from the right side of carcasses and stored at −80 °C. Muscle samples were divided into three groups: fast pH decline (F) group (pH_{6h} < 5.75), moderate pH decline (M) group (5.75 < pH_{6h} < 6.20) and slow pH decline (S) group (pH_{6h} > 6.20) based on muscle pH values at 6 h (pH_{6h}) postmortem. Muscle pH values at 0.5 and 6 h PM were significantly different among the three groups ($P < 0.01$).

6.2.1 *Global Phosphorylation Level of Sarcoplasmic Protein with Different Glycolytic Rates*

Sarcoplasmic proteins were stained with Pro-Q Diamond and SYPRO Ruby after separating by SDS-PAGE to quantify protein phosphorylation by densitometric analysis (Fig. 6.1). Individual protein bands differences in phosphorylation levels were observed among the three groups. As shown in Fig. 6.2, the global phosphorylation level of sarcoplasmic proteins in all the three groups of muscles increased early postmortem and then decreased afterwards. The highest phosphorylation level in the F group was observed at 2 h PM, which was not determined in the M and the S groups until 6 h PM. The global phosphorylation level of sarcoplasmic proteins was significantly lower in the F group than in the S and the M groups at 6 h PM ($P < 0.05$).

6.2.2 *Phosphorylation Analysis of Individual Protein Bands*

Totally 20 protein bands were detected via significance analysis. The phosphorylation levels of 8 protein bands were significantly different among the three groups (Fig. 6.3). The phosphorylation levels of band 4 and 8 were significantly different at 0.5 h PM, whereas band 8, 11, 13, and 18 were significantly different at 2 h PM among the three groups. In general, the phosphorylation level of band 4, 11, 13, and 18 was higher in the F group, while band 8 was higher in the S group.

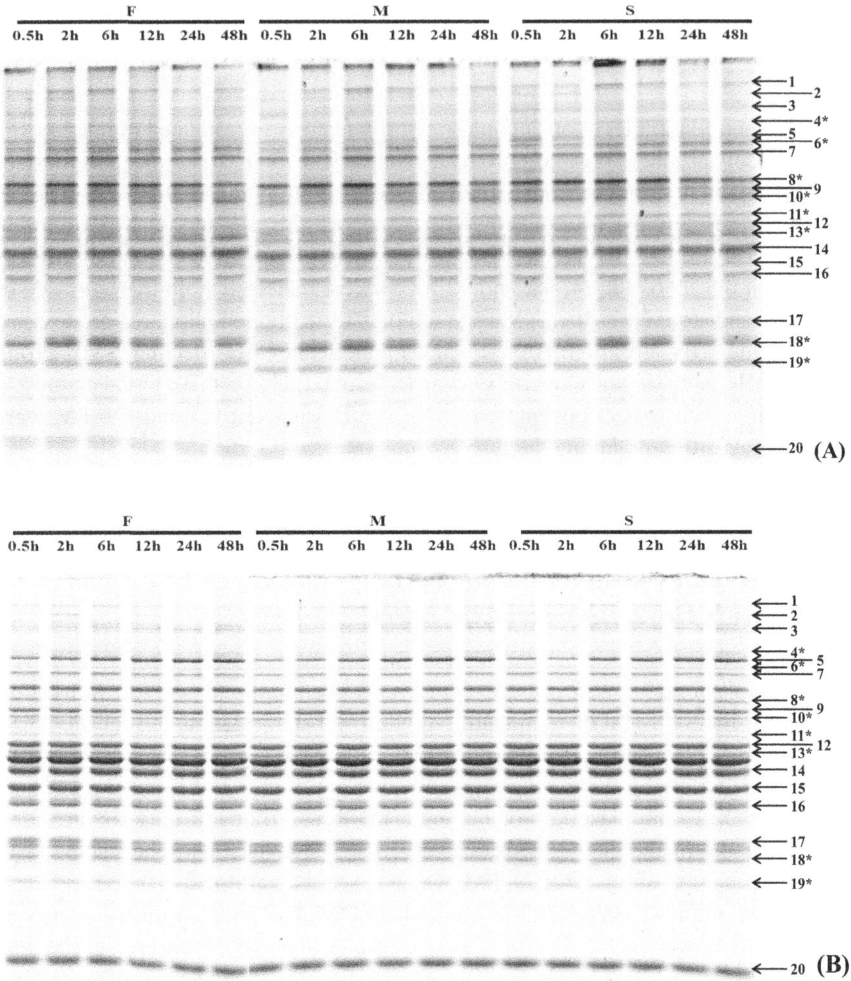

Fig. 6.1 Gel-base analysis of sarcoplasmic protein phosphorylation. (**a**) Images of gels stained with Pro-Q Diamond. (**b**) Images of gels stained with SYPRO Ruby. Bands marked with * were selected for protein identification by LC-MS/MS (Chen et al. 2018)

Correlation analysis between glycolytic rate attributes and phosphorylation level was conducted and the results were shown in Table 6.1. Band 6, 8, 11, 13, and 19 were found to correlate with the glycolytic rate (pH, glycogen, lactate acid). The glycolytic rate was negatively correlated with band 6, 11, 13, and 19, and positively correlated with band 8.

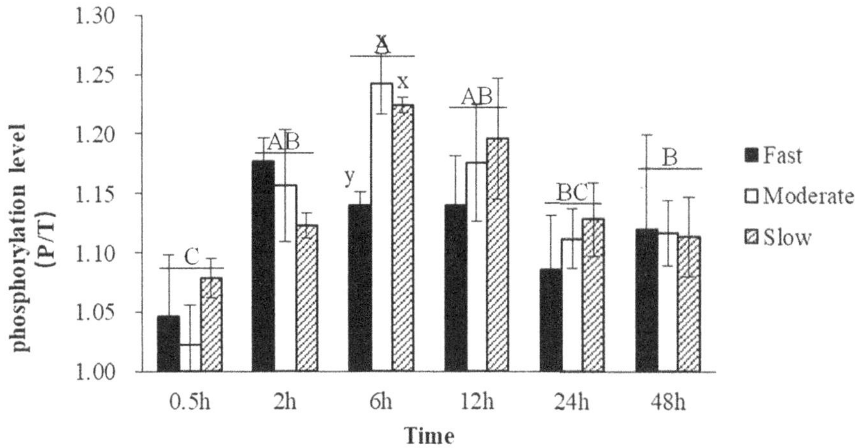

Fig. 6.2 Quantification of global phosphorylation levels of sarcoplasmic proteins. Different letters (x, y) at the same time points are significantly different between groups ($P < 0.05$). Different letters (A–C) are significantly different at different PM time ($P < 0.05$) (Chen et al. 2018)

6.2.3 Sarcoplasmic Protein Identification with Different Phosphorylation Level of the Three Glycolytic Rate Groups

The eight protein bands significantly different in phosphorylation levels were excised and identified by LC-MS/MS. Protein peptides were matched with Ovis aries database, and the proteins with a very high percentage of sequence coverage or a very high score were selected. In total, 17 unique proteins, molecular weight ranging from 21.75 to 118.18 kDa, were identified (Table 6.2). The name of all identified proteins was confirmed using Protein Knowledgebase (UniProtKB in www.uniprot.org). Most of the sarcoplasmic proteins identified were clustered in glycometabolism. Glycogen phosphorylase, creatine kinase, 6-PFK, enolase 3, phosphoglucomutase 1, glucose-6-phosphate isomerase, pyruvate kinase, phosphoglycerate kinase, triosephosphate isomerase, and AKI 1 were confirmed in the eight bands with significantly different phosphorylation level, which might be associated with the glycolytic rate.

There were significant correlations between the glycolytic rate and the phosphorylation levels of band 6, 8, 11, 13, and 19, where band 8 was positively correlated, and the others had a negative correlation.

PFK, identified in band 6, was the first committing step in the glycolytic pathway. The phosphorylation of PFK altered the kinetic behavior of the enzyme and regulated the compartmentalization of the enzyme which was observed in vitro and vivo studies (Luther and Lee 1986). Some studies have demonstrated that PFK was activated after phosphorylation by AMPK, contributing to maintain high glycolysis (Marsin et al. 2000).

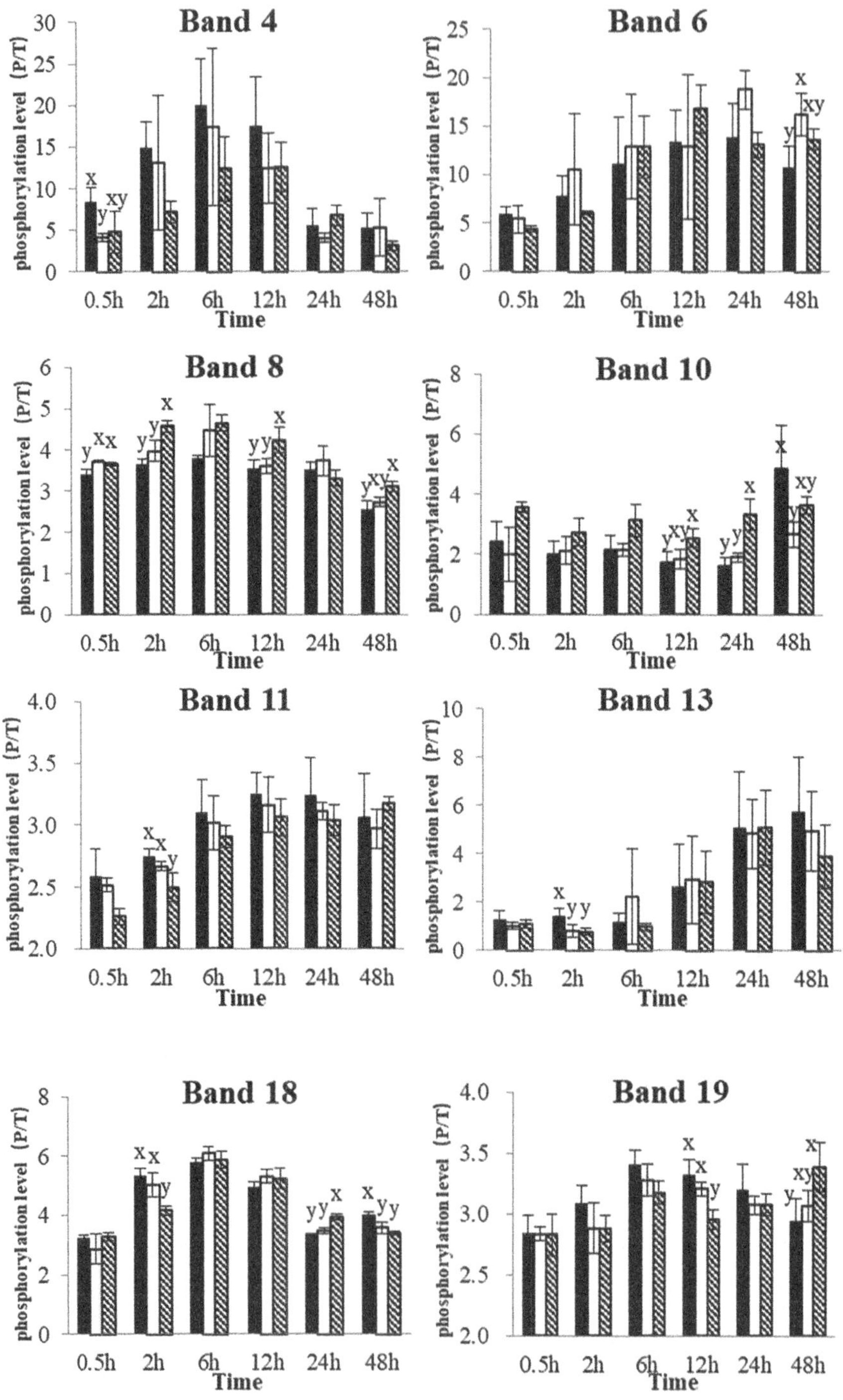

Fig. 6.3 The phosphorylation levels of 8 glycolytic rate related protein bands. The black, white, and dashed bars represent the fast glycolytic rate group, moderate glycolytic rate group, and slow

Table 6.1 Pearson correlation coefficients between glycolytic rate attributes and the phosphorylation level of total sarcoplasmic proteins or individual protein bands (Chen et al. 2018)

	pH value	Glycogen content	Lactate acid
Global phosphorylation level	−0.17039	−0.17906	−0.05086
Band 1	−0.46893*	−0.42422	0.56162*
Band 2	0.23692	0.32326	−0.25137
Band 3	−0.20943	−0.18129	0.28011
Band 4	−0.03318	0.07177	−0.38464
Band 5	0.28724	0.13035	−0.35938
Band 6	−0.80322**	−0.83210**	0.70452**
Band 7	−0.09155	−0.13035	−0.10880
Band 8	0.52808*	0.49561*	−0.58264*
Band 9	−0.24468	−0.33500	0.30962
Band 10	0.01289	−0.06243	0.23078
Band 11	−0.93183**	−0.91226**	0.69681**
Band 12	−0.13858	−0.08518	−0.14148
Band 13	−0.82957**	−0.85672**	0.94551**
Band 14	−0.38669	−0.38515	0.40643
Band 15	0.23815	0.27530	−0.21473
Band 16	0.49102*	0.37941	−0.23656
Band 17	−0.18796	−0.33037	0.24943
Band 18	−0.06107	−0.01097	−0.28540
Band 19	−0.64045**	−0.53918*	0.31309
Band 20	0.19653	0.21542	−0.37442

Note: Significant levels: *, $0.01 < p < 0.05$; **, $P < 0.01$

Enolase, identified in band 11 and 13, was a key enzyme in the glycolysis pathway. Enzyme activity assay demonstrated that the hyper-phosphorylation of a-enolase reduced its activity (Jin et al. 2008). The studies on phosphorylated enolase showed that the forward reaction was activated and the backward reaction was inhibited (Nettelblad and Engström 1987).

AKI, identified in band 19, was a key enzyme that could maintain muscle energy homeostasis. As the phosphotransferase, several isoforms of Adenylate kinase existed in mammalian tissues were well documented (Dzeja and Terzic 2009). Marjan Amiri reported that AKI catalyzed the phosphorylation of AMP and CMP with GTP as the phosphate donor, while AMP, dAMP, CMP, and dCMP were all phosphorylated when ATP was the phosphate donor (Amiri et al. 2013). Three phosphorylation sites, including S38, S178, and S181, have been identified in adenylate kinase in human skeletal muscle (Højlund et al. 2009). Therefore, AKI

Fig. 6.3 (continued) glycolytic rate group, respectively. Data are presented as mean ± standard deviation. Different letters (x, y) at the same time points are significantly different between groups ($p < 0.05$). (Chen et al. 2018)

Table 6.2 Identified proteins in the 8 glycolytic rate related protein bands by LC-MS/MS (Chen et al. 2018)

No.	Accession no[a]	Protein name	Score[b]	Mass[c]	Matches[d]	Sequence coverage[e]	Uniprot[f]
Band 4	W5NVR5	Uncharacterized protein	9308	118,179	555(345)	48	Calcium-transporting ATPase
	O18751	Glycogen phosphorylase	9284	97,702	524(334)	80	Glycogen phosphorylase, muscle form
	W5Q3Y4	Calcium-transporting ATPase	4030	107,971	260(177)	43	Calcium-transporting ATPase
	A8DR93	Heat shock protein	4742	85,077	249(166)	56	Submitted name: Heat shock protein 90 alpha family class A member 1
Band 6	W5QAA9	Uncharacterized protein (fragment)	11,188	85,981	604 (444)	56	Aconitate hydratase, mitochondrial
	O18751	Glycogen phosphorylase	6324	97,702	365 (262)	67	Glycogen phosphorylase, muscle form
	W5QDD4	6-phosphofructokinase	5696	95,200	311 (238)	52	ATP-dependent 6-phosphofructokinase
Band 8	W5PJB6	Uncharacterized protein (fragment)	38,705	65,544	1730 (1397)	87	Submitted name: Phosphoglucomutase 1
	W5QC41	Pyruvate kinase (fragment)	6407	62,180	297 (220)	65	Pyruvate kinase
	W5P323	Glucose-6-phosphate isomerase	4885	63,079	227 (164)	71	Glucose-6-phosphate isomerase
Band 10	W5P323	Glucose-6-phosphate isomerase	18,487	63,079	975 (695)	90	Glucose-6-phosphate isomerase
	P51977	Retinal dehydrogenase 1	5724	55,417	275 (206)	66	Retinal dehydrogenase 1
	W5QC41	Pyruvate kinase (fragment)	3667	62,180	210 (136)	67	Pyruvate kinase
Band 11	W5P663	Uncharacterized protein	10,328	47,382	502 (384)	73	Submitted name: Enolase 3
	W5PIG7	Uncharacterized protein	8853	48,565	415 (307)	69	Submitted name: Uncharacterized protein
	W5PIG6	Uncharacterized protein	8823	49,727	405 (305)	69	Submitted name: Uncharacterized protein
	W5PJ69	Uncharacterized protein	8209	43,213	469 (324)	82	Submitted name: Creatine kinase, M-type
	B7TJ13	Phosphoglycerate kinase	7134	44,936	436 (313)	87	Phosphoglycerate kinase

	W5PN24	Elongation factor 1-alpha	7057	56,244	413 (289)	58	Submitted name: Uncharacterized protein
	W5PEP7	ATP synthase subunit beta	6340	56,174	261 (202)	70	ATP synthase subunit beta
	W5PD15	Elongation factor 1-alpha	4149	50,451	267 (186)	53	Elongation factor 1-alpha
Band 13	B7TJ13	Phosphoglycerate kinase	23,359	44,936	1256 (926)	92	Phosphoglycerate kinase
	W5NYJ1	Uncharacterized protein	14,982	42,366	716 (558)	90	Submitted name: Uncharacterized protein
	W5PG09	Phosphoglycerate kinase	13,225	45,071	620 (478)	37	Phosphoglycerate kinase
	W5PJ69	Uncharacterized protein	11,124	43,213	641 (425)	88	Submitted name: Creatine kinase, M-type
	W5PZK7	Uncharacterized protein	10,977	42,381	617 (479)	84	Submitted name: Uncharacterized protein
	W5P663	Uncharacterized protein	7971	47,382	391 (289)	62	Submitted name: Enolase 3
	D7RIF5	Beta-actin variant 2	6885	42,052	450 (347)	64	Submitted name: Beta-actin variant 2
	W5Q0P3	Uncharacterized protein	5943	45,905	371 (265)	74	Submitted name: Creatine kinase, mitochondrial 2
Band 18	W5P5W9	Triosephosphate isomerase (fragment)	13,232	23,021	575 (413)	91	Triosephosphate isomerase
Band 19	C5IJA8	Adenylate kinase isoenzyme 1	8047	21,750	671 (425)	84	Adenylate kinase isoenzyme 1

[a]Accession numbers were derived from the UniProt database.
[b]For the proteins identified in more than one band, the highest score was presented.
[c]Theoretical molecular weight (recorded in UniProt database).
[d]Number of matched peptides.
[e]Percentage of coverage of the entire amino acid sequence.
[f]The result searching on the website, http://www.uniprot.org/.

was critical in glycolysis not only for the phosphorylation itself, but also for catalyzing the phosphorylation of the other phosphates.

Therefore, protein phosphorylation may be one of the reasons for the difference in glycolysis rate at 0.5 h PM. Further changes of phosphorylation level at different PM time, especially the regulation of kinase and phosphatase, will provide more clues to explore its precise mechanism.

6.3 Quantitative Phosphoproteomic Analysis of Muscle with Different Postmortem Glycolytic Rate

Protein phosphorylation of ovine muscle at 0.5 h postmortem was profiled in relationship to glycolytic rate. Quantitative proteomic tools isobaric Tags for Relative and Absolute Quantitation (iTRAQ) and bioinformatics analysis were used to further understand the mechanisms regulating postmortem changes.

6.3.1 Phosphoprotein Identification and Motif Analysis

A total of 1905 phosphopeptides were identified, which were assigned to 704 phosphoproteins. As to significant difference, 97 phosphopeptides were determined to be different in abundance between F and M groups, of which 67 peptides were downregulated and 30 were upregulated in F group compared to the M group. Forty one phosphopeptides had a significant difference in abundance between M and S groups, of which 15 peptides were downregulated and 26 were upregulated compared to S group. Eighty nine phosphopeptides were differently expressed in F and S groups, of which 51 peptides were downregulated and 38 were upregulated compared to S group. A total of 116 unique phosphopeptides, matching to 98 phosphoproteins and containing 188 phosphorylation sites, were significantly different in abundance among the three groups after one-way ANOVA analysis. Generally, 160 phosphoserine, 26 phosphothreonine and 2 phosphotyrosine residues (ratios of 85.11%, 13.83% and 1.06% respectively) were identified. Sixteen putative phosphorylation motifs were identified after analyzing with the Motif-X software, 13 serine motifs PS and 3 threonine motifs PT were included (Fig. 6.4). It is logical to deduce that proteins phosphorylated at these sites may play a role in the glycolysis pathway.

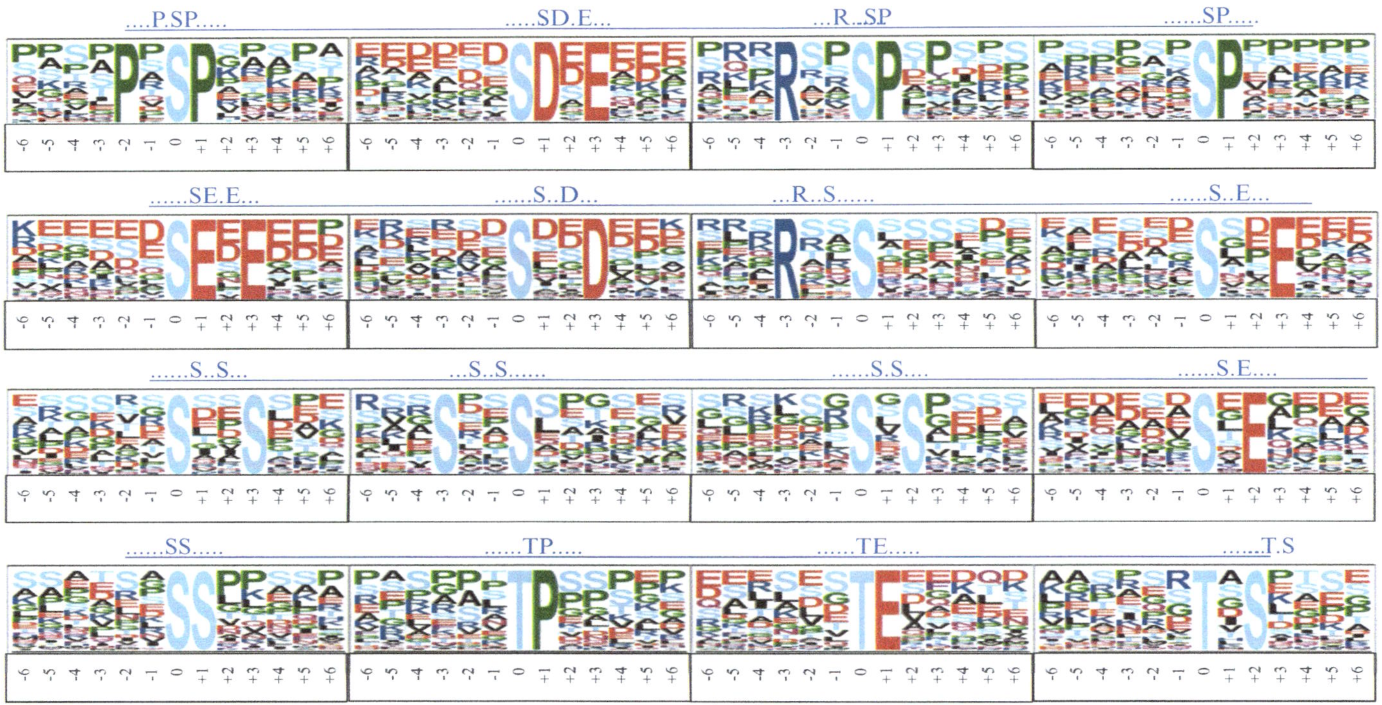

Fig. 6.4 Motif analysis among the three glycolytic rate groups. Note: The Logo-like representations of putative motifs are identified from all unambiguous phosphorylation sites. The height of the residues represents the frequency with which they occur at the respective positions. The color of the residues represents their physicochemical properties

6.3.2 Hierarchical Clustering Analysis

The 116 phosphopeptides with a significant difference in phosphorylation level were applied to hierarchical clustering analysis in Fig. 6.5. All 116 phosphopeptides were shown in 116 rows. Different colors in the same row represented phosphorylation levels of one phosphopeptide among the three groups. Red color represented a high abundance and a high phosphorylation level, while the green color meant a low abundance and a low phosphorylation level. The phosphorylation level of all the 116 phosphopeptides were significantly different among the three groups, especially between the F and S groups. The phosphopeptides with high phosphorylation levels in the F group (upper part in Fig. 6.5) had a low phosphorylation level in the S group,

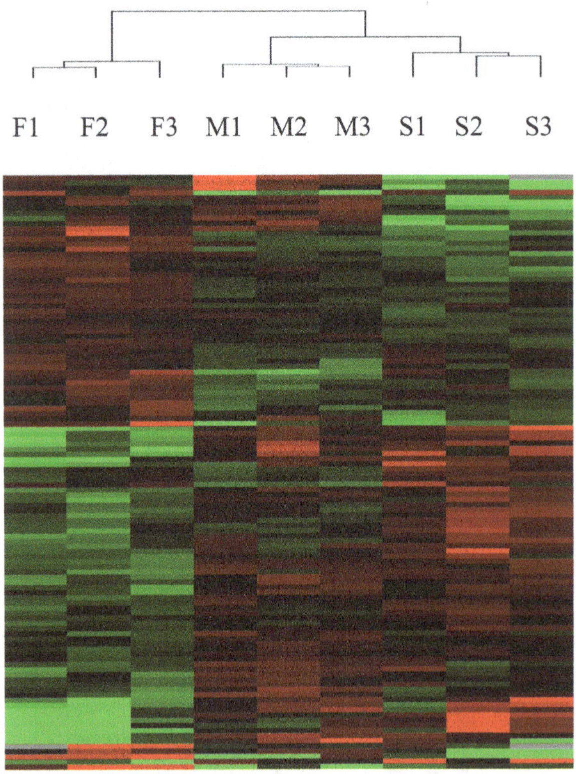

Fig. 6.5 Hierarchical clustering analysis among three glycolytic rate groups. Note: Muscle samples are displayed in columns and classified by phosphoproteomic subtypes as indicated by different glycolytic rates. Two random samples in the same group were mixed as a replication. F1, F2, and F3 are the three replications in the fast glycolytic rate group. M1, M2, and M3 are the three replications in the moderate glycolytic rate group. S1, S2, and S3 are the three replications in the slow glycolytic rate group. The same row represents one phosphopeptide, and different color represent phosphorylation levels. A redder color means a higher phosphorylation level and a greener color means a lower phosphorylation level of the phosphopeptide (Chen et al. 2019b)

while the phosphopeptides with low phosphorylation level in the F group (lower part of Fig. 6.5) had a high phosphorylation level in the S group. At the bottom, a few phosphopeptides were expressed irregularly among the three groups. Hierarchical cluster analysis visualized the specificity and reproducibility of the experiment.

6.3.3 Functional Enrichment Analysis

Gene Ontology (GO) terms enrichment and Kyoto Encyclopedia of Genes and Genomes (KEGG) pathway enrichment were conducted to obtain the important information about the regulation mechanism of glycolysis (Fig. 6.6). All the differentially expressed phosphoproteins were used to perform protein–protein interaction network analysis and three different clusters were showed in Fig. 6.7. The largest one was the muscle contraction related proteins, and the closest interaction one was the glycolytic enzymes. Pyruvate kinase, phosphoglucomutase 1, enolase2, enolase3, and fructose-bisphosphate aldolase were identified to be the glycolytic rate related phosphoproteins in the present study, which were detected to be phosphorylated at Thr157, Ser402, Ser177, Ser176, S124, and Ser127, respectively, whose phosphorylation level was significantly different among the three groups.

6.3.4 Quantitative Analysis of Phosphopeptides

More than one phosphopeptides and phosphosites were identified in the 24 glycolytic rate related phosphoproteins in the present study. Totally, 390 phosphopeptides were identified from 24 phosphoproteins, in which 32 phosphopeptides were different in phosphorylation levels among the three groups (Table 6.3). Moreover, the phosphopeptides and phosphosites revealed different phosphorylation levels among the three groups.

Glycolysis is a sequence of enzymatic reactions that are determined by the activities of glycolytic enzymes. Several studies revealed that the glycolytic enzymes influenced the transition process of muscle to meat for its activities changed after slaughter (Werner et al. 2010; Huang et al. 2011). Protein phosphorylation played a regulatory role in protein structure, function, signaling, and activity regulation. Most glycolytic enzymes have been reported to be phosphorylated and phosphorylation increase enzymes' activity or stability (Sale et al. 1987; Reiss et al. 1986). Pyruvate kinase, phosphoglucomutase 1, enolase2, enolase3, and fructose-bisphosphate aldolase were identified to be the glycolytic rate related phosphoproteins in the present study, which were detected to be phosphorylated at Thr157, Ser402, Ser177, Ser176, S124, and Ser127, respectively. Their phosphorylation level was significantly different among the three groups, with a significantly higher level in M and S groups than in the F group (Table 6.4).

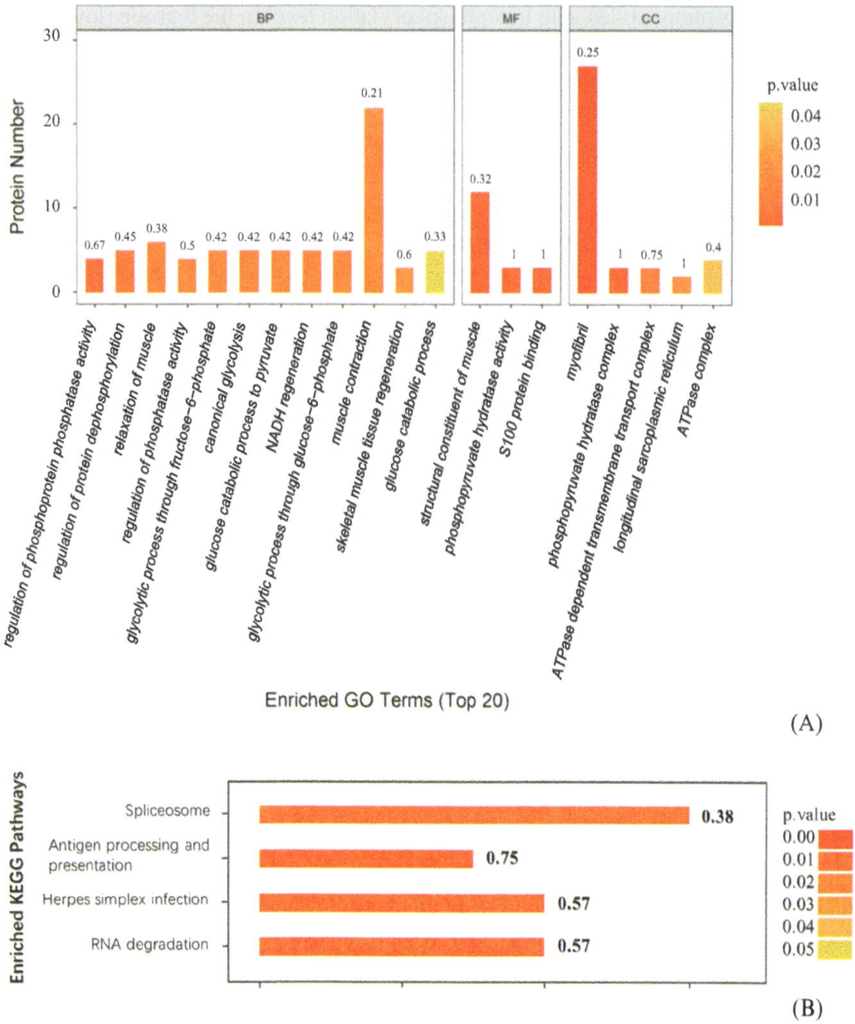

Fig. 6.6 GO terms enrichment and KEGG pathway enrichment of differently phosphorylated proteins in muscles of different glycolytic rate groups. (**a**) The enriched GO terms of differently expressed phosphoprotein among the three groups. (**b**) The enriched KEGG pathways of differently expressed phosphoprotein among the three groups. Note: The abscissa in the **a** indicates the enriched GO function, BP, MF, and CC represent the biological process, molecular function, and cellular component, respectively. The ordinate in the **b** indicates the significant KEGG pathway. The numbers above the bars called rich factor, it indicates the ratio of proteins corresponding to the phosphopeptides significantly different in phosphorylation level to all identified proteins (Chen et al. 2019b)

Pyruvate kinase, a rate-limited glycolytic enzyme, catalyzed the conversion of phosphoenolpyruvate to pyruvate irreversibly. Phosphorylation at Thr157 showed a significant difference among the three groups, with phosphorylation level being

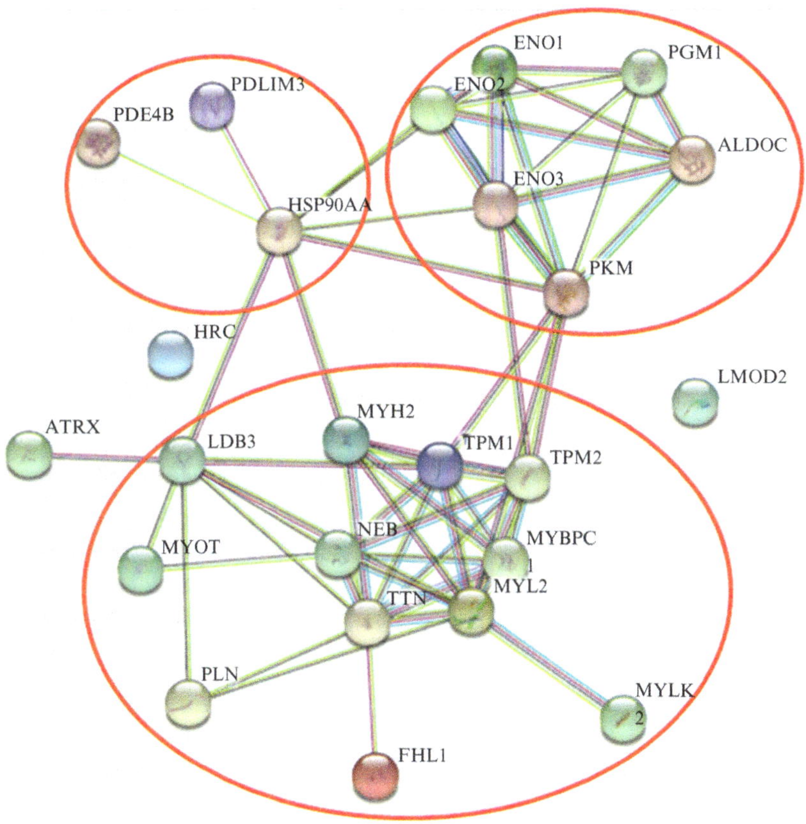

Fig. 6.7 Protein–protein interaction networks of identified glycolytic rate related phosphoproteins in ovine muscle. Different clusters of interacting proteins were identified using STRING to obtain a high confidence evidence network (Chen et al. 2019b)

higher in the S group and lower in the F group. The trend was consistent with Chen's result (Chen et al. 2018). It has been reported that pyruvate kinase had two isoforms in normal muscle. Isoforms 2 arose from isoform 1 through phosphorylation. Phosphorylation of pyruvate kinase could result in an additional, more acid-stable enzyme isoform, and maintain high activity in PSE meat (Zhang and Liu 2017). Therefore, the low phosphorylation level in the F group at early postmortem indicated the high activity of pyruvate kinase.

The activity of phosphoglucomutase 1 may alter after phosphorylation, which can stimulate the drop of pH value (Gururaj et al. 2004). Fructose 1,6-bisphosphate aldolase was one of the phosphoproteins that may be directly linked to postmortem pH decline (D'alessandro and Zolla 2013). The activity of enolase 1 was reduced at hyper-phosphorylation condition (Jin et al. 2008). Based on these reports, we supposed that the phosphorylation level of these glycolytic enzymes was related to the glycolytic rate at early postmortem.

Table 6.3 The 32 phosphopeptides and their phosphosites of 24 glycolytic rate related phosphoproteins (Chen et al. 2019b)

No	Accessions	Phosphoproteins	Sequence (significantly different among groups)	Phosphosite in peptide	Location in phosphoproteins	F	M	S
1	W5NQP9	Fructose-bisphosphate aldolase	gILAADEsVGsMAk	S(8): 100.0; S(11): 100.0	S124,S127			
2	W5QC41	Pyruvate kinase	gPEIRtGLIk	T(6): 100.0	T157			
3	W5PJB6	Phosphoglucomutase 1	fFGNLMDAsk	S(9): 100.0	S402			
4	W5P5C0	Enolase 2	lAMQEFmILPVGAEsFR	S(15): 100.0	S177			
5	W5P663	Enolase 3	lAMQEFmILPVGAsSFR	S(14): 97.2; S(15): 2.8	S176			
6	W5PT09	Myosin heavy chain 2	gQTVEQVtNAVGALAk	T(3): 0.0; T(8): 100.0	T422			
7	W5Q0I1	Myosin binding protein C	sGEGQDDAGELDFSGLLk	S(1): 100.0; S(14): 0.0	S151			
			fDScSFDLEVHESTGtPNIDIR	S(3): 0.0; S(5): 0.0; S(13): 13.3; T(14): 13.3; T(16): 86.7; T(17): 86.7	T138,T139			
8	W5Q754	Titin	tSEMEASSsVR	T(1): 0.0; S(2): 0.0; S(7): 2.2; S(8): 2.2; S(9): 95.6	S34313			
			aEFVcsISk	S(6): 97.4; S(8): 2.6	S14375			
			rsLGDIsDEELLLPIDDYLAmk	S(2): 100.0; S(7): 100.0; Y(18): 0.0	S33892,S33897			
			vTsDNLmSR	T(2): 2.3; S(3): 95.4; S(8): 2.3	S18570			
			tIVsTAQISETR	T(1): 0.0; S(4): 2.1; T(5): 95.8; S(9): 2.1; T(11): 0.0	T210			
9	W5PFV9	Nebulin	ftSVTDsLEQVLAk	T(2): 97.4; S(3): 2.5; T(5): 0.1; S(7): 99.9	T2938, S2943			
			mVGFQsLQDDPk	S(6): 100.0	S1099			
10	W5Q1Q5	Myotilin	sRssSRGDmSDQDAIQEk	S(1): 97.7; S(3): 67.3; S(4): 67.3; S(5): 67.3; S(10): 0.6	S230, S232, S233, S234			
11	B2LU28	TPM1	aIsEELDHALNDMTsI	S(3): 100.0; T(14): 50.0; S(15): 50.0	S271, T282/S283			
			lEksIDDLEDELYAQk	S(4): 100.0; Y(13): 0.0	S252			
12	W5QFE3	Cardiac phospholamban	astIEmPQQAR	S(2): 100.0; T(3): 100.0	S16, T17			
13	W5NU05	Myosin light chain kinase 2	iSsSGALMALGV	S(2): 1.3; S(3): 97.4; S(4): 1.3	S617			
14	B6VCA9	Myosin light chain 2	rAEGANsNVFSmFEQTQIQEFk	S(6): 100.0; S(10): 100.0; T(15): 100.0	S15,S19,T24			
15	A9P323	Heat shock protein 90 alpha	lGIHEDsQNR	S(7): 100.0	S443			
16	W5PDJ3	PDZ and LIM domain 3	lWsPQVTEDGk	S(3): 100.0; T(7): 0.0	S89			
17	W5PBB8	Phosphodiesterase	eWYQStIPQSPsPAPDDQEEGR	Y(3): 0.0; S(5): 0.0; T(6): 0.0; S(10): 100.0; S(12): 100.0	S585, S587			
18	C8BKC8	Four and a half LIM domains 1 protein	hSGPsSYk	S(2): 1.1; S(5): 48.9; S(6): 48.9; Y(7): 1.1	S10/S11			
19	W5NR35	LIM domain binding 3	dLTVDSAsPVYQAVIk	T(3): 0.0; S(6): 0.0; S(8): 100.0; Y(11): 0.0	S213			
20	W5QBA6	ATRX, chromatin remodeler	aTSsSNPsSPAPDWYk	T(2): 2.7; S(3): 2.7; S(4): 47.5; S(5): 47.5; S(8): 49.8; S(9): 49.8; Y(15): 0.0	S1945/S1946			
21	W5PT08	Histidine rich calcium binding protein	hAGHEDDDDGDDAVsTER	S(15): 100.0; T(16): 0.0	S245			
			eSDsEEDEEEkEEDRssHEEANEGSEEGGEGTR	S(2): 3.6; S(4): 96.4; S(16): 50.0; S(17): 50.0; S(25): 100.0; T(32): 0.0	S517, S538			
			hQGHEEEADDEDDDDIVsTEHR	S(18): 50.0; T(19): 50.0	S298, T299			
			eEEDEDEGEENVsTEYGQQVHR	S(13): 98.5; T(14): 1.5; Y(16): 0.0	S196			
22	W5NS29	Leiomodin 2	yEsIDEDELLAsLSAEELk	Y(1): 0.2; S(3): 99.8; S(12): 96.4; S(14): 3.6	S15, S24			
23	W5PIG7	Enolase 1	-	-	-		-	
24	W5PQL7	Tropomyosin 2	-	-	-		-	

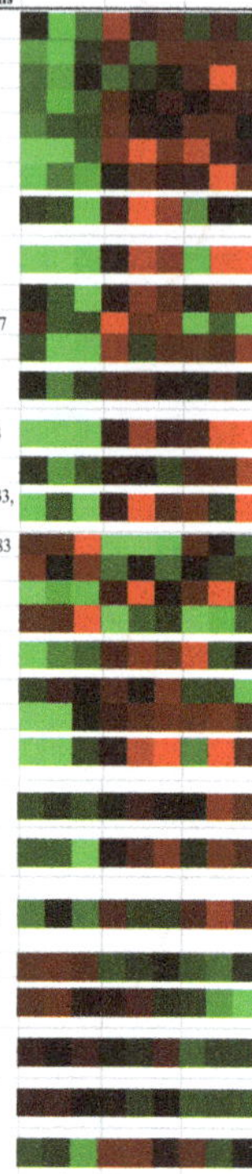

Note: Location in peptide indicates phosphosite in peptide derived in UniProt database. F, M and S represent three groups. The colour bar represents the phosphorylation level of phosphopeptide from Hierarchical clustering analysis. The phosphorylation levels were marked from red to green in abundance from high to low. A redder colour means a higher phosphorylation level and a greener colour means a lower phosphorylation level of the phosphopeptides. "-" represents there is no significantly different phosphopeptide in the phosphoprotein.

Glycolysis was a dynamic physiological and developmental process. In the glycolysis pathway, the metabolism of one molecule of glucose to two molecules of pyruvate had a net yield of two molecules of ATP, with the concomitant production of lactate and muscle pH decline. Protein phosphorylation was regulated by protein kinase which transferred a phosphate group from a nucleoside triphosphate (usually ATP) and covalently attached it to amino acids. Phosphorylation of pyruvate kinase could maintain high activity in low pH value. Therefore, ATP and low pH could also affect phosphorylation.

Further research with different postmortem time will be beneficial to fully understand the regulation of protein phosphorylation on postmortem changes and meat quality development.

Table 6.4 Protein identification of 11 gel bands with significant difference (Chen et al. 2019a)

No.	Accession no[a]	Protein name	Mass[b]	Score[c]	Matches[d]	Sequence coverage[e]
Band 1	W5NZK9	Filamin C	278,503	5152	282(168)	59(45)
Band 5	O18751	Glycogen phosphorylase	97,702	1994	188(86)	38(25)
Band 7	W5QDD4	ATP-dependent 6-phosphofructokinase	95,200	1874	123(73)	21(16)
Band 8	W5PJB6	Phosphoglucomutase 1	65,544	3401	188(125)	24(21)
Band 9	W5QC41	Pyruvate kinase	62,180	3084	164(114)	22(19)
Band 10	W5P323	Glucose-6-phosphate isomerase	63,079	3466	220(151)	16(13)
Band 11	W5PIG6	Enolase 1	49,727	1096	86(52)	15(11)
	W5P663	Enolase 3	47,382	1006	86(44)	14(11)
Band 14	W5P1X9	Fructose-bisphosphate aldolase	39,925	1855	122(69)	20(18)
Band 15	W5PDG3	Glyceraldehyde-3-phosphate dehydrogenase	36,241	1502	117(51)	14(9)
Band 18	W5P5W9	Triosephosphate isomerase	23,021	2637	116(89)	10(10)
Band 19	C5IJA8	Adenylate kinase isoenzyme 1	21,750	2140	142(99)	11(8)

Note: The ovine *Longissimus thoracis* were crushed and incubated at 4 °C, and then snap frozen in liquid nitrogen after collection at different time. The 11 bands selected were significantly different in the phosphorylation level among the three groups

[a]Accession numbers were derived from the UniProt database.

[b]Theoretical molecular weight (recorded in UniProt database).

[c]For the proteins identified in more than one band, the highest score was presented.

[d]Number of matched peptides, total matched peptides (credible matched peptides).

[e]Number of matched amino acid sequence, total matched sequence (credible matched sequence).

6.4 Validation of Protein Phosphorylation on Glycolysis and the Regulation Mechanism of Enzyme Activity

To verify the effect of protein phosphorylation on glycolysis and elucidate the regulatory mechanism from the perspective of enzyme activity, the ovine muscle was treated with a kinase inhibitor, dimethyl sulfoxide, or a phosphatase inhibitor, and the activity of glycogen phosphorylase, pyruvate kinase, and PFK was measured.

6.4.1 Global Phosphorylation of Sarcoplasmic Protein with the Regulation of Kinase or Phosphatase Inhibitors

The global phosphorylation level of the sarcoplasmic protein was evaluated by densitometry analysis of phosphoproteins and total proteins with gel stain. The global phosphorylation level of the sarcoplasmic protein increased early PM and then decreased ($P < 0.05$). These results were consistent with Huang's data about protein phosphorylation in muscle with intermediate pH decline rates (Huang et al. 2011). After incubation for 12 h, the phosphorylation level of the kinase inhibitor group was significantly lower than that in the other two groups, and the phosphorylation level of the phosphatase inhibitor group was significantly higher at 48 h PM compared to the control group ($P < 0.05$) (Fig. 6.8).

Protein phosphorylation and dephosphorylation had diverse effects in cellular regulation and signalling, which were regulated by the competing activities of protein kinases and phosphatases (Manning et al. 2002). Protein kinases catalyzed the transfer of phosphate groups from ATP to molecules, whereas phosphatases removed phosphate groups from molecules. P0300, as a kinase inhibitor, was a

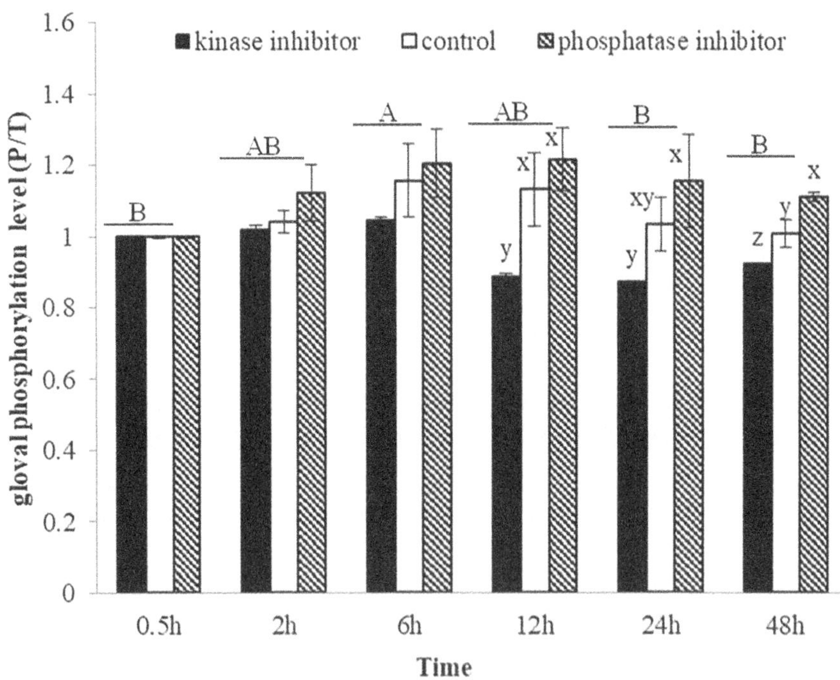

Fig. 6.8 Global phosphorylation level of sarcoplasmic protein. Different letters (x, y, z) at the same time points represent significant difference among the three groups ($P < 0.05$). Different letters (A–B) in the same group represent significant difference at different PM time ($P < 0.05$). (Chen et al. 2019a)

synthetic peptide used for studying the cAMP-dependent protein kinase and could inhibit this protein kinase competitively. PhosStop Roche, used as a phosphatase inhibitor, had a broad spectrum of action against phosphatases, inhibiting protein phosphatases competitively. The samples were "intact" meat, which involved a series of physic-biochemical changes. The inhibitors showed a significant difference after 12 h of incubation, maybe because they needed time to enter the muscle cell to function. The significant difference among the three groups revealed that the inhibitors were effective in modulating the level of protein phosphorylation.

6.4.2 Glycolytic Rate

The pH value and lactic acid content of the three groups were shown in Fig. 6.9. Generally, the ovine muscle pH decreased dramatically within the first 12 h, but remained stable afterwards. Meanwhile, the lactic acid corresponded well with the pH value, which increased within the first 12 h but remained stable afterwards. As to the three groups, the pH value in the phosphatase inhibitor group was significantly lower at 2 h and 6 h than those in the control group, and the lactic acid in the phosphatase inhibitor group was significantly higher than that in the kinase inhibitor group during the PM time except at 6 h ($p < 0.05$). There was no significant difference in pH values and lactic acid between the control and kinase inhibitor groups throughout the whole PM period ($p < 0.05$).

The pH value was negatively correlated with lactate during the PM time. Lactate was the ultimate product of anaerobic metabolism. With the formation of lactate, one hydrogen from Nicotinamide adenine dinucleotide (NADH) and one hydrogen from the solution were removed from the cytoplasm (Ferguson and Gerrard 2014). Ferguson claimed there was utility in lactate formation for extending anaerobic muscle metabolism and suggested that lactate accumulation was a good indicator for the extent and rate of glycolysis.

Ovine muscle in the phosphatase inhibitor group with a high phosphorylation level had higher lactate and a lower pH value compared to the control group. Li et al. (2017) reported that the muscle of the kinase inhibitor group with a low phosphorylation level had lower lactate and a higher pH value compared to the control group. Therefore, it is logical to conclude that a high protein phosphorylation is one of the reasons for the fast decline in the pH value.

6.4.3 Gel Band Identification

Nineteen protein bands were detected on the gels, among which 11 protein bands that significantly differed in phosphorylation level were excised and identified (Table 6.4). In total, 12 unique proteins were identified and most of them were clustered into the pathways of glycogenolysis and glycolysis. They were PFK,

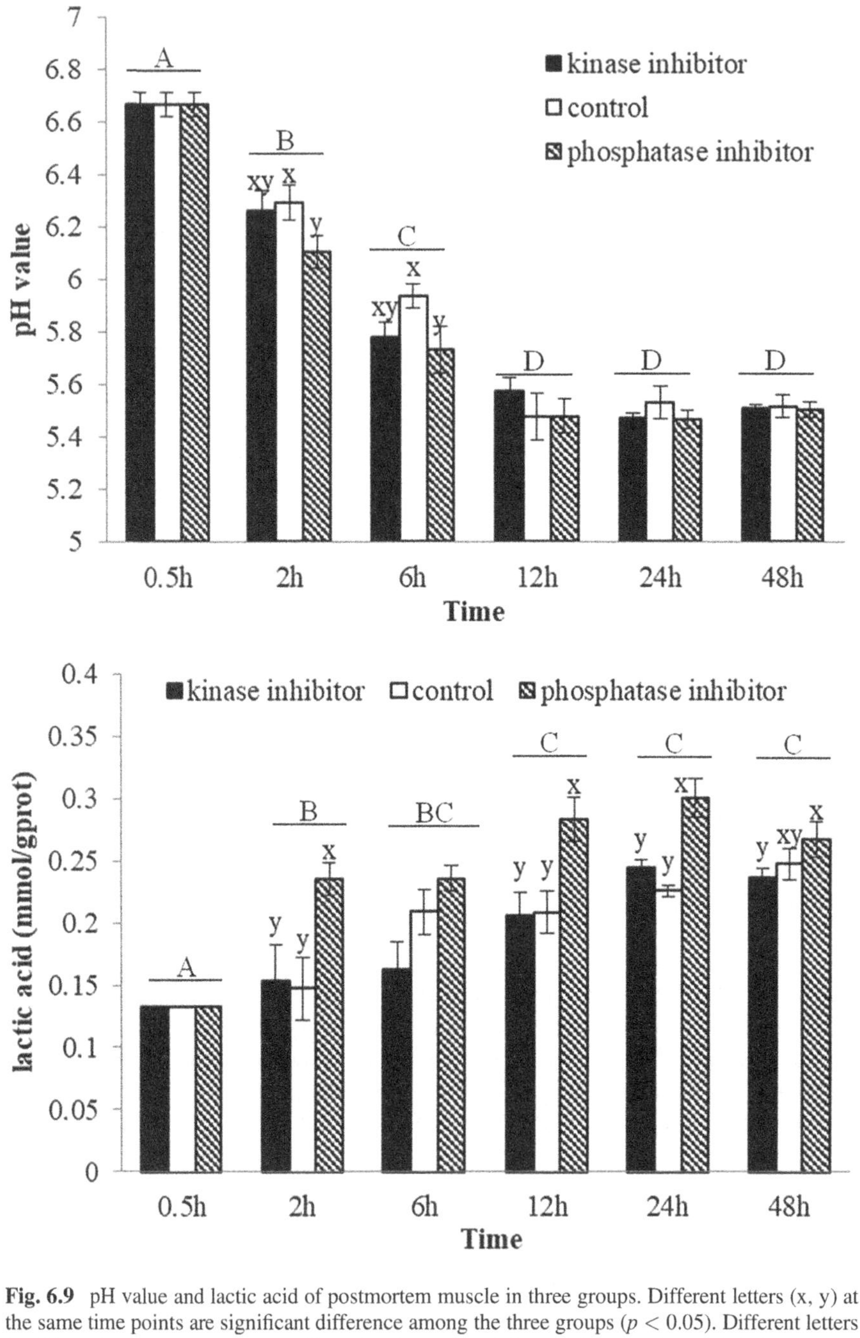

Fig. 6.9 pH value and lactic acid of postmortem muscle in three groups. Different letters (x, y) at the same time points are significant difference among the three groups ($p < 0.05$). Different letters (A–D) in the same group are significantly different at different PM time ($p < 0.05$) (Chen et al. 2019a)

Filamin C, GP, Phosphoglucomutase 1, PK, Glucose-6-phosphate isomerase, Fructose-bisphosphate aldolase, Enolase 1, Enolase 3, Triosephosphate isomerase, Glyceraldehyde-3-phosphate dehydrogenase, and AKI 1. Among these enzymes, PFK and PK were rate-limiting enzymes in glycolysis, while GP catalyzed the rate-limiting step in glycogenolysis. To explain glycolysis, the properties of glycogen phosphorylase, PFK, and pyruvate kinase have been studied (Scheffler and Gerrard 2007). Therefore, these enzymes should be given more attention.

6.4.4 Glycolytic Enzymes Activities and Phosphorylation Level After Regulation

6.4.4.1 Glycogen Phosphorylase

The phosphorylation level and activity of glycogen phosphorylase (GP) were shown in Fig. 6.10. The activity of GP was basically stable during the postmortem (PM) time. As to the three groups, the activity of GP in the phosphatase inhibitor group was significantly higher than in the other two groups ($P < 0.05$). The phosphorylation level of GP in the phosphatase inhibitor group at 2 h PM was significantly higher than that in the control group, which was not observed at the other PM times ($P < 0.05$). The correlation analysis given in Table 6.5 showed that the activity of GP had a positive correlation with the phosphorylation level of GP during the PM time.

GP catalyzed the breakdown of glycogen to glucose-1-phosphate for glycolysis and was crucial in controlling glycogenolysis (Aizawa et al. 2017). GP existed in phosphorylated GP a and dephosphorylated GP b forms. When GP b was phosphorylated at serine 14, its structure changed, transforming it into the active GP a form, which represented the first step in enzyme activation (Schwägele et al. 1996; Johnson 1992; Sprang et al. 1988). The change in structure involved of the amino- and carboxyl-terminal domains of GP rotating apart by 5°, which increased the access of substrates to the catalytic site (Sprang et al. 1991). The activity of GP was significantly higher in the phosphatase inhibitor group, theoretically, more GP a form and a higher phosphorylation level should exist in the phosphatase inhibitor group. Although without a statistical difference, there was a numerical difference between the control and phosphatase inhibitor groups. The increased activity of GP further accelerated the rate of glycogenolysis, resulting in low glycogen content and pH value. Therefore, phosphorylation of GP was important in the regulation of glycolysis.

6.4.4.2 Pyruvate Kinase

The phosphorylation level and the activity of pyruvate kinase (PK) was shown in Fig. 6.11. There was no significant difference in phosphorylation level among the

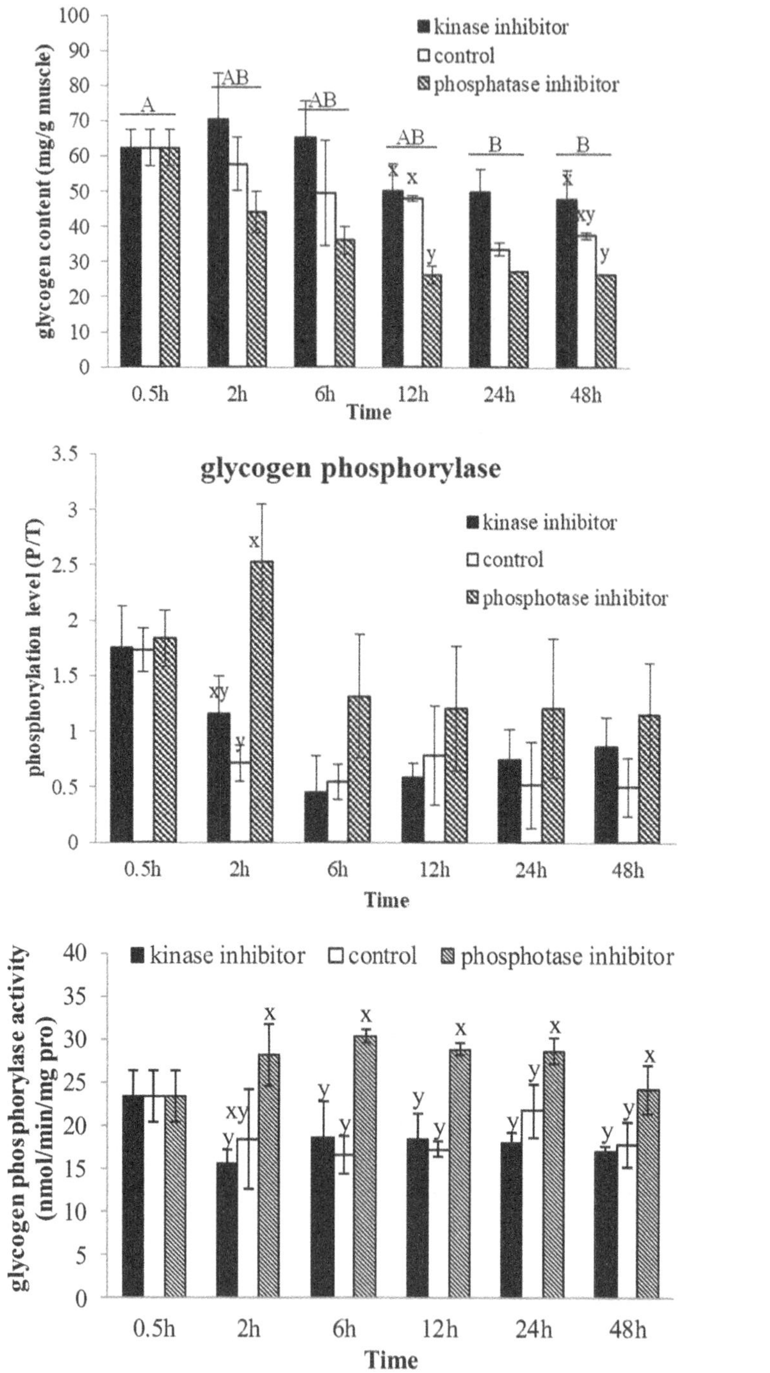

Fig. 6.10 Glycogen content, the phosphorylation level, and activity of glycogen phosphatase in postmortem muscle. Different letters (x, y) at the same time points are significantly different among

three groups except at 12 h PM, where the phosphorylation level of PK in the phosphatase inhibitor group was significantly higher than that in the control group ($P < 0.05$). As for the enzymatic activity, there was a significant difference among the three groups from 6 h onwards, with the activity in the phosphatase inhibitor group being significantly higher than in the other two groups ($P < 0.05$).

PK was a critical rate-limiting enzyme that catalyzed the irreversible conversion of phosphoenolpyruvate to pyruvate (Gupta and Bamezai 2010). It has been reported that PK had two isoforms in muscle: isoform 2 arose from isoform 1 through phosphorylation (Schwägele et al. 1996). Phosphorylation could cause pyruvate kinase to retain higher activity under acidic conditions (Schwägele et al. 1996). Heiden et al. (2010) reported that PK was found to be more active after enzyme phosphorylation. In the phosphatase inhibitor group, the phosphorylation level and the activity had the same trend in PM time. In terms of the three groups, the phosphatase inhibitor group had a relatively higher phosphorylation level and activity. Since protein phosphorylation altered the structure, activity, and stability of the protein, it was likely that phosphorylation of PK had a positive effect on the activity of PK.

6.4.4.3 Phosphofructokinase

The phosphorylation level and activity of PFK were shown in Fig. 6.12. There was no significant difference in the phosphorylation level between the kinase inhibitor and control groups PM ($P < 0.05$). After 24 h incubation, the phosphorylation level of PFK in the phosphatase inhibitor group was significantly higher than that in the control group ($P < 0.05$). The activity of PFK was significantly higher in the phosphatase inhibitor group at 6 h and 12 h than in the other two groups ($P < 0.05$). The activity of PFK was positively correlated with the phosphorylation level (Table 6.5).

PFK catalyzed the conversion of fructose 6-phosphate and ATP to fructose 1,6-bisphosphate and ADP, which was the first key step in the glycolytic pathway. Phosphorylation of PFK altered the kinetic behavior of the enzyme and regulated the compartmentalization of the enzyme in vitro and in vivo (Luther and Lee 1986). Some studies have demonstrated that PFK was activated after phosphorylation by AMPK, thus promoting glycolysis (Marsin et al. 2000). The kinase inhibitor P0300 used in the present research was a specific kinase inhibitor, which inhibited the cAMP-dependent protein kinase, specially protein kinase A, contributing no significant difference in PFK phosphorylation level between the kinase inhibitor and control groups. AMPK could phosphorylate PFK-2 which catalyzed the formation of fructose 2,6-bisphosphate. This product was an allosteric activator of PFK-1, a

Fig. 6.10 (continued) the three groups ($P < 0.05$). Different letters (A, B) in the same group are significantly different at different PM time ($P < 0.05$). Adapted from (Chen et al. 2019a)

Table 6.5 Pearson correlation coefficients of glycolytic rate, glycolytic enzymes phosphorylation level and activities (Chen et al. 2019a)

	Glycogen	GP activity	GP p-level	PK activity	PK p-level	PFK activity	PFK p-level	Global p-level	pH	lactate
Glycogen	1	**−0.527***	0.1006	−0.0577	−0.4174	−0.2105	−0.38	**−0.496***	**0.7199****	**−0.912****
		0.0246	0.6912	0.82	0.0848	0.4019	0.1197	**0.036**	**0.0008**	**<0.0001**
GP activity	**−0.52704***	1	**0.6326****	**0.7325****	**0.5842***	**0.7592****	**0.771****	**0.588***	0.0352	0.3638
	0.0246		**0.0048**	**0.0005**	**0.0109**	**0.0003**	**2E-04**	**0.01**	0.8896	0.1378
GP p-level	0.10061	**0.6326****	1	**0.619****	0.1543	**0.5528***	**0.561***	0.2009	**0.5842***	−0.171
	0.6912	**0.0048**		**0.0062**	0.5409	**0.0174**	**0.016**	0.4241	**0.0109**	0.4962
PK activity	−0.05773	**0.7325****	**0.619****	1	0.2837	**0.7711****	**0.594****	0.3455	0.4599	−0.105
	0.82	**0.0005**	**0.0062**		0.254	**0.0002**	**0.009**	0.1603	0.0548	0.6797
PK p-level	−0.41739	**0.5842***	0.1543	0.2837	1	0.3872	**0.492***	**0.492***	−0.3325	0.4838
	0.0848	**0.0109**	0.5409	0.254		0.1124	**0.038**	**0.038**	0.1776	0.0419
PFK activity	−0.21046	**0.7592****	**0.5528***	**0.7711****	0.3872	1	**0.503***	0.2336	0.1623	0.0682
	0.4019	**0.0003**	**0.0174**	**0.0002**	0.1124		**0.033**	0.351	0.52	0.7881
PFK p-level	−0.38014	**0.7709****	**0.5605***	**0.5938****	**0.4918***	**0.503***	1	**0.622****	0.0353	0.3472
	0.1197	**0.0002**	**0.0155**	**0.0094**	**0.0382**	**0.0333**		**0.006**	0.8895	0.158
Global p-level	**−0.49599***	**0.5884***	0.2009	0.3455	**0.4919***	0.2336	**0.622****	1	−0.137	0.3682
	0.0363	**0.0102**	0.4241	0.1603	**0.0381**	0.351	**0.006**		0.5878	0.1327
pH	**0.71992****	0.0352	**0.5842***	0.4599	−0.3325	0.1623	0.0353	−0.137	1	**−0.848****
	0.0008	0.8896	**0.0109**	0.0548	0.1776	0.52	0.8895	0.5878		**<0.0001**
Lactate	**−0.9123****	0.3638	−0.1715	−0.1046	**0.4838***	0.0682	0.3472	0.3682	**−0.848****	1
	<0.0001	0.1378	0.4962	0.6797	**0.0419**	0.7881	0.158	0.1327	**<0.0001**	

Abbreviations: *GP* glycogen phosphorylase, *PK* pyruvate kinase, *PFK* phosphofructokinase, *p-level* phosphorylation level. The ovine Longissimus thoracis were crushed and incubated at 4 °C, and then snap frozen in liquid nitrogen after collection at different times. The first row of each indicator represents correlation coefficient, and the second row represents p value. Significant levels: *0.01 $<p<$ 0.05; **$p<$ 0.01

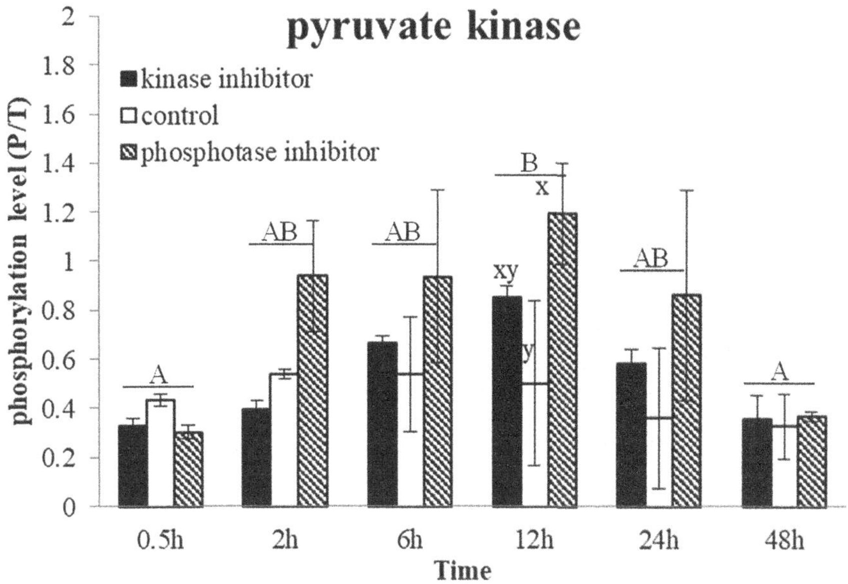

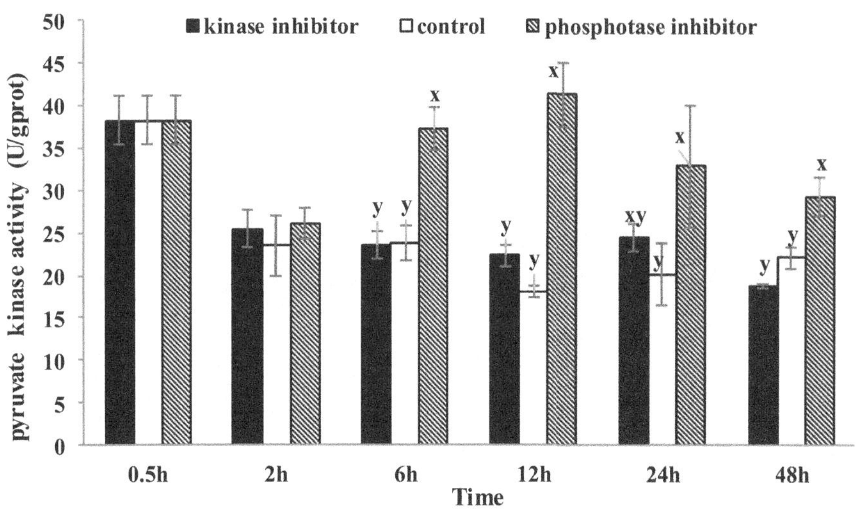

Fig. 6.11 The phosphorylation level and activity of pyruvate kinase in postmortem muscle. Different letters (x, y) at the same time points are significantly different among the three groups ($P < 0.05$). Different letters (A, B) in the same group are significantly different at different PM time ($P < 0.05$). Adapted from (Chen et al. 2019a)

key rate-limiting enzyme of glycolysis (Scheffler and Gerrard 2007). In the present study, the activity of PFK was positively correlated with the phosphorylation level.

Glycolysis was a sequence of enzymatic reactions that were determined by the activity of glycolytic enzymes. GP, PK, and PFK were rate-limiting enzymes that

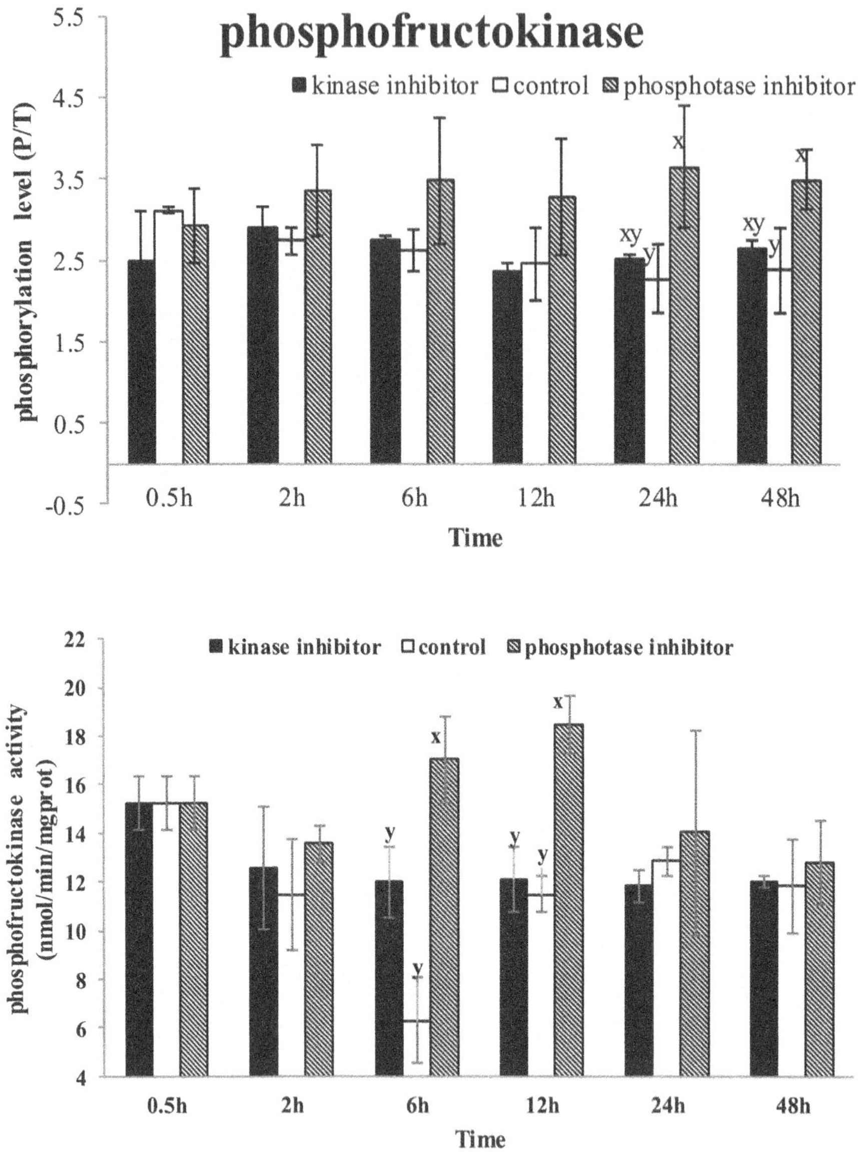

Fig. 6.12 The phosphorylation level and activity of PFK in postmortem muscle. Different letters (x, y) at the same time points are significantly different among the three groups ($P < 0.05$). Adapted from (Chen et al. 2019a)

controlled the rate of PM metabolism. The activity of these enzymes was significantly higher in the phosphatase inhibitor group with high phosphorylation level, which revealed protein phosphorylation might play a positive role in the regulation of glycolysis. Shen and Du (2005) reported that the glycolysis and pH decline were indirectly affected by the phosphorylation status of AMPK in PM muscle. Therefore, we could infer that glycolysis was indirectly affected by protein phosphorylation through the regulation of enzyme activity.

6.5 Conclusions

The present research focused on the effects of protein phosphorylation on postmortem glycolysis and glycolytic enzymes in ovine muscle. The global phosphorylation level of sarcoplasmic proteins increased early postmortem and then decreased afterwards in general. The results of fast, moderate, and slow glycolytic rate groups revealed that the phosphorylation of sarcoplasmic proteins was related to muscle pH decline at early postmortem. 24 phosphoproteins were identified to be related to the glycolytic rate via Quantitative proteomic iTRAQ. Phosphorylation of pyruvate kinase at Thr157, negatively correlated with the glycolytic rate, deserved more attention in future research. To validate the effect of protein phosphorylation on glycolysis, ovine muscle was treated with a kinase inhibitor, dimethyl sulfoxide, or a phosphatase inhibitor, therefore resulted in different phosphorylation levels. The pH value and lactate content revealed that a high phosphorylation level was one of the reasons for the fast pH decline. The activity of glycogen phosphorylase, pyruvate kinase, and PFK was positively correlated with their phosphorylation level, from which we could infer that protein phosphorylation played a role in postmortem glycolysis through the regulation of enzyme activity in ovine muscle. In summary, protein phosphorylation was related to postmortem glycolysis and glycolytic enzymes, and the phosphorylation level of glycolytic enzymes may influence its activities, further regulating postmortem glycolysis.

Acknowledgments Parts of this chapter are reprinted from Food Chemistry, 280, Chen, L., et al., quantitative phosphoproteomic analysis of ovine muscle with different postmortem glycolytic rates, 203-209; Food Chemistry, 293, Chen, L., et al., Effects of protein phosphorylation on glycolysis through the regulation of enzyme activity in ovine muscle, 537-544; International Journal of Food Science and Technology, 53(12), Chen, L., et al., The effect of sarcoplasmic protein phosphorylation on glycolysis in postmortem ovine muscle, 2714-2722. Copyright (2020), with permission from Elsevier.

References

Aizawa, H., Yamada, S.-I., Xiao, T., Shimane, T., Hayashi, K., Qi, F., et al. (2017). Difference in glycogen metabolism (glycogen synthesis and glycolysis) between normal and dysplastic/ malignant oral epithelium. *Archives of Oral Biology, 83*, 340–347.

Amiri, M., Conserva, F., Panayiotou, C., Karlsson, A., & Solaroli, N. (2013). The human adenylate kinase 9 is a nucleoside mono- and diphosphate kinase. *The International Journal of Biochemistry & Cell Biology, 45*, 925–931.

Bell, R. A. V., & Storey, K. B. (2014). P06: Posttranslational modification of glyceraldehyde-3-phosphate dehydrogenase from a hibernating mammal: Insight into cold-adaptation and structural diversity of a housekeeping enzyme. *Cryobiology, 69*, 196–197.

Chen, L., Li, X., Ni, N., Liu, Y., Chen, L., Wang, Z., et al. (2016). Phosphorylation of myofibrillar proteins in post-mortem ovine muscle with different tenderness. *Journal of the Science of Food and Agriculture, 96*, 1474–1483.

Chen, L., Li, Z., Li, X., Chen, J., Everaert, N., & Zhang, D. Q. (2018). The effect of sarcoplasmic protein phosphorylation on glycolysis in postmortem ovine muscle. *International Journal of Food Science and Technology, 53*, 2714–2722.

Chen, L., Bai, Y., Everaert, N., Li, X., Tian, G., Hou, C., et al. (2019a). Effects of protein phosphorylation on glycolysis through the regulation of enzyme activity in ovine muscle. *Food Chemistry, 293*, 537–544.

Chen, L., Li, Z., Everaert, N., Lametsch, R., & Zhang, D. (2019b). Quantitative phosphoproteomic analysis of ovine muscle with different postmortem glycolytic rates. *Food Chemistry, 280*, 203–209.

D'alessandro, A., & Zolla, L. (2013). Meat science: From proteomics to integrated omics towards system biology. *Journal of Proteomics, 78*, 558–577.

Dzeja, P., & Terzic, A. (2009). Adenylate kinase and AMP signaling networks: Metabolic monitoring, signal communication and body energy sensing. *International Journal of Molecular Sciences, 10*, 1729–1772.

Ferguson, D. M., & Gerrard, D. E. (2014). Regulation of post-mortem glycolysis in ruminant muscle. *Animal Production Science, 54*, 464.

Graves, J. D., & Krebs, E. G. (1999). Protein phosphorylation and signal transduction. *Pharmacology & Therapeutics, 82*, 111–121.

Gupta, V., & Bamezai, R. N. (2010). Human pyruvate kinase M2: a multifunctional protein. *Protein Science, 19*, 2031–2044.

Gururaj, A., Barnes, C. J., Vadlamudi, R. K., & Kumar, R. (2004). Regulation of phosphoglucomutase 1 phosphorylation and activity by a signaling kinase. *Oncogene, 23*, 8118.

Heiden, M. G. V., Locasale, J. W., Swanson, K. D., Sharfi, H., Heffron, G. J., Amador-Noguez, D., et al. (2010). Evidence for an alternative glycolytic pathway in rapidly proliferating cells. *Science, 329*, 1492–1499.

Højlund, K., Bowen, B. P., Hwang, H., Flynn, C. R., Madireddy, L., Geetha, T., et al. (2009). In vivo phosphoproteome of human skeletal muscle revealed by phosphopeptide enrichment and HPLC– ESI– MS/MS. *Journal of Proteome Research, 8*, 4954–4965.

Huang, H., Larsen, M. R., Karlsson, A. H., Pomponio, L., Costa, L. N., & Lametsch, R. (2011). Gel-based phosphoproteomics analysis of sarcoplasmic proteins in postmortem porcine muscle with pH decline rate and time differences. *Proteomics, 11*, 4063–4076.

Immonen, K., & Puolanne, E. (2000). Variation of residual glycogen-glucose concentration at ultimate pH values below 5.75. *Meat Science, 55*, 279–283.

Jin, X., Wang, L.-S., Xia, L., Zheng, Y., Meng, C., Yu, Y., et al. (2008). Hyper-phosphorylation of α-enolase in hypertrophied left ventricle of spontaneously hypertensive rat. *Biochemical and Biophysical Research Communications, 371*, 804–809.

Johnson, L. (1992). Glycogen phosphorylase: Control by phosphorylation and allosteric effectors. *The FASEB Journal, 6*, 2274–2282.

Lametsch, R., Larsen, M. R., Ess N-Gustavsson, B., Jensen-Waern, M., Lundström, K., & Lindahl, G. (2011). Postmortem changes in pork muscle protein phosphorylation in relation to the RN genotype. *Journal of Agricultural and Food Chemistry, 59*, 11608–11615.

Li, M., Li, X., Xin, J., Li, Z., Li, G., Zhang, Y., et al. (2017). Effects of protein phosphorylation on color stability of ground meat. *Food Chemistry, 219*, 304–310.

Lindahl, G., Henckel, P., Karlsson, A. H., & Andersen, H. J. (2006). Significance of early postmortem temperature and pH decline on colour characteristics of pork loin from different crossbreeds. *Meat Science, 72*, 613–623.

Luther, M. A., & Lee, J. C. (1986). The role of phosphorylation in the interaction of rabbit muscle phosphofructokinase with F-actin. *Journal of Biological Chemistry, 261*, 1753–1759.

Manning, G., Whyte, D. B., Martinez, R., Hunter, T., & Sudarsanam, S. (2002). The protein kinase complement of the human genome. *Science, 298*, 1912–1934.

Marsin, A. S., Bertrand, L., Rider, M. H., Deprez, J., Beauloye, C., Vincent, M. F., et al. (2000). Phosphorylation and activation of heart PFK-2 by AMPK has a role in the stimulation of glycolysis during ischaemia. *Current Biology, 10*, 1247–1255.

Nettelblad, F. A., & Engström, L. (1987). The kinetic effects of in vitro phosphorylation of rabbit muscle enolase by protein kinase C: A possible new kind of enzyme regulation. *FEBS Letters, 214*, 249–252.

Paredi, G., Raboni, S., Bendixen, E., de Almeida, A. M., & Mozzarelli, A. (2012). "Muscle to meat" molecular events and technological transformations: The proteomics insight. *Journal of Proteomics, 75*, 4275–4289.

Pösö, A. R., & Puolanne, E. (2005). Carbohydrate metabolism in meat animals. *Meat Science, 70*, 423–434.

Reiss, N., Kanety, H., & Schlessinger, J. (1986). Five enzymes of the glycolytic pathway serve as substrates for purified epidermal-growth-factor-receptor kinase. *Biochemical Journal, 239*, 691–697.

Sale, E. M., White, M. F., & Kahn, C. R. (1987). Phosphorylation of glycolytic and gluconeogenic enzymes by the insulin receptor kinase. *Journal of Cellular Biochemistry, 33*, 15–26.

Scheffler, T. L., & Gerrard, D. E. (2007). Mechanisms controlling pork quality development: The biochemistry controlling postmortem energy metabolism. *Meat Science, 77*, 7–16.

Schwägele, F., Buesa, P. L. L., & Honikel, K. O. (1996). Enzymological investigations on the causes for the PSE-syndrome, II. Comparative studies on glycogen phosphorylase from pig muscles. *Meat Science, 44*, 41–53.

Shen, Q. W., & Du, M. (2005). Role of AMP-activated protein kinase in the glycolysis of postmortem muscle. *Journal of the Science of Food and Agriculture, 85*, 2401–2406.

Shen, Q., Means, W., Thompson, S., Underwood, K., Zhu, M., McCormick, R., et al. (2006a). Pre-slaughter transport, AMP-activated protein kinase, glycolysis, and quality of pork loin. *Meat Science, 74*, 388–395.

Shen, Q. W., Means, W. J., Underwood, K. R., Thompson, S. A., Zhu, M. J., McCormick, R. J., et al. (2006b). Early post-mortem AMP-activated protein kinase (AMPK) activation leads to phosphofructokinase-2 and -1 (PFK-2 and PFK-1) phosphorylation and the development of pale, soft, and exudative (PSE) conditions in porcine longissimus muscle. *Journal of Agricultural and Food Chemistry, 54*, 5583–5589.

Sprang, S., Acharya, K., Goldsmith, E., Stuart, D., Varvill, K., Fletterick, R., et al. (1988). Structural changes in glycogen phosphorylase induced by phosphorylation. *Nature, 336*, 215.

Sprang, S. R., Withers, S. G., Goldsmith, E. J., Fletterick, R. J., & Madsen, N. B. (1991). Structural basis for the activation of glycogen phosphorylase b by adenosine monophosphate. *Science, 254*, 1367.

Werner, C., Natter, R., & Wicke, M. (2010). Changes of the activities of glycolytic and oxidative enzymes before and after slaughter in the longissimus muscle of Pietrain and Duroc pigs and a Duroc-Pietrain crossbreed. *Journal of Animal Science, 88*, 4016–4025.

Zhang, B., & Liu, J. Y. (2017). Serine phosphorylation of the cotton cytosolic pyruvate kinase GhPK6 decreases its stability and activity. *FEBS Open Bio, 7*, 358–366.

Chapter 7
Mechanism of the Effect of Protein Phosphorylation on Myofibril Protein Degradation

Abstract Myofibrillar proteins degradation contributes to meat tenderization during postmortem. Protein phosphorylation has been revealed to be associated with meat tenderness in recent years. This chapter aimed to investigate the effect of myofibrillar protein phosphorylation on the degradation susceptibility by μ-calpain. Protein kinase A (PKA) and alkaline phosphatase (AP) were added to myofibrillar proteins, extracted from ovine *longissimus lumborum* (LL) muscles, to modify the phosphorylation level for 30 min at 30 °C. The samples were then incubated with μ-calpain at 4 °C at different Ca^{2+} concentrations. The results showed that more degradation products of myosin heavy chain, actin, titin, desmin, and troponin T were detected with the increase of Ca^{2+} concentration. The protein degradation was higher in AP treated group compared to the control group. This research indicated that phosphorylation prevented proteolytic susceptibility of myofibrillar proteins to degradation by μ-calpain. Protein phosphorylation made a major contribution in meat tenderization during postmortem.

Keywords Phosphorylation · Protein degradation · Myosin heavy chain · Actin · Titin · Desmin · Troponin T

7.1 Introduction

Tenderness is considered to be one of the most important quality attributes. The degradation of myofibrillar proteins and disruption of myofibril structure by proteolytic enzymes are reported to be the key mechanisms for tenderization during postmortem (Huff Lonergan et al. 2010). The calpain system is the primary endogenous enzyme (Du et al. 2017). The stability and proteolysis of myofibrillar proteins are also regulated by protein modifications. For example, oxidation and nitrosylation affect the proteolytic activity of calpain (Li et al. 2018).

Protein phosphorylation, the most common and important post-translational modification, has been reported to be associated with meat tenderness in recent years (Chen et al. 2016; Gannon et al. 2008). High phosphorylation of Z-disc related proteins has been detected in tough bovine meat by D'alessandro and Zolla (2013),

D. Zhang et al., *Protein Phosphorylation and Meat Quality*,
https://doi.org/10.1007/978-981-15-9441-0_7

who found that phosphorylation of Z-disc proteins probably increased interactions of proteins at Z-line and reduced the accessibility of proteases, resulting in tougher meat (D'alessandro et al. 2012a, b; D'Alessandro and Zolla 2013). In addition, Chen et al. (2016) observed that the global phosphorylation level of tough meat was significantly higher than tender meat at 4, 12, and 24 h postmortem. All these studies have shown that protein phosphorylation could regulate protein stability and degradation protein, which contributed to the meat tenderization. In addition, another research founded out that phosphorylation protected proteins from calpain-mediated proteolysis by decreasing the substrate recognition or improving the protein stability (Zhang et al. 2012), while phosphorylation enhanced the degradation of some protein by μ-calpain (Shen et al. 2001).

However, the mechanism by which protein phosphorylation influences meat tenderness is still unclear, especially the impact of phosphorylation of myofibrillar proteins on substrate recognition and degradation by μ-calpain remains unknown. Therefore, we investigated the effect of the phosphorylation on myofibrillar protein (myosin, action, titin, Troponin T and desmin) degradation in vitro, with the addition of protein kinase A (PKA) and alkaline phosphatase (AP) to regulate phosphorylation level which affects degradation of these proteins. Moreover, the 3D structure of titin fragment was simulated and molecular dynamics trajectory of the constructed proteins was analyzed using the Discovery Studio™. The outcome of this study may further improve our understanding of the mechanism of phosphorylation on degradation of titin to regulate postmortem meat tenderization.

7.2 The Effect of Protein Phosphorylation on Myosin and Actin Degradation

Proteolytic degradation of muscle occurs simultaneously with postmortem cell death and meat aging, resulting in the production of protein fragments. Myosin and actin are the first two abundant proteins in skeletal muscle and the main components of skeletal muscle thin filaments. Protein degradation is primarily responsible for postmortem meat tenderization, which might be affected by protein phosphorylation. The purpose of this research was to investigate the effect of protein phosphorylation on myosin and actin degradation during postmortem. Here, we regulated the phosphorylation pattern of protein by protein kinase inhibitors and phosphatase inhibitors, and analyzed the effect of protein phosphorylation on myosin and actin degradation.

7.2.1 Global Phosphorylation Level of Myofibrillar Protein

The phosphorylation level of myofibrillar proteins was defined as the ratio of phosphoproteins to total proteins. After electrophoresis, as shown in Fig. 7.1, phosphorylated proteins and total proteins were stained with Pro-Q Diamond and SYPRO Ruby fluorescence staining agents, respectively. All the 21 visible myofibrillar protein bands were selected for total phosphorylation level quantification. The global phosphorylation level of myofibrillar proteins significantly reduced during storage for all the three groups (Fig. 7.2). Compared with the control group, the total

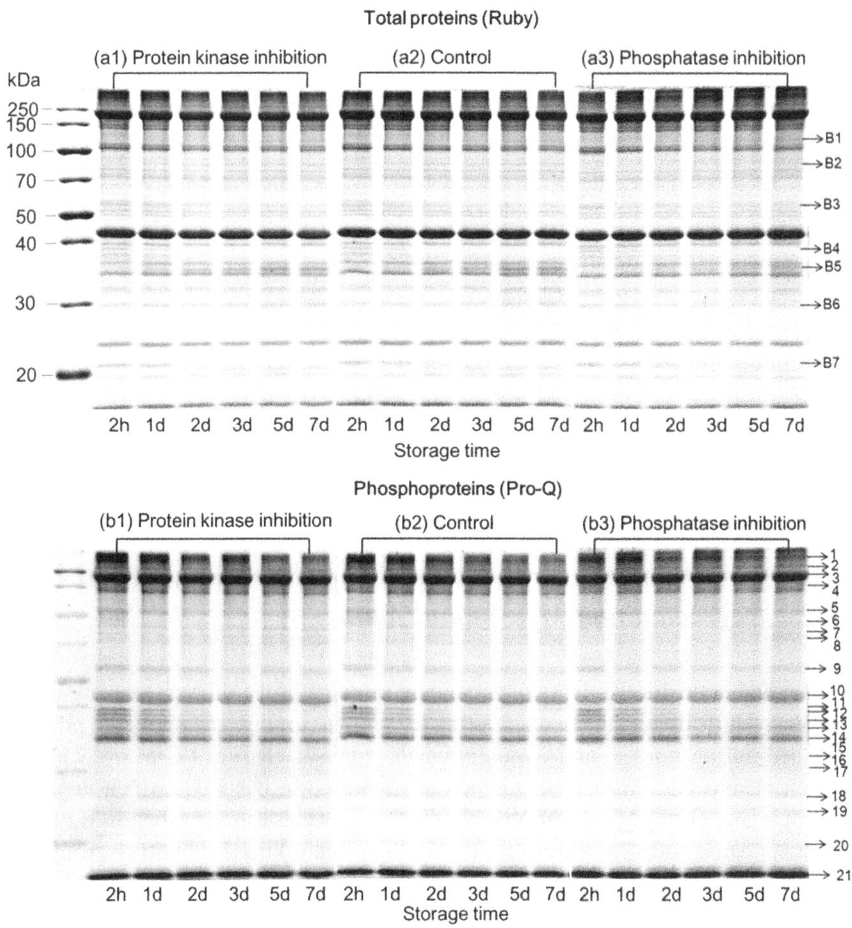

Fig. 7.1 Phosphorylation level of myofibrillar proteins (protein kinase inhibition, control, phosphatase inhibition, ovine *longissimus*). (**a**) Gel image of total (T) proteins stained with SYPRO® Ruby. (**b**) Gel image of phosphoproteins (P) stained with Pro-Q®. Protein bands B1–B7 were selected for LC-MS/MS identification. Protein bands 1–21 were selected for relative intensity and phosphorylation level analysis (Li et al. 2018)

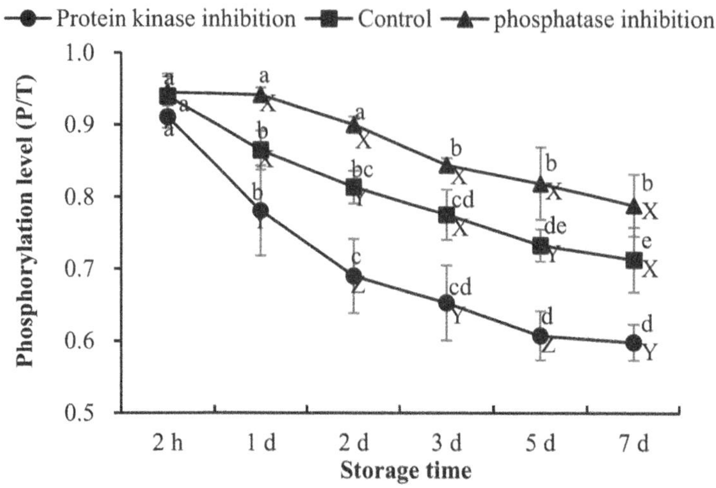

Fig. 7.2 Global phosphorylation level of myofibrillar proteins. Values lacking common lowercase letters are significantly different within the same group at different storage time points ($P < 0.05$). Values lacking common capital letters are significantly different among the three groups at the same storage time point ($P < 0.05$) (Li et al. 2018)

phosphorylation level of myofibrillar proteins was significantly lower in the protein kinase inhibition group. Furthermore, the global phosphorylation level of myofibrillar proteins in muscle storage for 2 days and 5 days was significantly higher than that in the control group. The results showed that the inhibitors used in the study could effectively regulate the phosphorylation status of myofibrillar proteins in postmortem muscle (Li et al. 2018).

7.2.2 Degradation of Myosin and Actin

Myosin and actin are the two most abundant proteins in the thick filament and thin filament (Huff Lonergan et al. 2010). As shown in Fig. 7.1, several proteins degraded during storage, and 7 protein bands were selected for LC-MS/MS identification (Table 7.1). Myosin fragment was identified in band B1, which might be myosin heavy chain. Moreover, myosin fragments were also identified in B2 band, which appeared during storage for 1 day in the protein kinase inhibitory group and the control group, but not detected in the phosphatase inhibitory group. Actin was identified in band B4, B5, B6, and B7. As shown in Fig. 7.3, myosin heavy chain and actin were not observed with significant degradation. This was consistent with Huff Lonergan et al. (2010), which reported no significant changes in myosin heavy chain and actin during postmortem. However, myosin degradation fragments were identified in bands B1, B2, B3, B4 and B5, and actin degradation fragments were identified in bands B4, B5, B6, and B7 (Table 7.1) by LC-MS/MS. This was

Table 7.1 LC-MS/MS identified proteins in 7 bands from SDS-PAGE gel (Li et al. 2018)

No.	Accession	Protein name	Score	Mass	Matches	Sequences	Sequence coverage
B1	Q9BE40	Myosin-1	223,764	30,521	1471(942)	191(166)	67
	F1MRC2	Myosin-2	224,106	27,110	1325(841)	181(155)	64
	F1MJ28	Alpha-1,4 glucan phosphorylase	8215	97,674	527(317)	69(60)	70
B2	Q9BE40	Myosin-1	223,764	19,750	965(589)	169(133)	65
	F1MRC2	Myosin-2	224,106	18,130	894(545)	163(123)	62
	Q0IIG5	ATP-dependent 6-phosphofructokinase	86,095	7613	432(261)	39(33)	56
	Q9BE39	Myosin-7	223,889	5450	347(191)	112(73)	53
B3	O62654	Desmin	8878	53,556	555(336)	61(51)	88
	P19483	ATP synthase subunit alpha	2325	59,797	110(65)	30(23)	57
	P48616	Vimentin	2060	53,752	162(87)	38(28)	65
	Q9BE40	Myosin-1	1823	223,764	105(60)	54(35)	30
	P00829	ATP synthase subunit beta	1769	56,249	108(57)	23(21)	63
	F1MRC2	Myosin-2	1747	224,106	108(55)	53(33)	29
B4	P68138	Actin	42,366	9493	509(362)	33(29)	85
	Q8MKI0	Troponin T fast skeletal muscle type	30,682	8488	524(328)	31(23)	64
	Q8MKH9	Troponin T fast skeletal muscle type	30,684	8266	473(302)	31(22)	64
	Q3ZC07	Actin	42,334	7087	446(311)	32(27)	85
	Q9BE40	Myosin-1	223,764	7086	351(204)	94(57)	43
	F1MRC2	Myosin-2	224,106	6917	351(200)	91(51)	41
B5	Q5KR49	Tropomyosin alpha-1 chain	32,732	11,777	726(454)	59(50)	92
	Q5KR48	Tropomyosin beta chain	32,931	9175	621(355)	58(51)	95
	Q5KR48–2	Isoform 2 of Tropomyosin beta chain	33,049	8783	590(341)	53(45)	85
	Q5KR47	Tropomyosin alpha-3 chain	32,856	6298	412(233)	54(41)	94
	P68138	Actin	42,366	4623	266(158)	31(29)	91
	Q8MKH8	Troponin T fast skeletal muscle type	29,798	4073	269(149)	25(19)	52

(continued)

Table 7.1 (continued)

No.	Accession	Protein name	Score	Mass	Matches	Sequences	Sequence coverage
	Q8MKH7	Troponin T fast skeletal muscle type	29,800	3959	248(137)	25(18)	52
	Q9BE40	Myosin-1	223,764	3842	217(106)	74(40)	34
B6	P68138	Actin	42,366	6672	420(270)	31(28)	87
	Q8MKH7	Troponin T fast skeletal muscle type	29,800	4176	270(144)	27(19)	58
	Q8MKH8	Troponin T fast skeletal muscle type	29,798	4170	289(155)	27(20)	58
	P02722	ADP/ATP translocase 1	33,174	3402	237(150)	26(24)	62
B7	A5PJM2	TNNI2 protein (fragment)	23,707	7452	669(332)	29(22)	74
	P68138	Actin	42,366	1561	117(54)	25(17)	76
	Q9XSC6	Creatine kinase M-type	43,190	1411	92(48)	23(12)	58
	E1BEL7	Heat shock protein beta-1	22,621	1102	85(42)	15(13)	68

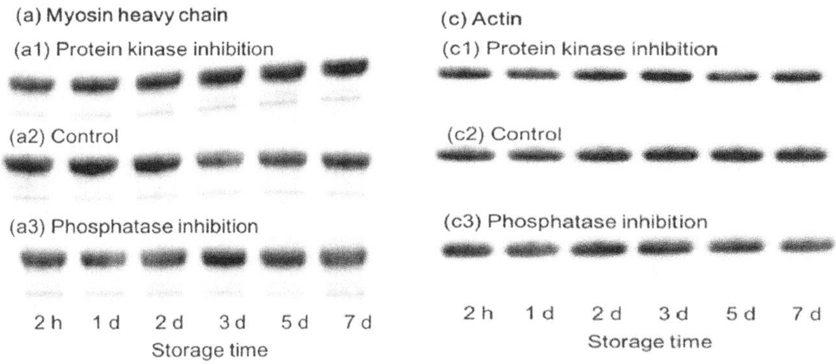

Fig. 7.3 Immunoblot analysis of myosin heavy chain (**a**) and actin (**c**) in ovine *longissimus* during storage (Li et al. 2018)

consistent with the research of Lametsch et al. (2002), which reported that myosin and actin degradation fragments could be detected by sensitive methods (LC-MS/MS) in postmortem muscle. In general, myosin heavy chain and actin showed limited degradation in postmortem muscle. This study showed that dephosphorylation can partly promote the degradation of myosin heavy chain and actin, suggesting that protein phosphorylation may play an important role in postmortem meat tenderization.

7.3 Phosphorylation Prevents Myosin and Actin Degradation by μ-Calpain In Vitro

Degradation of myofibrillar proteins is conducive to meat tenderization during postmortem. Gannon et al. (2008) reported that protein phosphorylation was associated with meat tenderness. The aim of this research was to determine the effect of myofibrillar protein phosphorylation on the degradation sensitivity by μ-calpain. Myofibrillar proteins were incubated with PKA or AP to increase or decrease protein phosphorylation levels, respectively, and the effect of protein phosphorylation on myosin and actin degradation was analyzed.

7.3.1 Phosphorylation Level of Myosin Heavy Chain and Actin

The ratio of phosphorylated protein to total protein was used to assess protein phosphorylation level. As shown in Fig. 7.4, the global phosphorylation level of myofibrillar proteins in the PKA group was significantly higher than that in the

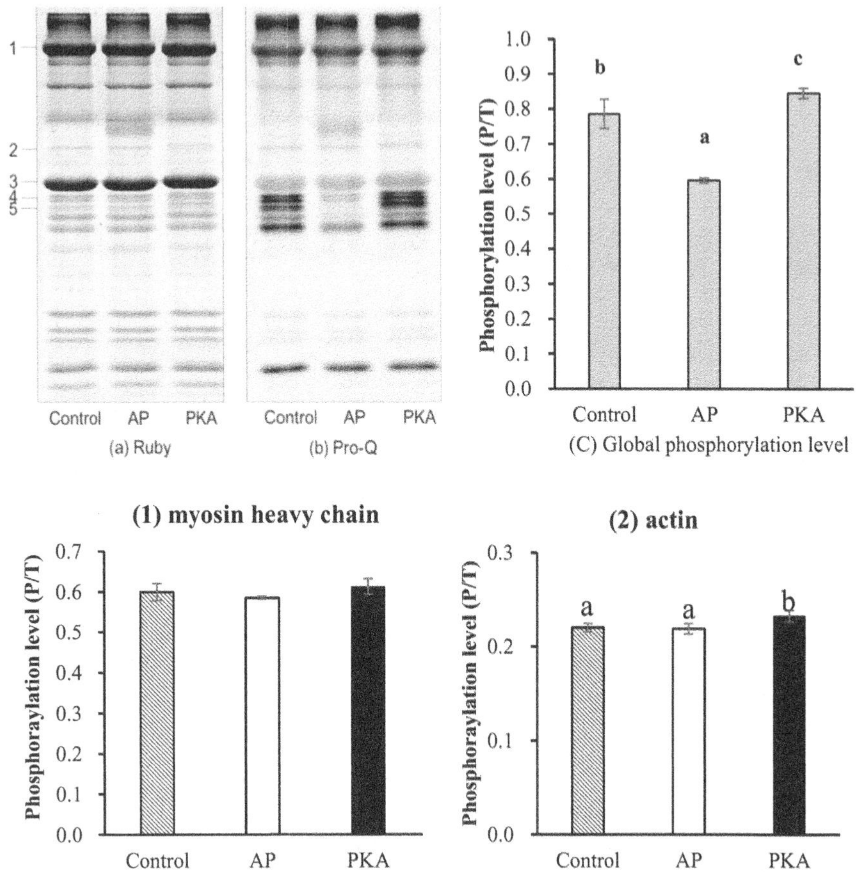

Fig. 7.4 Assessment of the phosphorylation level of myofibril proteins (Control, AP, and PKA) by 1-DE. (**a**) Gel image of total (T) proteins stained with SYPRO Ruby. (**b**) Gel image of phosphoproteins (P) stained with Pro-Q. (**c**) Global phosphorylation level (P/T). (1–2) protein phosphorylation level (P/T) in two individual protein bands labeled in Fig. 7.1a, (1) Band 1, mainly contains myosin heavy chain, (2) Band 2, mainly contains actin. Values with different letters are significantly different ($P < 0.05$) (Li et al. 2017)

control group. And the global phosphorylation level of myofibrillar proteins in the AP group was significantly lower than that in the control group. These results indicated that PKA and AP can significantly increase and decrease the overall phosphorylation level of myofibrillar proteins in vitro, respectively. The results also indicated that PKA and AP had an obvious regulatory effect on protein phosphorylation in vitro.

In order to further figure out which type of protein phosphorylation was influenced by PKA and AP, the phosphorylation level of myosin heavy chains (MHC, band 1 in Fig. 7.4a) and actin (band 3 in Fig. 7.4a) was analyzed. As shown in Fig. 7.4, PKA and AP had no significant influence on MHC phosphorylation. However, PKA incubation increased the phosphorylation of actin, but AP did not change the phosphorylation level of actin.

7.3.2 *Myosin and Actin Degradation by μ-Calpain at Different Ca^{2+} Concentration*

In a bid to evaluate the effect of protein phosphorylation on myofibrillar protein degradation by μ-calpain, myofibrillar proteins treated with PKA or AP were incubated with μ-calpain at different Ca^{2+} concentrations.

Myosin is the most abundant protein in skeletal muscle and the main component of muscle filaments in sarcomere. Four bands of MHC in untreated samples of μ-calpain were detected by immunoblotting, the molecular weight of four bands was approximately 180, 130, 110, and 70 kDa, respectively (Fig. 7.5). Approximately 95 KDa new degradation bands were detected in the control and AP group (1 mM Ca^{2+}) (Fig. 7.5: a3), indicating the degradation of MHC. The gray value of band 1 in control group did not change significantly, but increased significantly in AP group after incubation for 30 min (Fig. 7.5: a3 and b). Furthermore, the relative gray value of band 1 in the AP group was significantly higher than that in the control group at incubation for 5 and 30 min.

These results indicated that MHC phosphorylation by PKA prevented the hydrolysis of μ-calpain, but the dephosphorylation by AP promoted its degradation.

Actin is the second most abundant protein in skeletal muscle and the main component of skeletal muscle thin filaments. Degradation of actin and myosin disrupts the formation of actomyosin, resulting in impaired muscle structural integrity (Wang et al. 2013). Degradation of actin in the PKA-treated group was also measured. As shown in Fig. 7.6, no degradation fragments were detected in PKA-treated group after incubation with μ-calpain at any Ca^{2+} concentrations used in this study. Unlike the PKA-treated group, the control group was detected with two actin fragments at $[Ca^{2+}] = 0.1$ mM, while three degradation fragments were detected at $[Ca^{2+}] = 1.0$ mM. Three actin fragments were detected in the AP treatment group at 0.1 mM and 1.0 mM Ca^{2+}. At $[Ca^{2+}] = 1.0$ mM. Quantitative analysis of actin fragments showed that the density of these three bands increased with incubation time. The fragments of the three bands in the AP treatment group were higher than that in the control group, indicating that the degradation of actin increased after incubation with AP. These results showed that PKA and AP pretreatment reduced and increased actin degradation by μ-calpain, respectively. SDS-PAGE analysis showed that MHC and actin did not significantly degrade after incubation with μ-calpain, but immunoblotting identified degradation products with lower molecular weight in new bands, demonstrating that they can be hydrolyzed by

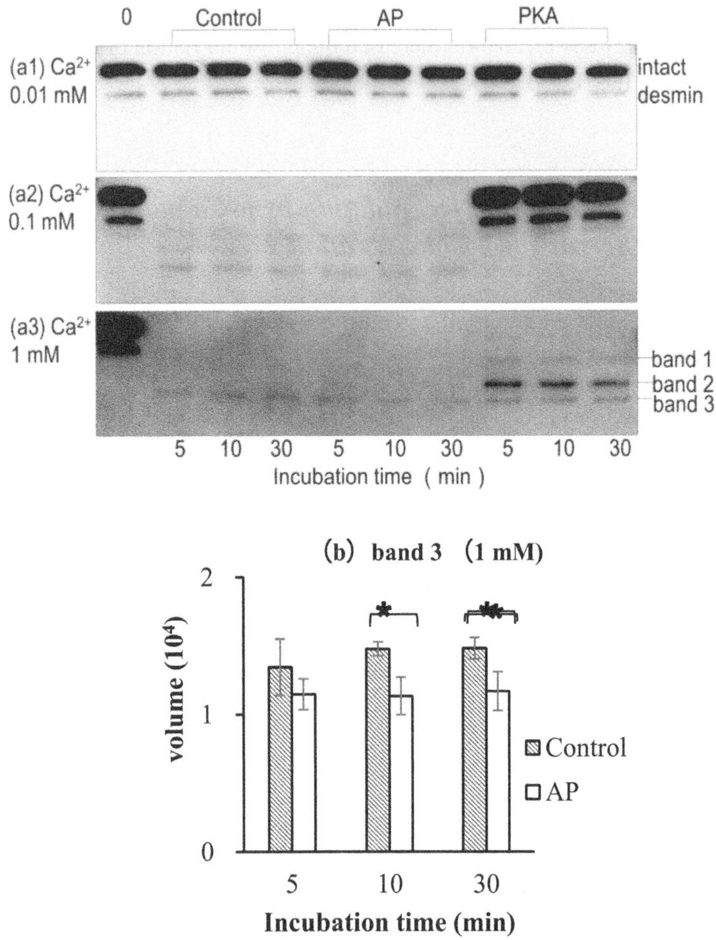

Fig. 7.5 Analysis of the degradation of myosin heavy chain after incubation with μ-calpain for 30 min. (al–3) Western blot analysis of myosin heavy chain after myofibrillar proteins were incubated with μ-calpain for 30 min at 0.01, 0.1, and 1 mM Ca^{2+}. (**b**) Densitometric analysis of band 1 in (**a3**). Values with/show significant difference between Control and AP group in the same incubation time ($P < 0.05$); values with different letters are significantly different within the AP group at different incubation time ($P < 0.05$) (Li et al. 2017)

μ-calpain. MHC and actin did not change significantly during postmortem. However, scholars have reported that they were all substrates of μ-calpain, and two protein degradation products were detected in postmortem muscle (Lametsch et al. 2002; Lametsch et al. 2004).

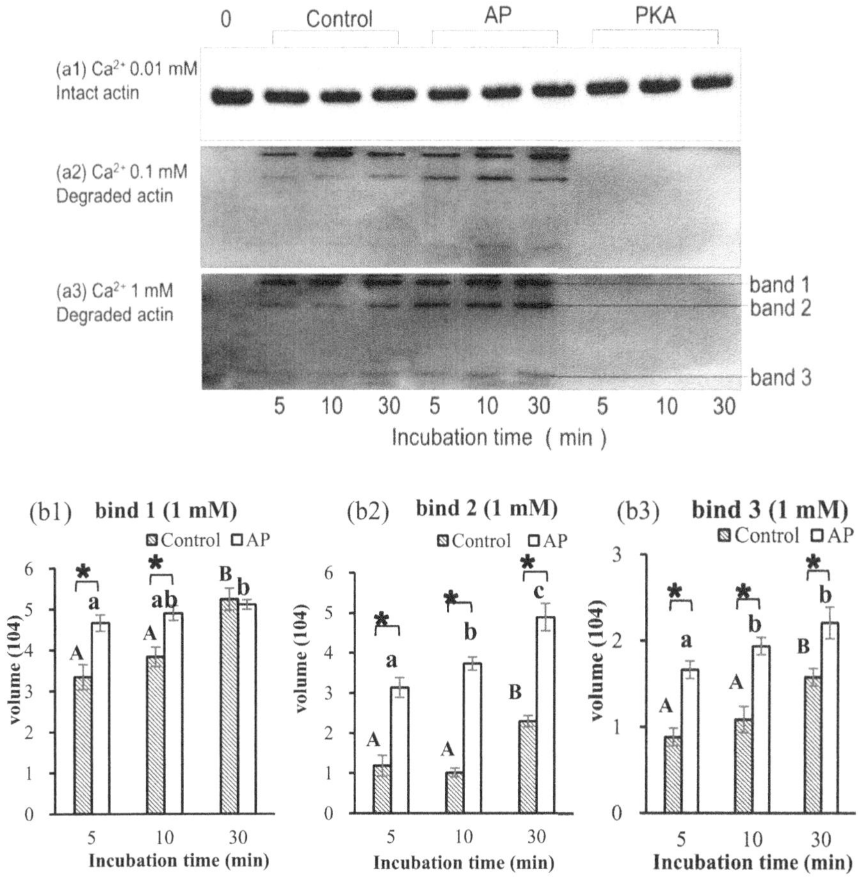

Fig. 7.6 Analysis of the degradation of actin after incubation with μ-calpain for 30 min. (**a1–3**) Western blot analysis of actin after myofibrillar proteins were incubated with μ-calpain for 30 min at 0.01, 0.1, and 1 mM Ca^{2+}. (**b1–3**) Densitometric analysis of band 1–3 in (a3). Values with/show significant difference between Control and AP group at the same incubation time (*P < 0.05*); values with different letters are significantly different within the same group at different incubation time (*P < 0.05*) (Li et al. 2017)

7.4 Influence Mechanism of Protein Phosphorylation on Titin Degradation

Titin is the third most abundant protein in muscle that spans half of a sarcomere with its C-terminal end localizing in the M-line and the N-terminal forming an integral part of Z-line in striated muscles. Its degradation weakens the longitudinal structure of myofibrillar sarcomere and the integrity of muscle. The objective of this part was to investigate the degradation of titin and its phosphorylation level in three positions of sheep muscles. The pH of *longissimus lumborum* (LL), *semimembranosus* (SM), and *psoas major* (PM) muscles were measured at 30 min, 1, 2, 7, 14, 21, and 28days postmortem. Myofibrillar proteins were extracted, separated by SDS-PAGE, and quantified by phosphor-specific staining. Phosphorylation of titin was evaluated by Pro-Q Diamond-SYPRO Ruby staining.

7.4.1 Changes in pH Value of the Three Ovine Muscles

Muscle pH values were shown in Fig. 7.7. The pH of all the three muscles decreased rapidly within 1 day postmortem ($P < 0.05$) and then did not change afterwards. The pH of the PM muscle was significantly lower than that of the other two muscles at 0.5 h postmortem, but higher for the rest time ($P < 0.05$).

The pH decline was majorly due to the accumulation of lactate produced through glycolysis during postmortem. Bouton et al. (1971) reported that pH_u (ultimate pH) in ovine muscles was achieved on the third day postmortem at 1–2 °C. In the present research, all the three muscles reached pH_u after 1 day postmortem. It might because of vacuum packing, which resulted in the quick accumulation of lactate (Napravnikova et al. 2002). In addition, pH_u of the PM muscle was significantly higher than that of the LL and SM muscles ($P < 0.05$). This might be caused by the various amount of glycogen, which determined pH_u (Ferguson et al. 2008; Honikel 2014a, b).

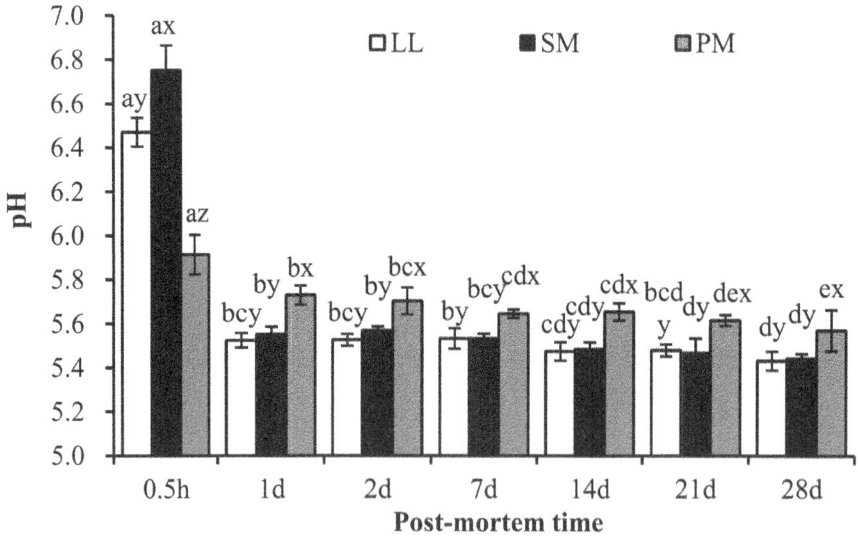

Fig. 7.7 pH values of LL, SM, and PM muscles stored at 4 °C for 28 days. Values with x-y-z differ among the three muscle types ($P < 0.05$). Values with a-b-c are significantly different among aging time points within the same group ($P < 0.05$) (Wang et al. 2018)

7.4.2 Changes in Phosphorylation Level of Titin

Phosphorylation of titin was evaluated by SDS-PAGE and fluorescence staining. As shown in Fig. 7.8, the phosphorylation level of titin from PM muscle was significantly higher at 0.5 h, 1 and 2 days postmortem than that from the same muscle in other time points ($P < 0.05$). The phosphorylation of titin from LD and SM muscles was higher at 1 day postmortem than that of LD and SM at 7, 14, 21, and 28 days postmortem ($P < 0.05$). After 2 days postmortem, the phosphorylation level of titin from all muscles decreased. The results may be caused by the content of ATP in cold (2–4 °C) stored muscles. Protein phosphorylation was a common post-translational modification.

Kinases catalyze phosphorylation of proteins, and this reaction occurs almost exclusively with the existence of ATP (Macek et al. 2009). Mattio et al. (1992) found out that a maximum content of ATP occurred 6 h after death, and then, there were no significant difference in the slopes of ATP decrease during 1–2 days postmortem. Finally, ATP showed a decreasing tendency after 3 days postmortem. With the exhaustion of ATP, accumulation of different metabolic by-products and the development of rigor mortis, the phosphorylation status of myofibrillar proteins from three muscles (LL, SM, and PM) would be changed with postmortem time (Huang et al. 2012). In addition, the phosphorylation of titin from PM muscle was significantly higher than that of the SM and LL muscles within 14 days postmortem ($P < 0.05$), with LL muscle being lowest in titin phosphorylation. The reason might be that there was a higher content of ATP in the PM muscles. Linear fitting analysis has been conducted for comparison of the rate of phosphorylation level of titin decrease. Fitting functions were shown in Fig. 7.8f. The absolute value of the slope of the equation from PM ($|K_{PM}| = 0.2476$) was higher than the other two absolute values of the slopes. The result indicated titin from PM muscles degraded faster.

7.4.3 Degradation of Titin

Titin is the largest protein in nature with a molecular weight approximately of 3000 kDa (Labeit et al. 1990). When titin is degraded, a major degradation product is observed that migrates only slightly faster under SDS-PAGE conditions than intact titin. This product migrates at approximately 2400 kDa (T1) (Kurzban and Wang 1988; Huff-Lonergan et al. 2010). Another titin degradation product that has been observed migrates at approximately 1200 kDa (T2) by SDS-PAGE analysis (Huff-Lonergan et al. 2010). Titin from PM muscle migrated as three closely spaced bands (T1, T1–2, T2, in the increasing order of migration) (Fig. 7.8b), which matched with the results of Huff-Lonergan et al. (1996). Intact titin (T1) decreased and degraded titin (T2) increased in three samples with postmortem time. However, there were two bands (T1, T1–2) from PM muscles at 0.5 h postmortem, while there

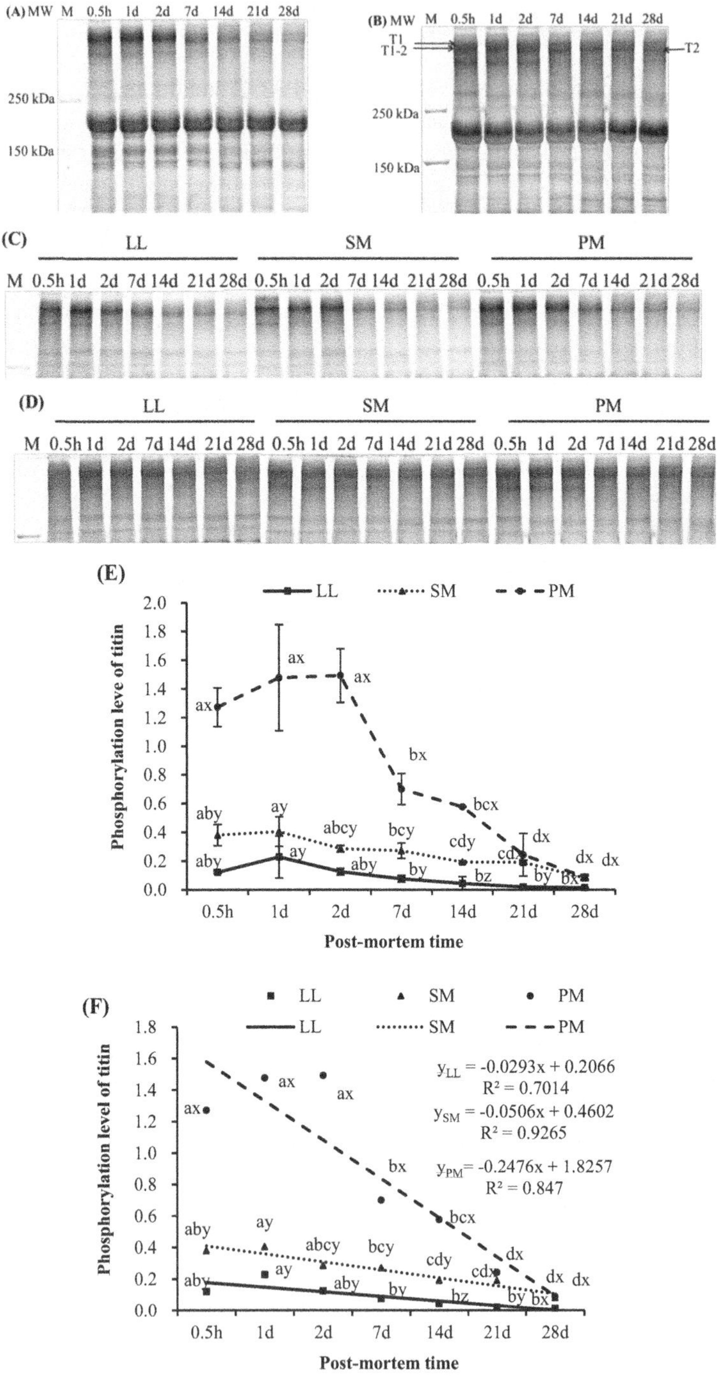

Fig. 7.8 (**a**) Gel images of phosphoproteins and total proteins and quantification of protein phosphorylation level of titin from three ovine muscles stored in refrigerator for 28 days. (**a**) and (**b**) Gel images of phosphoproteins and total proteins from PM muscle. (**a**) and (**c**) Gels stained with

was only one band (T1) from LL and SM muscles at the same time (Fig. 7.8d). The rate of conversion of T1 to T2 was faster in PM muscle than in LL and SM muscles. The slope of fitting functions (Fig. 7.8f) also indicated the rate of titin from PM degraded faster. This was consistent with literature in which Melody et al. (2004) reported that the degradation of titin occurred earlier in the PM than in the other two muscles. The degradation of beef large structural proteins, including titin, was associated with inherent muscles pH_u (Wu et al. 2014a, b). High-pH_u ($\geq$6.2) muscles had faster rate of titin degradation than muscles of lower pH_u ($\leq$5.79). In this study, the fastest degradation of titin was detected in PM muscle of higher pH compared to that of LL and SM muscles. The reason might be that μ-calpain was most active near neutral pH (Koohmaraie and Geesink 2006). Higher pH_u could lead to faster autolysis of the enzyme to the 76 kDa isoform and thus high l-calpain activity in muscle. In addition, Fig. 7.8e showed that the phosphorylation level of titin in PM, which had the faster degradation of titin, was determined to be higher than in the other two muscles. The reason might be that the phosphorylation of titin had a positive effect on its degradation. Titin has been known to be phosphorylatable and contained phosphorylation sites (Gautel et al. 2001). Protein phosphorylation was a common post-translational modification in regulating the structure and function of protein. The described series of steps were progressive phosphorylation, the structural change and then degradation (Chiu et al. 2008). The PEST sequence was rich in proline (P), glutamic acid (E), serine (S), and threonine (T). The amino acid sequences appeared to be signals for degradation. After being phosphorylated, the structure of the protein may change. The hidden PEST may be unfolded. The inferred conformational change allowed the casein kinases to access a previous titin easily and led to the degradation of protein.

7.5 Effects of Phosphorylation on Titin Degradation at Different Ca^{2+} Concentrations Incubation In Vitro

Titin was extracted from ovine *longissimus lumborum* muscles. PKA and AP were individually added to titin in vitro to regulate its phosphorylation level. The samples were then incubated with μ-calpain at 4 °C for 2 days at Ca^{2+} concentrations of either 0.05 or 2 mmol/L. The pH of the systems was measured and the level of phosphorylation and degradation of titin were determined.

Fig. 7.8 (continued) Pro-Q Diamond. (**b**) and (**d**) Gels stained with SYPRO Ruby. (**f**) Linear fitting analysis for protein phosphorylation level of titin. Values with x-y-z differ among the three muscle types ($P < 0.05$). Values with a-b-c are significantly different among aging time points within the same group ($P < 0.05$) (Wang et al. 2018)

7.5.1 Phosphorylation and Dephosphorylation of Titin

The effects of PKA and AP on the phosphorylation of titin were determined (Fig. 7.9). The bands of titin were shown on Pro-Q Diamond (Figure 7.9a) and SYPRO Ruby stained gels (Figure 7.9b). The phosphorylation level of titin from the PKA group, control group, and AP group was shown in Figure 7.9c. The result showed that the phosphorylation level of titin from the PKA group was significantly higher than that in the control group, which had a higher phosphorylation level than the AP group ($P < 0.05$). It indicated that the phosphorylation of titin in vitro was effectively altered by the treating of PKA and AP. Skeletal titin contained the N2 region (Tskhovrebova and Trinick 2003). The kinase domain in the sequence of titin and potential phosphorylation site of titin could be phosphorylated by PKA (Gautel et al. 1993; Sebestyen et al. 1995). In addition, the phosphorylation level of titin from the AP group was the lowest among the three groups (Figure 7.9c, $P < 0.05$). This may due to the significant effect of AP on titin. AP was a phosphohydrolase that could catalyze the hydrolysis of monophosphate to release free inorganic phosphate

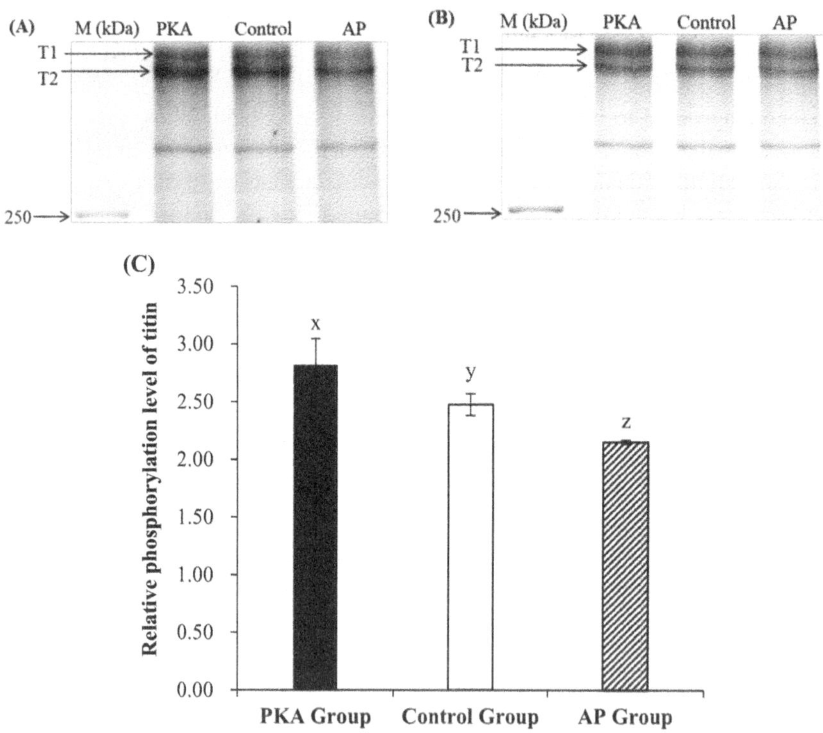

Fig. 7.9 Phosphorylation levels of titin from PKA group, Control group, and AP group. (**a**) Image of gel stained with Pro-Q Diamond. (**b**) Image of gel stained with SYPRO Ruby. (**c**) Quantification of phosphorylation level of titin. Values with different letters differ among three groups ($P < 0.05$) (Wang et al. 2019)

or to transfer the phosphoryl group to other alcohols (Deng et al. 2015; Nguyen et al. 2017). AP not only inhibited the phosphorylation of the kinase domain of titin, but also promoted the dephosphorylated of other domains in the present research.

7.5.2 pH Values of Incubation Systems

The pH values of reaction systems before and after incubation were shown in Table 7.2. The results showed that the pH value was lower after incubation than before in each group ($P < 0.05$). In addition, there was no significant difference of the pH among three treatments in this study (Table 7.2, $P > 0.05$). Many previous studies have proved that pH played the vital role in the activity of μ-calpain (Du et al. 2017; Lomiwes et al. 2014; Pulford et al. 2009). The result of this research excluded the possible different activity of μ-calpain caused by pH value.

7.5.3 Degradation of Titin

When intact titin (the T1 band) was degraded, a major degradation product T2 (approximately 2400 kDa) and another degradation product (about 1200 kDa) were observed by using SDS-PAGE analysis Huff-Lonergan et al. 1995; Huff Lonergan et al. 2010). Degradation of titin from the PKA group, control group, and AP group at different concentrations of Ca^{2+} were present in Fig. 7.10. At the Ca^{2+} concentrations of 0.05 mM, T1 and T2 bands were detected in the PKA and control groups (Fig. 7.10a), while besides T1 and T2 bands, there was another band weighted 1200 kDa at the Ca^{2+} concentration of 2 mM (Fig. 7.10b). The band weighted 1200 kDa after 12 h incubation occurred at the Ca^{2+} concentration of 0.05 mM (Fig. 7.10b), while occurred at 0.5 h at the Ca^{2+} concentration of 2 mM in AP group (Fig. 7.10). The degradation rate of titin became faster as the Ca^{2+} concentration increased (Fig. 7.10). Titin was the substrate of μ-calpain (Wu et al. 2014a, b), which was a calcium dependent protease and required micro-molar of calcium for its activity (Biswas et al. 2016). The increased concentration of Ca^{2+} promoted the degradation of μ-calpain and activated μ-calpain at the same time (Du et al. 2017).

Table 7.2 Changes in pH of incubated system before and after incubation. (Wang et al. 2019)

Treatment	pH(before)	pH(after)
PKA group	6.85 ± 0.02^{ax}	6.72 ± 0.01^{bx}
Control group	6.82 ± 0.02^{ax}	6.73 ± 0.03^{bx}
AP group	6.81 ± 0.02^{ax}	6.72 ± 0.02^{bx}

Notes: Values with x-y-z differ among three groups at the same aging time ($P < 0.05$). Values with a-b are significantly different among aging time points within group ($P < 0.05$)

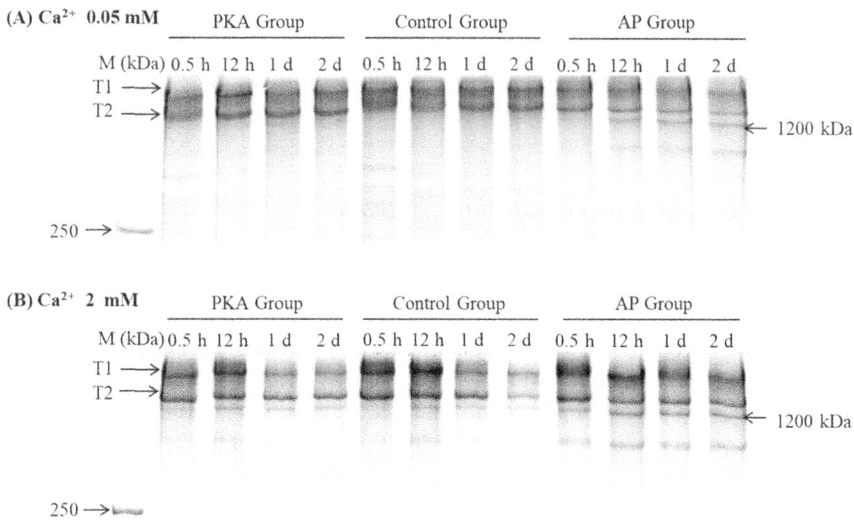

Fig. 7.10 Degradation of titin from PKA group, Control group, and AP group at different Ca^{2+} concentrations (Wang et al. 2019)

This present study also found out that the band with a molecular weight of 1200 kDa was obtained at the Ca^{2+} concentrations of 0.05 in the AP group, while it was not detected in the PKA and control groups (Fig. 7.10a, b). At the Ca^{2+} concentration of 2 mM, the band weighted 1200 kDa showed up in all three groups. However, there was the band weighted 1200 kDa in the AP group at 0.5 h incubation, at 12 h incubation in the PKA group and control group (Fig. 7.10b). The results indicated that titin in the AP group degraded faster than that in the PKA and control groups at the same Ca^{2+} concentration. Previous studies reported that pH of muscle affected the degradation of titin. In their research, they discovered that titin degraded faster in the meat muscles with the high ultimate pH, which had a higher activity of μ-calpain (Lomiwes et al. 2014; Wu et al. 2014a, b). However, there was no significant difference in pH among three groups in this study (Table 7.2, $P < 0.05$). It indicated that μ-calpain activity was not differently affected by pH in the three treatments (Du et al. 2017). In the present work, titin degraded faster in the AP group which had a lowest relative phosphorylation level (Figs. 7.9 and 7.10, $P < 0.05$). The result demonstrated the dephosphorylation of titin caused by AP accelerated the degradation of titin. The effect of AP may release free inorganic phosphate or transfer the phosphoryl group to cause the change of structure of titin. It was likely that dephosphorylation of titin influenced the stability of the protein. However, the structure of the dephosphorylated titin was not measured in the present study. Therefore, how the stability dephosphorylation of titin in vitro contributes to the degradation of titin needs further research.

7.6 Influence of Protein Phosphorylation on Other Myofibril Proteins Degradation

Previous researches have reported that the degradation of Troponin T (TnT) and desmin contributed to the meat tenderization. Degradation of TnT has been proved be highly related to the postmortem tenderization of meat (Anderson et al. 2012; Cruzen et al. 2014). Desmin is associated with Z-disc to connect Z-disc with other cytoskeletal elements. Desmin degrades rapidly during postmortem meat maturation and serves as an indicator of overall postmortem proteolysis (Huff-Lonergan et al. 2010). Further, we found out Tnt and desmin could also be phosphorylated (Li et al. 2017).

7.6.1 Troponin T (TnT)

A main TnT degradation product (30 kDa, band 2) was detected for control and AP samples at both $[Ca^{2+}] = 0.1$ and 1.0 mM, which was only detected in PKA samples at the highest Ca^{2+} concentration used in this study (Fig. 7.11: a2 and a3). In addition, two other products less in abundance were also detected in this research. Quantification of intact TnT showed that PKA-treated samples contained the most unchanged TnT among the three treatments at any incubation time (Fig. 7.11: b1 and b2). When comparing the control and AP treatments, less or the same amount of intact TnT was measured in AP samples. All these data showed that increased phosphorylation of TnT by PKA prevented its degradation, while dephosphorylation by AP increased its susceptibility to μ-calpain.

After incubation with μ-calpain, all desmin and TnT were degraded in AP and control samples at $[Ca^{2+}]$ 0.1 mM. Pretreatment of proteins with PKA reduced, while AP increased the degradation of all the four proteins by μ-calpain. In particular, the phosphorylation level of TnT showed a good inverse relationship with the degradation of the proteins, suggesting that phosphorylation may play an important role in protein degradation. High phosphorylation of Z-disc related proteins, including titin and TnT, were detected in tough bovine meat. Phosphorylation of Z-disc protein probably led to increased interactions of myotilin, myozenin 1, and troponin at Z-discs and reduced accessibility by proteases, resulting in tougher meat (D'Alessandro and Zolla 2013; D'Alessandro et al. 2012a, b). In addition, higher level of phosphorylation of structural proteins was related to lower myofibrillar degradation index in the *Longissimus dorsi* of Chianina and Maremmana cattle (D'Alessandro et al. 2012a, b). Troponin I with lower phosphorylation was more susceptible to degradation by calpain compared with Troponin I with higher phosphorylation (Gomes et al. 2005), which was consistent with our study. Another possibility was that phosphorylation may change the conformation of the protein, leading to changed stability (Tornavaca et al. 2011) (Fig. 7.11).

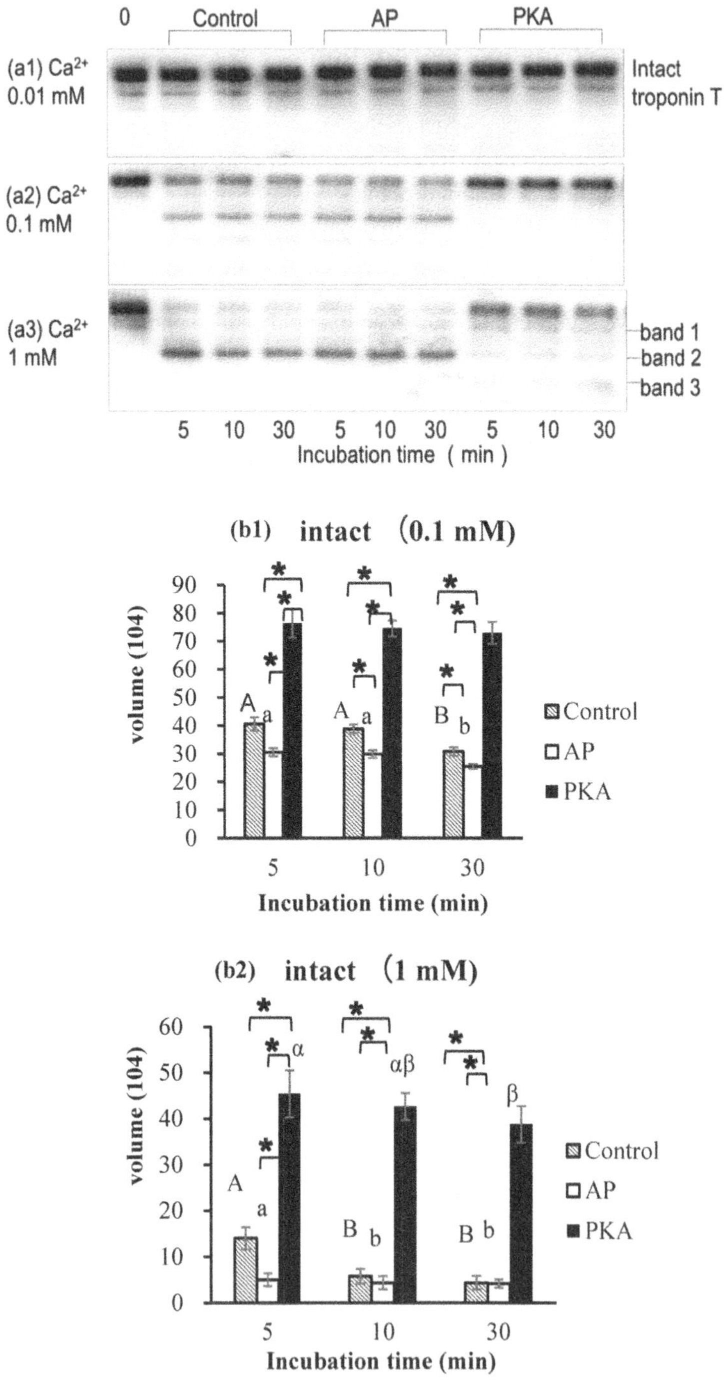

Fig. 7.11 Analysis of the degradation of troponin T after incubation with μ-calpain for 30 min. (**a1–4**) Western blot analysis of troponin T after myofibrillar proteins were incubated with μ-calpain for 30 min at 0.01, 0.1 and 1 mM Ca^{2+}. (**b1**) Densitometric analysis of troponin T in (**a2**). (**b2**)

7.6.2 Desmin

Degradation of desmin by μ-calpain was shown in Fig. 7.12. Before incubation with μ-calpain, two bands of intact desmin were detected in isolated myofibrillar proteins. Almost all desmin was degraded and no intact desmin was detectable in control and AP samples after incubation with μ-calpain when [Ca^{2+}] increased to 0.1 mM. Three bands were detected for desmin hydrolysis. Different with the control and AP treatments, desmin in the PKA-treated samples was unchanged and no fragments were detected (Fig. 7.12: a2), indicating the increased stability of PKA phosphorylated desmin. When [Ca^{2+}] was further increased to 1.0 mM, only one band of desmin fragment (band 3) was visible in control and AP samples. The other two bands (band 1 and 2) were probably further digested to smaller fragments that could not be detected by the anti-desmin antibody used in this study (Fig. 7.12: a3). The increased proteolysis of desmin was also measured in the PKA sample at [Ca^{2+}] = 1.0 mM, for which three products were determined. To compare the difference between control and AP samples, band 3 was quantified by densitometric analysis. As shown in Fig. 7.12, the relative density of band 3 was lower for AP samples than for control, showing relatively higher level of protein degradation in AP samples. All these data showed that dephosphorylation of desmin by AP enhanced its degradation, while phosphorylation by PKA reduced its degradation by μ-calpain.

Two isoforms of desmin (54 and 52 kDa) and three degraded fragments (50, 47 and 39 kDa) were reported in a previous study (Muroya et al. 2010). Desmin was susceptible to proteolysis and purified desmin was thoroughly degraded after incubation with μ-calpain for 1 h at 25 °C (Baron et al. 2004). Consistent with the literature, two bands of desmin and three degradation products were determined in this study. Also, the intact desmin was thoroughly degraded in control and AP samples at 0.1 and 1 mM Ca^{2+}.

Structural studies have confirmed that phosphorylation induced conformational changes in peptides corresponding to the hinge region (Penrose et al. 2004). Kim reported that the conformation of C/EBPb changed after dual phosphorylation, which facilitated dimerisation through its leucine zipper domains, leading to reduced degradation (Kim et al. 2007).

Fig. 7.11 (continued) Densitometric analysis of troponin T in (**a3**). /means significant difference between treatments at the same incubation time (*$P < 0.05$*); values with different letters are significantly different within the same group at different incubation time (*$p < 0.05$*) (Li et al. 2017)

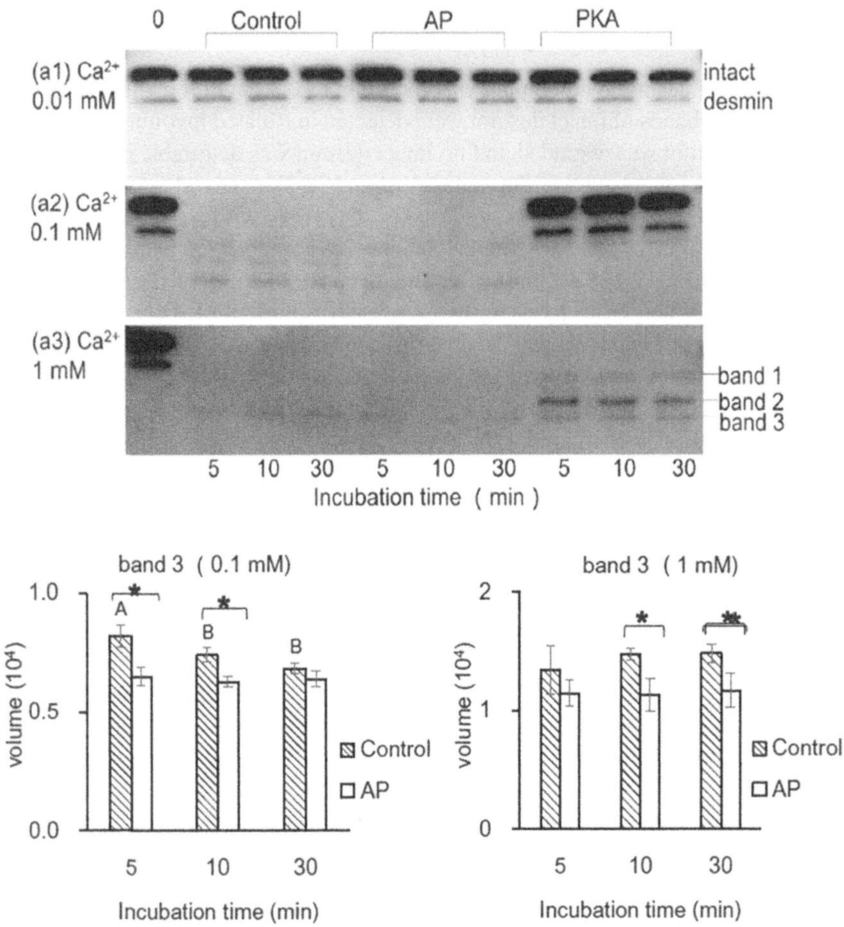

Fig. 7.12 Analysis of the degradation of desmin after incubation with μ-calpain for 30 min. (**a1–3**) Western blot analysis of desmin after myofibrillar proteins were incubated with μ-calpain for 30 min at 0.01, 0.1 and 1 mM Ca^{2+}. (**b1**) Densitometric analysis of band 3 in (**a2**). (**b2**) Densitometric analysis of band 3 in (**a3**). Values with/show significant difference between Control and AP group at the same incubation time ($P < 0.05$); values with different letters are significantly different within the Control group at different incubation time ($P < 0.05$) (Li et al. 2017)

7.7 Conclusions

PKA and AP can effectively alter the phosphorylation of isolated myofibrillar proteins and individual proteins, such as titin, desmin, and TnT. The phosphorylation of isolated proteins was significantly decreased by AP and increased proteolysis of the three proteins by μ-calpain. In this present study, we used DS software to carry out the simulation and make the molecular dynamics trajectory analysis in order to give the hypothesis. The hypothesis was that AP changed the dihedral angle of serine small and made the structure of titin unstable to promote its degradation. Additional research is needed to determine the structural changes of the intact titin after dephosphorylation to verify the hypothesis. The outcome of this study may improve our understanding of the mechanism of phosphorylation on degradation of myofibrillar proteins to regulate postmortem meat tenderization.

Acknowledgments Parts of this chapter are reprinted from Food Chemistry, 245, Li, Z., et al., Dephosphorylation enhances postmortem degradation of myofibrillar proteins. Food Chemistry, 233–239; Food Chemistry, 218, Li, Z., et al., Phosphorylation prevents in vitro myofibrillar proteins degradation by μ-calpain. Food Chemistry, 455–462; International Journal of Food Science and Technology, 53, Wang, Y., et al., Changes in phosphorylation levels and degradation of titin in three ovine muscles during postmortem aging. 913–920, Copyright (2020), with permission from Elsevier. Parts of this chapter are translated from Food Science, 40, Wang, Y., et al., Effects of Phosphorylation on titin Degradation at different Ca^{2+} concentrations in Vitro (Chinese). 52–57. Copyright (2020), with permission from Journal of Food Science.

References

Anderson, M. J., Lonergan, S. M., Fedler, C. A., Prusa, K. J., Binning, J. M., & Huff-Lonergan, E. (2012). Profile of biochemical traits influencing tenderness of muscles from the beef round. *Meat Science, 91*, 247–254.

Baron, C. P., Jacobsen, S., & Purslow, P. P. (2004). Cleavage of desmin by cysteine proteases: Calpains and cathepsin B. *Meat Science, 68*, 447–456.

Biswas, A. K., Tandon, S., & Beura, C. K. (2016). Identification of different domains of calpain and calpastatin from chicken blood and their role in post-mortem aging of meat during holding at refrigeration temperatures. *Food Chemistry, 200*, 315–321.

Bouton, P. E., Harris, P. V., & Shorthose, W. R. (1971). Effect of ultimate pH upon the water-holding capacity and tenderness of mutton. *Journal of Food Science, 36*, 435–439.

Chen, L. J., Li, X., Ni, N., Liu, Y., Chen, L., Wang, Z. Y., et al. (2016). Phosphorylation of myofibrillar proteins in post-mortem ovine muscle with different tenderness. *Journal of Science and Food Agricultral, 96*, 1474–1483.

Chiu, J. C., Vanselow, J. T., Kramer, A., & Edery, I. (2008). The phospho-occupancy of an atypical SLIMB-binding site on period that is phosphorylated by doubletime controls the pace of the clock. *Genes Development, 22*, 1758–1772.

Cruzen, S. M., Paulino, P. V. R., Lonergan, S. M., & Huff-Lonergan, E. (2014). Postmortem proteolysis in three muscles from growing and mature beef cattle. *Meat Science, 96*, 854–861.

D'Alessandro, A., Marrocco, C., Rinalducci, S., Mirasole, C., Failla, S., & Zolla, L. (2012b). Chianina beef tenderness investigated through integrated Omics. *Journal of Proteomics, 75*, 4381–4398.

D'Alessandro, A., Rinalducci, S., Marrocco, C., Zolla, V., Napolitano, F., & Zolla, L. (2012a). Love me tender: An Omics window on the bovine meat tenderness network. *Journal of Proteomics, 75*, 4360–4380.

D'Alessandro, A., & Zolla, L. (2013). Meat science: From proteomics to integrated omics towards system biology. *Journal of Proteomics, 78*, 558–577.

Deng, J., Yu, P., Wang, Y., & Mao, L. (2015). Real-time ratiometric fluorescent assay for alkaline phosphatase activity with stimulus responsive infinite coordination polymer nanoparticles. *Analytical Chemistry, 87*, 3080–3086.

Du, M. T., Li, X., Li, Z., Li, M., Gao, L. L., & Zhang, D. Q. (2017). Phosphorylation inhibits the activity of μ-calpain at different incubation temperatures and Ca2+ concentrations in vitro. *Food Chemistry, 228*, 649–655.

Ferguson, D. M., Daly, B. L., Gardner, G. E., & Tume, R. K. (2008). Effect of glycogen concentration and form on the response to electrical stimulation and rate of post- mortem glycolysis in ovine muscle. *Meat Science, 78*, 202–210.

Gannon, J., Staunton, L., O'connell, K., Doran, P., & Ohlendieck, K. (2008). Phosphoproteomic analysis of aged skeletal muscle. *International Journal of Molecular Medicine, 22*, 33–42.

Gautel, M., Leonard, K., & Labeit, S. (1993). Phosphorylation of KSP motifs in the C-terminal region of titin in differentiating myoblasts. *EMBO Journal, 12*, 3827–3834.

Gautel, M., Leonard, K., & Labeit, S. (2001). Phosphorylation of KSP motifs in the C-terminal region of titin in differentiating myoblasts. *The EMBO Journal, 12*, 3827–3384.

Gomes, A. V., Harada, K., & Potter, J. D. (2005). A mutation in the N-terminus of troponin I that is associated with hypertrophic cardiomyopathy affects the Ca^{2+}-sensitivity, phosphorylation kinetics and proteolytic susceptibility of troponin. *Journal of Molecular and Cellular Cardiology, 39*, 754–765.

Honikel, K. O. (2014a). Chemical and physical characteristics of meat pH measurement. *Encyclopedia of Meat Science, 1*, 262–266.

Honikel, K. O. (2014b). Conversion of muscle to meat|glycolysis. In *Encyclopedia of meat sciences* (2nd ed., pp. 353–357). Oxford: Academic Press.

Huang, H., Larsen, M. R., & Lametsch, R. (2012). Changes in phosphorylation of myofibrillar proteins during postmortem development of porcine muscle. *Food Chemistry, 134*, 1999–2006.

Huff Lonergan, E., Zhang, W., & Lonergan, S. M. (2010). Biochemistry of postmortem muscle - lessons on mechanisms of meat tenderization. *Meat Science, 86*, 184–195.

Huff-Lonergan, E., Mitsuhashi, T., Beekman, D. D., Parrish Jr., F. C., Olson, D. G., & Robson, R. M. (1996). Proteolysis of specific muscle structural proteins by μ-calpain at low pH and temperature is similar to degradation in postmortem bovine muscle. *Journal of Animal Science, 74*, 993–1008.

Huff-Lonergan, E., Parrish, F. C., & Robson, R. M. (1995). Effects of postmortem aging time, animal age, and sex on degradation of titin and nebulin in bovine longissimus muscle. *Journal of Animal Science, 73*, 1064–1073.

Huff-Lonergan, E., Zhang, W., & Lonergan, S. M. (2010). Biochemistry of postmortem muscle—Lessons on mechanisms of meat tenderization. *Meat Science, 86*, 184–195.

Kim, J., Tang, Q., Li, X., & Lane, M. D. (2007). Effect of phosphorylation and S-S bond-induced dimerization on DNA binding and transcriptional activation by C/EBPb. In *Proceedings of the National Academy of Sciences of the United States of America* (Vol. 104, pp. 1800–1804).

Koohmaraie, M., & Geesink, G. H. (2006). Contribution of postmortem muscle biochemistry to the delivery of consistent meat quality with particular focus on the calpain system. *Meat Science, 74*, 34–43.

Kurzban, G. P., & Wang, K. (1988). Giant polypeptides of skeletal-muscle titin — Sedimentation equilibrium in guanidine-hydrochloride. *Biochemical and Biophysical Research Communications, 150*, 1155–1161.

Labeit, S., Barlow, D. P., Gautel, M., Gibson, T., Holt, J., Hsieh, C. L., et al. (1990). A regular pattern of two types of 100-residue motif in the sequence of titin. *Nature, 345*, 273–276.

Lametsch, R., Roepstorff, P., & Bendixen, E. (2002). Identification of protein degradation during post-mortem storage of pig meat. *Journal of Agricultural and Food Chemistry, 50*, 5508–5512.

Lametsch, R., Roepstorff, P., Moller, H. S., & Bendixen, E. (2004). Identification of myofibrillar substrates for mu-calpain. *Meat Science, 68*, 515–521.

Li, Z., Li, M., Du, M., Shen, Q. W., & Zhang, D. (2018). Dephosphorylation enhances postmortem degradation of myofibrillar proteins. *Food Chemistry, 245*, 233–239.

Li, Z., Li, X., Gao, X., Shen, Q. W., Du, M. T., & Zhang, D. Q. (2017). Phosphorylation prevents in vitro myofibrillar proteins degradation by μ-calpain. *Food Chemistry, 218*, 455–462.

Lomiwes, D., Farouk, M. M., Wu, G., & Young, O. A. (2014). The development of meat tenderness is likely to be compartmentalized by ultimate pH. *Meat Science, 96*, 646–651.

Macek, B., Mann, M., & Olsen, J. V. (2009). Global and site-specific quantitative phosphoproteomics: Principles and applications. *Annual Review of Pharmacology and Toxicology, 49*, 199–223.

Mattio, N. D., Paredi, M. E., & Crupkin, M. (1992). Postmortem changes in glycogen, ATP, hypoxanthine and ja:Math absorbance ratio in extracts of adductor muscles from Aulacomya ater ater (Molina) at different biological conditions. *Comparative Biochemistry and Physiology Part A: Physiology, 103*, 605–608.

Melody, J. L., Lonergan, S. M., Rowe, L. J., Huiatt, T. W., Mayes, M. S., & Huff Lonergan, E. (2004). Early postmortem biochemical factors influence tenderness and water-holding capacity of three porcine muscles. *Journal of Animal Science, 82*, 1195–1205.

Muroya, S., Ertbjerg, P., Pomponio, L., & Christensen, M. (2010). Desmin and troponin T are degraded faster in type IIb muscle fibers than in type I fibers during postmortem aging of porcine muscle. *Meat Science, 86*, 764–769.

Napravnikova, E., Vorlova, L., & Malota, L. (2002). Changes in hygienic quality of vacuum-packed pork during storage. *Acta Veterinaria Brno, 71*, 255–262.

Nguyen, H., Wu, S., Su, C., & Hwang, T. (2017). Biochemical, biophysical, and thermal properties of alkaline phosphatase from thermophile Thermus sp. NTU-237. *Biotechnology, Agronomy, Society and Environment, 21*, 117–126.

Penrose, K. J., Garcia-Alai, M., de Prat-Gay, G., & Mcbride, A. A. (2004). Casein kinase II phosphorylation-induced conformational switch triggers degradation of the papillomavirus E2 protein. *Journal of Biological Chemistry, 279*, 22430–22439.

Pulford, D. J., Dobbie, P., Fraga Vazquez, S., Fraser-Smith, E., Frost, D. A., & Morris, C. A. (2009). Variation in bull beef quality due to ultimate muscle pH is correlated to endopeptidase and small heat shock protein levels. *Meat Science, 83*, 1–9.

Sebestyen, M. G., Wolf, J. A., & Greaser, M. L. (1995). Characterization of a 5.4 kb cDNA fragment from the Z-line region of rabbit cardiac titin reveals phosphorylation sites for proline-directed kinase. *Journal of Cell Science, 108*, 3029–3037.

Shen, J., Channavajhala, P., Seldin, D. C., & Sonenshein, G. E. (2001). Phosphorylation by the protein kinase CK2 promotes calpain-mediated degradation of IkappaBalpha. *Journal of Immunology, 167*, 4919–4925.

Tornavaca, O., Sarro, E., Pascual, G., Bardaji, B., Montero, M. A., & Salcedo, M. T. (2011). KAP degradation by calpain is associated with CK2 phosphorylation and provides a novel mechanism for cyclosporine A-induced proximal tubule injury. *PLoS One, 6*, e25746.

Tskhovrebova, L., & Trinick, J. (2003). Titin: Properties and family relationships. *Nature Reviews Molecular Cell Biology, 4*, 679–689.

Wang, D., Dong, H., Zhang, M., Liu, F., Bian, H., Zhu, Y., et al. (2013). Changes in actomyosin dissociation and endogenous enzyme activities during heating and their relationship with duck meat tenderness. *Food Chemistry, 141*, 675–679.

Wang, Y., Li, X., Li, Z., Li, M., Zhu, J., & Zhang, D. Q. (2018). Changes in phosphorylation levels and degradation of titin in three ovine muscles during post-mortem ageing. *International Journal of Food Science & Technology, 53*, 913–920.

Wang, Y., Li, X., Li, Z., Zhu, J., Zhang, S., & Zhang, D. (2019). Effects of phosphorylation on titin degradation at different Ca^{2+} concentrations *in vitro*. *Journal of Food Science, 40*, 52–57.

Wu, G., Clerens, S., & Farouk, M. M. (2014a). LC-MS/MS identification of large structural proteins from bull muscle and their degradation products during postmortem storage. *Food Chemistry, 150*, 137–144.

Wu, G., Farouk, M. M., Clerens, S., & Rosenvold, K. (2014b). Effect of beef ultimate pH and large structural changes with ageing on meat tenderness. *Meat Science, 98*, 637–645.

Zhang, Y., Li, S., Qian, S., Zhang, Y., Liu, Y., Tang, Q. Q., et al. (2012). Phosphorylation prevents C/EBPβ from the calpain-dependent degradation. *Biochemical and Biophysical Research Communications, 419*, 550–555.

Chapter 8
Mechanism of the Effect of Protein Phosphorylation on Calpain Activity

Abstract μ-Calpain is the key enzyme in postmortem meat tenderization. Studies have shown that μ-calpain can be phosphorylated and phosphorylation influences the activity of μ-calpain. However, whether μ-calpain can be phosphorylated in postmortem muscles and the possible relative influencing mechanism remains unknown. The *longissimus lumborum* (LL) muscles were used in this research. The relationship between phosphorylation level of sarcoplasmic proteins, μ-calpain activity, and calpastatin degradation in mutton with different tenderness was analyzed. Alkaline phosphatase and phosphatase inhibitor were used to modulate the phosphorylation level of sarcoplasmic proteins, and the effect of phosphorylation of sarcoplasmic proteins on μ-calpain activity was studied. The phosphorylation level of μ-calpain was regulated in vitro, and the influence of dephosphorylation and phosphorylation of μ-calpain on μ-calpain activity was investigated. The changes in the secondary structure of μ-calpain were studied and the phosphorylation sites of μ-calpain were identified to illustrate the directly regulatory mechanism of μ-calpain activity by phosphorylation. The effects of calpastatin on phosphorylated μ-calpain and the effects of phosphorylated calpastatin on μ-calpain were also analyzed. The study of the effects of phosphorylation on the interaction between calpastatin and μ-calpain clarified the indirectly regulatory mechanism of phosphorylation on μ-calpain activity.

Keywords Tenderness · μ-calpain · Calpastatin · Phosphorylation · Dephosphorylation

8.1 Introduction

Tenderness is an important sensory quality attribute of meat, especially for beef and lamb. The calpain system, which includes μ-calpain, m-calpain, and calpastatin, is considered to be primarily responsible for the postmortem proteolytic tenderization of meat (Pomponio and Ertbjerg 2012; Kemp et al. 2010). The μ-calpain is a heterodimer composed of two subunits of 80 kDa and 28 kDa. The 80 kDa subunit degrades to a 76 kDa peptide through a 78 kDa intermediate by autolysis for its

D. Zhang et al., *Protein Phosphorylation and Meat Quality*,
https://doi.org/10.1007/978-981-15-9441-0_8

activation (Goll et al. 2003). In addition, the activity of μ-calpain is regulated by many factors, such as Ca^{2+}, temperature, calpastatin, etc. (Koohmaraie 1992; Geesink et al. 2006).

The calpains are calcium activated cysteine proteases. The 80 kDa subunit has EF hand domains which bind Ca^{2+} to activate calpain. Therefore, the concentration of Ca^{2+} in sarcoplasm directly influences μ-calpain activation (Goll et al. 2003; Ohno et al. 1986). The μ-calpain is very sensitive to temperature. In postmortem bovine muscle, μ-calpain activity is primarily regulated by muscle temperature (Mohrhauser et al. 2014). In addition, accelerated μ-calpain activation is detected in both lamb and beef when muscles are incubated at elevated temperatures (Geesink et al. 2000; Hwang et al. 2004; Pomponio and Ertbjerg 2012). As one of the well-characterized parts in calpain system, the endogenous calpastatin is the only known protein inhibitor specific for μ-calpain and m-calpain (Crawford 1990). Each calpastatin can inhibit four calpain molecules to limit postmortem muscle proteolysis and meat tenderization (Goll et al. 2003; Kent et al. 2004).

Protein phosphorylation, one of the most common post-translational modifications, has intrigued scientists' great interest in recent years for its regulation of postmortem energy metabolism and probably meat quality (Huang et al. 2012). It is also reported that phosphorylation of both sarcoplasmic and myofibrillar proteins in beef carcass is altered by electrical stimulation (Li et al. 2015), indicating protein phosphorylation may be related to meat tenderness. However, the exact biochemical mechanism is not well understood. Biomedical studies reveal that μ-calpain is phosphorylated by protein kinase A (PKA) and protein kinase C (PKC) (Storr et al. 2011). There are nine phosphorylation sites in the four domains of the 80 kDa subunit of μ-calpain, including four Ser residues, three Thr residues, and two Tyr residues (Vazquez et al. 2008). Phosphorylation of μ-calpain alters its proteolysis ability and the location within cells (Vazquez et al. 2008; Xu and Deng 2006a, b). However, whether the μ-calpain function affected by phosphorylation in postmortem muscle remains unknown.

8.2 Relationship Between Protein Phosphorylation and Calpain Activity

Muscle samples were grouped into three according to myofibril fragmentation index (MFI) at 24 h and 48 h postmortem. The relationship between phosphorylation level of sarcoplasmic proteins, μ-calpain activity, and calpastatin degradation in postmortem mutton with different tenderness was investigated. In the present study, the MFI of 50 sheep at 24 h and 48 h postmortem was detected and the results were ranked. Sheep with highest, middle, and lowest (5 of each) MFI value were chosen as tender, middle, and tough groups. The results showed that there was a significant difference between the three groups, which could be used for further analysis.

8.2.1 Changes in pH in Mutton with Different Tenderness at Postmortem

The changes in pH values in the three groups were consistent during postmortem (Fig. 8.1). There was no significant difference in pH values among the three groups at each time point ($P > 0.05$).

The decline rate of pH during postmortem affected meat tenderness (Marsh et al. 1987). pH was also an important factor that influences the activity and autolysis degradation of μ-calpain (Maddock et al. 2005). Many studies have shown that the inactivation rate of μ-calpain was faster in muscle with faster pH decline rate at postmortem (Hwang and Thompson 2001; Rhee et al. 2006; Whipple et al. 1990). Muscle with low pH presented low μ-calpain activity (Claeys et al. 2001). In this study, there was no significant difference in pH value in three groups at each time point, indicating that the activity of μ-calpain and meat tenderness was not affected by pH value.

8.2.2 Phosphorylation Level of Sarcoplasmic Proteins During Postmortem

The relative phosphorylation level of sarcoplasmic proteins in tender, middle, and tough groups during postmortem is shown in Fig. 8.2. The changes in phosphorylation level of sarcoplasmic proteins in three groups were consistent, but tender

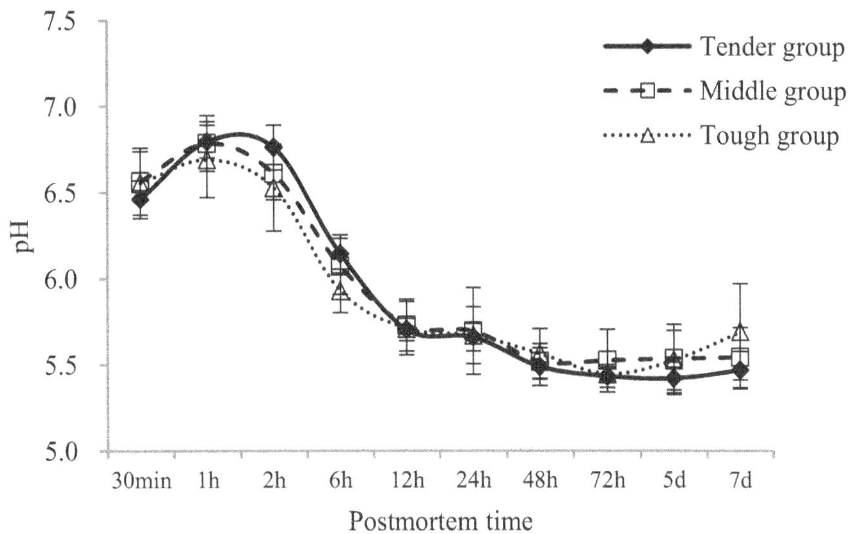

Fig. 8.1 Changes in pH in mutton with different tenderness at postmortem (Du 2018)

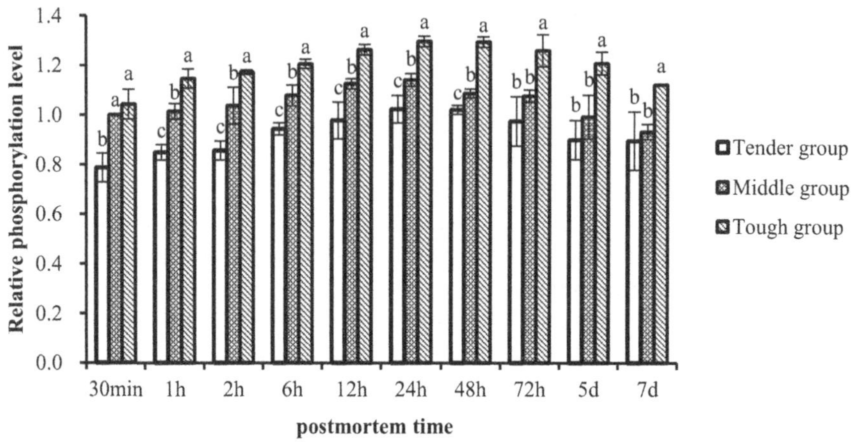

Fig. 8.2 Phosphorylation level of sarcoplasmic proteins during postmortem (Du 2018)

group had a significantly lower phosphorylation level than middle group, than that of the tough group (*P* < *0.05*).

Protein phosphorylation, one of the most common post-translational modifications, has intrigued scientists' great interest in recent years for its regulation of postmortem energy metabolism and probably meat quality (Huang et al. 2012). It is also reported that phosphorylation of both sarcoplasmic and myofibrillar proteins in beef carcass is altered by electrical stimulation (Li et al. 2015), indicating protein phosphorylation may be related to meat tenderness. Our team has studied the phosphorylation level of proteins in mutton with different tenderness and found out that tender meat showed lower phosphorylation level of both sarcoplasmic and myofibrillar protein, which was the same as the present study. Phosphorylation may affect meat tenderness by influencing rigor mortis and glycolysis (Chen et al. 2016).

8.2.3 *Casein Zymography Analysis of μ-/m-Calpain*

In the detection of μ-calpain and m-calpain by casein zymography in the present study, casein gels were incubated in the presence of 4 mM Ca^{2+}. This concentration of Ca^{2+} can activate both μ-calpain and m-calpain to degrade casein at their locations in the gel. Thus, the content of detected μ-calpain and m-calpain remained proteolysis ability. The results showed that the content of unautolyzed μ-calpain in all three groups decreased during postmortem (Fig. 8.3). The proteolysis activity of unautolyzed μ-calpain in tender group decreased faster at 6 h postmortem. At 24 h postmortem, μ-calpain in tender group autolyzed totally and deactivated at 72 h postmortem. In the middle group, μ-calpain deactivated at 5 d postmortem, but still remained activity at 7 d in the tough group. During postmortem, the proteolysis

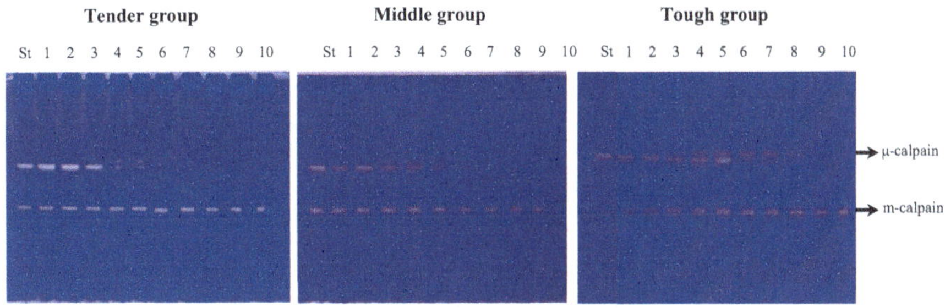

Fig. 8.3 Casein zymography of μ-/m-calpain. Sarcoplasmic protein of middle group at 30 min postmortem was regard as St. (Du 2018)

activity of m-calpain did not change in three groups. M-Calpain did not autolyze during 7 d aging.

The content of μ-calpain in tender, middle, and tough groups decreased during postmortem aging, indicating that μ-calpain was responsible for meat tenderization by autolysis and degradation. Studies have shown that the proteolysis ability of μ-calpain mainly concentrated on the first 3 days after slaughter. μ-calpain activated because of its self-autolysis and deactivated after degrading substrate proteins. In most cases, the concentration of Ca^{2+} in postmortem muscle was too low to activate m-calpain, thus, m-calpain might not have an effect on meat tenderization postmortem (Koohmaraie and Geesink 2006; Geesink et al. 2006; Koohmaraie 1992). O'Halloran et al. (1997) studied the relationship between μ-calpain and meat tenderness and found out that muscle with lower μ-calpain activity at 3 h postmortem had higher protein degradation level, which meant that muscle was tender with faster autolysis of μ-calpain. At 7 d postmortem, the activity of m-calpain showed no obvious change in all three groups, indicating that the differences in meat tenderness in three groups were not induced by m-calpain.

8.2.4 The Degradation of μ-Calpain 80 kDa Subunit

The 80 kDa subunit of μ-calpain degraded slowly in all the three groups during postmortem (Fig. 8.4). The degradation rate of μ-calpain in tender group was significantly higher than the middle and tough groups at 1, 6, 12, and 72 h postmortem ($P < 0.05$). Overall, the degradation rate of μ-calpain was: tender group > middle group > tough group.

Studies have shown that the autolysis and activation of μ-calpain were related to the degradation of myofibrillar proteins at early postmortem (Melody et al. 2004). Most of the protein degradation and tenderization was induced by μ-calpain activation. μ-calpain with faster autolysis rate showed earlier activation and faster deactivation (Killefer and Koohmaraie 1994). Kim et al. (2013) studied the autolysis of μ-calpain in mutton during postmortem aging by western blotting and discovered

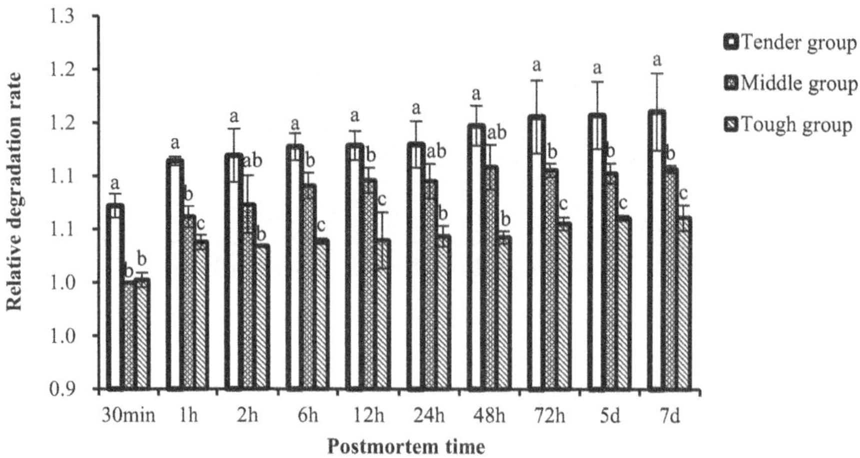

Fig. 8.4 Degradation rate of μ-calpain 80 kDa subunit (Du 2018)

that the autolysis and activation degree of μ-calpain was positively related to protein degradation. Muscle with higher content of μ-calpain intact 80 kDa subunit presented lower proteolysis degradation and lower tenderness (Cruzen et al. 2014). In the present study, the degradation rate of μ-calpain in tender group was higher than middle group, than that of in tough group, indicating that the activation degree and activity of μ-calpain were the highest in tender group and the lowest in tough group, which was consistent with the results of casein zymography. Muscle with different tenderness had different phosphorylation level of sarcoplasmic proteins. Tender muscle presented lower phosphorylation level of sarcoplasmic proteins but higher activity of μ-calpain, which demonstrated that phosphorylation of sarcoplasmic proteins may have negatively regulated μ-calpain activity and finally influenced meat tenderness during postmortem.

8.2.5 The Degradation of Calpastatin

During postmortem, the 130 kDa calpastatin degraded gradually in all three groups (Fig. 8.5). The 130 kDa calpastatin in tender and middle groups entirely degraded at 6 h and 5 d postmortem, respectively. However, calpastatin could still be detected in tough group at 7 d postmortem. Calpastatin in tender group degraded faster than middle group, than that of in the tough group.

As one of the well-characterized parts in calpain system, the endogenous calpastatin was the only known protein inhibitor specific for μ-calpain and m-calpain (Crawford 1990). Each calpastatin could inhibit four calpain molecules to limit postmortem muscle proteolysis and meat tenderization (Goll et al. 2003; Kent et al. 2004). Although the degradation of calpastatin did not equal to the loss of

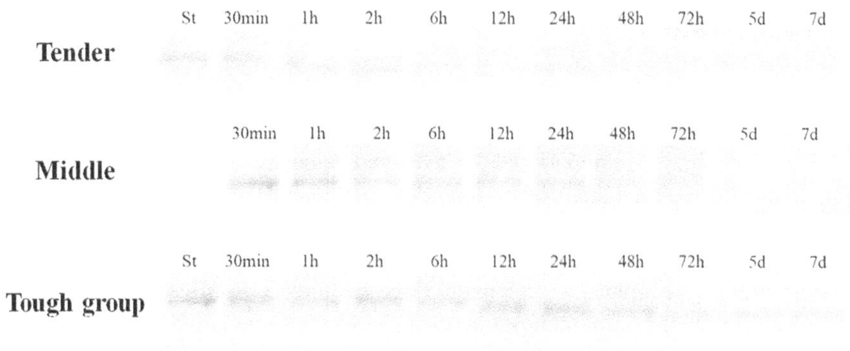

Fig. 8.5 Degradation of calpastatin during postmortem (Du 2018)

inhibitory activity, the inhibitory activity of calpastatin was weakened by degradation (Huang et al. 2014). Following the degradation of calpastatin, the muscle became tender during postmortem (Geesink et al. 2005; Boehm et al. 1998). Tough meat showed higher level of calpastatin, because of the inhibitory ability of calpastatin to the proteolysis activity of μ-calpain (Kemp et al. 2010). Tender group had a faster degradation rate of calpastatin, indicating the faster deactivation rate of calpastatin. Therefore, the inhibitory ability of calpastatin to μ-calpain was lowest in tender group. Salamino et al. (1994) studied the phosphorylation of calpastatin and found out that, compared to dephosphorylated calpastatin, phosphorylated calpastatin showed greater inhibition to m-calpain. In the present study, tough group had higher phosphorylation level of sarcoplasmic proteins and higher degradation level of calpastatin, which demonstrated that phosphorylation might play negative roles in meat tenderization by enhancing the inhibition of calpastatin to μ-calpain activity and then inhibiting protein proteolysis at postmortem.

8.3 Effects of Phosphorylation of Sarcoplasmic Proteins on μ-Calpain Activity at Different Incubation Temperature

The indirectly effects of phosphorylation on the activity of μ-calpain and the sensitivity of μ-calpain to temperature were investigated in this part. Sarcoplasmic proteins were extracted from three Fat Tail Han sheep (8 months) at 30 min postmortem, adjusted to 12 μg/μL and subjected to three treatments: (1) 25 U/100 μg protein alkaline phosphatase (AP); (2) One tablet/10 mL supernatant phosphatase inhibitor (PI); (3) Control. For temperature sensitivity analysis, sarcoplasmic protein was treated with AP and phosphatase inhibitor (PI) at 4, 25, and 37 °C, respectively.

8.3.1 Phosphorylation Level Analysis of Sarcoplasmic Proteins

To modulate the phosphorylation level of proteins, prepared sarcoplasmic proteins were treated with or without AP. As expected, incubation of proteins with AP remarkably decreased the phosphorylation of sarcoplasmic proteins, since the intensity of protein bands of the AP group on the Pro-Q Diamond stained gels was obviously lower than that of the other two groups (Fig. 8.6a, c, g, h). The phosphorylation level of sarcoplasmic proteins in three treatments at different temperatures was evaluated by the ratio of phosphoproteins to total proteins (Fig. 8.6). The results showed that the intensity of Pro-Q Diamond stained proteins in AP group was significantly lower compared to other groups during the 72 h incubation at both 4 and 25 °C ($P < 0.05$) (Fig. 8.6a, c, g, h). AP dephosphorylated proteins, while phosphatase inhibitors inhibited protein dephosphorylation both in vivo and in vitro (Sharma et al. 2014; Chen et al. 2016). These two chemicals were often used to modulate protein phosphorylation status. At all incubation temperatures, PI group showed a greater intensity in Pro-Q Diamond stained proteins and a higher phosphorylation level compared to control group. Overall, AP group had the lowest intensity, indicating the lowest phosphorylation level of proteins, which meant that both AP and phosphatase inhibitor worked in this study. The high, middle, and low protein phosphorylation level of PI, control, and AP groups revealed that protein phosphorylation status could be effectively regulated by AP and PI.

8.3.2 μ- and m-Calpain Activity

Generally, the 80 kDa subunit of μ-calpain degraded slowly in all three groups during incubation (Fig. 8.7). Comparatively, the AP group showed a higher level and a faster rate of μ-calpain degradation than the control and PI groups. As the temperature increased, the degradation of μ-calpain increased due to the intensity of degraded μ-calpain bands increased in all three groups. It has been reported that earlier autolysis/degradation of μ-calpain was associated with earlier activation of μ-calpain in postmortem muscle. The degradation degree of μ-calpain was positively correlated with μ-calpain activity (Melody et al. 2004). Thus, the activity of μ-calpain was supposed to be higher in AP group compared to other two groups. At 4 °C and 37 °C, the 80 kDa subunit of μ-calpain in control group degraded relatively faster than in PI group, indicating that the activity of μ-calpain in control group was higher than in the PI group of high protein phosphorylation level. However, no difference in the degradation of μ-calpain was detected between control and PI samples incubated at 25 °C. The optimal temperature for μ-calpain activity was 25 °C (Boehm et al. 1998). At this temperature, μ-calpain had the best proteolysis ability and maximal activity. In other words, the degradation rate of μ-calpain was the fastest at this temperature. In this study, the AP was more effective than PI in

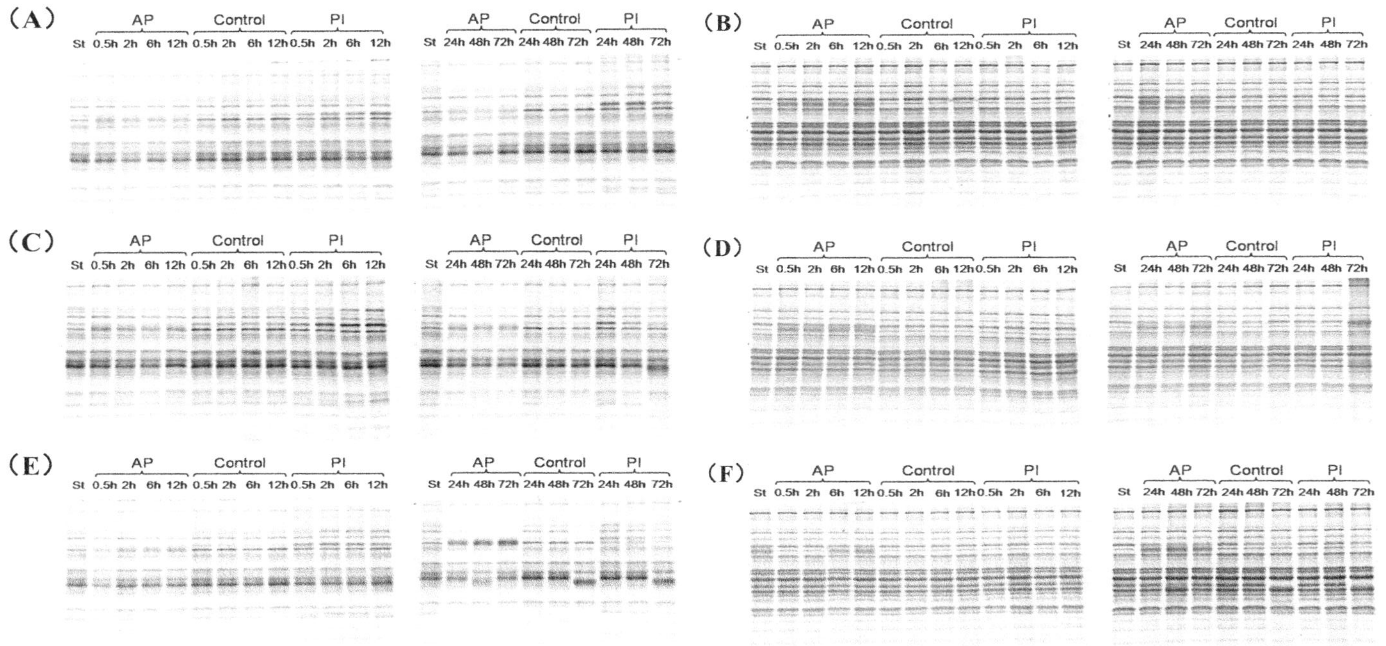

Fig. 8.6 Phosphorylation of sarcoplasmic proteins in three groups incubated at 4, 25, and 37 °C. (**a**), (**b**) Samples incubated at 4 °C. (**c**), (**d**) Samples incubated at 25 °C. (**e**), (**f**) Samples incubated at 37 °C. (**a**), (**c**), (**e**) SDS-PAGE gel stained by Pro-Q Diamond stains. (**b**), (**d**), (**f**) SDS-PAGE gel stained by Ruby stains. (**g**), (**h**), (**i**) Relative phosphorylation level of sarcoplasmic proteins in three treatments incubated at 4, 25, and 37 °C, respectively. Values with different letters show significant difference in the results in different groups at the same incubation time ($P < 0.05$). *AP* Alkaline phosphatase, *PI* Phosphatase inhibitor (Du et al. 2017a)

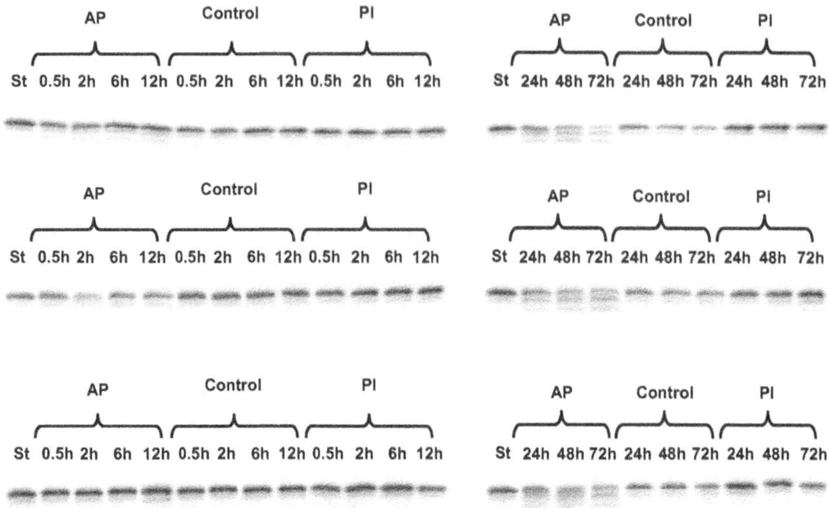

Fig. 8.7 Western blotting analysis during incubation at 4, 25, and 37 °C. (**a**), (**b**), (**c**) The degradation of μ-calpain 80 kDa subunit in three groups incubated at 4, 25, and 37 °C, respectively. *AP* Alkaline phosphatase, *PI* Phosphatase inhibitor (Du et al. 2017a)

regulating protein phosphorylation status, because the difference in protein phosphorylation levels between AP and control samples was greater than that between control and PI groups (Fig. 8.6g, h, i). Consequently, the difference in the degradation of 80 kDa μ-calpain subunit between AP sample and control was greater than that between control and PI groups when incubated at 4 °C and 37 °C. The degradation of 80 kDa subunit showed no difference between control and PI samples when incubated at 25 °C. This could be the optimal temperature and thus the high activity of μ-calpain eliminated the effect of the minute difference in protein phosphorylation on the autolysis of μ-calpain (Boehm et al. 1998).

Phosphorylation altered the physiological functions of proteins by changing their structure or biological activities. Phosphorylation of sarcoplasmic proteins in response to electrical stimulation increased the activity of metabolic enzymes (Li et al. 2015). The phosphorylation of μ-calpain catalyzed by extracellular protein kinases (ERK) not only enhanced its activity, but also promoted its secretion (Xu and Deng 2004). In contrast, the data obtained in this study showed that phosphorylation may negatively regulate μ-calpain degradation and activation. In addition, the influence of phosphorylation on μ-calpain degradation was greater at 37 °C, which reduced at 25 °C, indicating that temperature could influence the sensitivity of μ-calpain to phosphorylation.

In the detection of μ-calpain and m-calpain by casein zymography in the present study, casein gels were incubated in the presence of 4 mM Ca^{2+}. This concentration of Ca^{2+} can activate both μ-calpain and m-calpain to degrade casein at their locations in the gel. Thus, the content of detected μ-calpain remained 80 or 78 kDa subunit

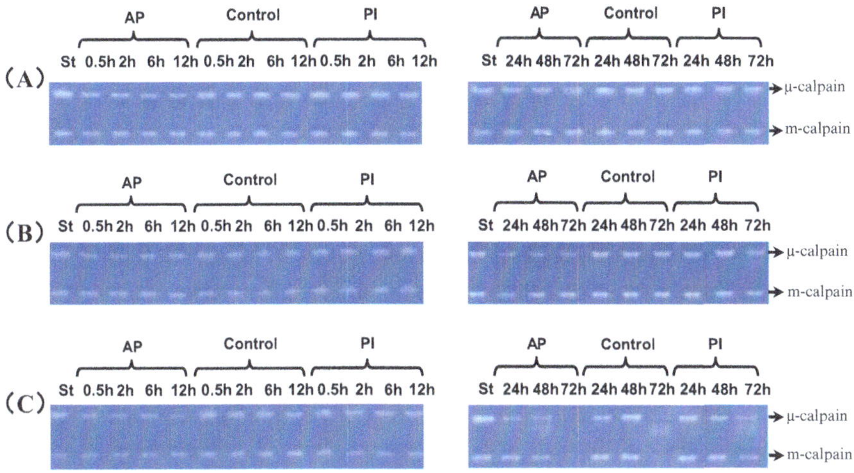

Fig. 8.8 Casein zymography gel analysis during incubation at 4, 25, and 37 °C. (**a**), (**b**), (**c**) μ- and m-calpain in three groups incubated at 4, 25, and 37 °C, respectively. *AP* Alkaline phosphatase; *PI* Phosphatase inhibitor (Du et al. 2017a)

which kept its proteolysis ability. The results showed that the autolyzed μ-calpain was only detected in AP samples incubated at 37 °C for 48 h and in both control and PI samples incubated at 37 °C for 72 h (Fig. 8.8c). After 48 h incubation at 37 °C, the content of autolyzed μ-calpain increased more enough to be activated in gels to hydrolyze casein. The less unautolyzed μ-calpain in AP samples incubated at any temperature in this study (Fig. 8.8) demonstrated that the autolysis degree of μ-calpain in this group was greater than that of the other two groups, which was consistent with the results of western blotting analysis of μ-calpain degradation.

The m-calpain, same as μ-calpain, consisted of an 80 kDa catalytic subunit and a 28 kDa regulatory subunit. m-calpain was also calcium dependent and could be activated by autolysis in the presence of calcium (Cong et al. 1989). In most cases, the concentration of Ca^{2+} in postmortem muscle was too low to activate m-calpain, thus, m-calpain may not be crucial in meat tenderization postmortem (Koohmaraie and Geesink 2006). However, autolyzed m-calpain was detected in pork stored in refrigerator for 6 days (Pomponio et al. 2008). Pomponio and Ertbjerg (2012) reported that m-calpain was activated by the high temperature in pork. In the present study, noticeable autolysis of m-calpain happened only in AP and control samples incubated at 37 °C for 72 h, indicating that high temperature activated m-calpain. Meanwhile, the content of unautolyzed m-calpain was much higher in higher phosphorylated sarcoplasmic proteins. m-calpain was phosphorylated by PKA or ERK. Phosphorylation of m-calpain increased its activity (Xu and Deng 2004; Xu and Deng 2006a, b; Leloup et al. 2010). Some literature reported that phosphorylation of m-calpain at serine 369 kept it in an inactive state (Shiraha et al. 2002). In our study, the autolysis degree of m-calpain was greater in AP samples than in the other two groups, showing that phosphorylation inhibited m-calpain activation in vitro.

8.4 Effects of Phosphorylation on μ-Calpain Activity at Different Incubation Temperature In Vitro

The effects of AP dephosphorylation and PKA phosphorylation on μ-calpain activity and sensitivity to temperature were investigated in this part. The purified μ-calpain was treated with AP or PKA for 30 min at 30 °C to regulate its phosphorylation level. Samples were then incubated at controlled freezing point (−1), 4, 25, and 37 °C, respectively. The results showed that PKA and AP had no influence on pH values of incubation solution. At −1 and 4 °C, the maximum degradation rate of μ-calpain was detected in AP group and the minimum was in control group. Low temperature of controlled freezing point prevented dephosphorylation and phosphorylation progression and delayed μ-calpain degradation. Higher incubation temperature of 4, 25, and 37 °C increased μ-calpain degradation. Two degradation products with about 50 kDa from μ-calpain were identified, of which the intensity was also lower in control group than in the other two groups. These observations demonstrated that AP dephosphorylation and PKA phosphorylation of μ-calpain promoted μ-calpain autolysis and activation.

8.4.1 The pH Values of μ-Calpain Solution Measured Before and After Treated with AP/PKA

The pH values were presented in Table 8.1. The results showed that there was no difference in pH values among the three treatments either before or after incubation ($P > 0.05$). pH was an important factor affecting μ-calpain activity in postmortem muscle (Koohmaraie et al. 1986). It has been reported that μ-calpain was optimally active at pH 7.5 and autolyzed rapidly in high pH muscle in which myofibrillar proteins degraded fast (Lomiwes et al. 2014a; Pulford et al. 2009). In the present study, no difference in pH values indicated that the effect of pH on μ-calpain activity was not significantly different between the three treatments.

Table 8.1 The pH values of μ-calpain solution measured before and after incubation at 30 °C for 30 min (Du et al. 2017b)

Treatment	0 min	30 min
AP	6.72 ± 0.06^{aA}	6.59 ± 0.02^{aB}
Control	6.77 ± 0.01^{aA}	6.60 ± 0.02^{aB}
PKA	6.74 ± 0.06^{aA}	6.63 ± 0.02^{aB}

a values with the same lowercase within each column do not differ ($P > 0.05$), *AB* within the same row, values with different uppercase letters differ ($P < 0.05$)

8.4.2 The Phosphorylation Level of μ-Calpain

To estimate the effect of PKA and AP on μ-calpain phosphorylation, the phosphorylation level of μ-calpain was measured after μ-calpain was incubated with or without PKA/AP. Figure 8.9a, b showed the phosphorylated and total μ-calpain, respectively. The intensity of phosphorylated μ-calpain was the highest in PKA group and the lowest in AP group, while there was no significant difference in the intensity of total μ-calpain among the three treatments. The phosphorylation level of μ-calpain was PKA group > control group > AP group (Fig. 8.9c). This result

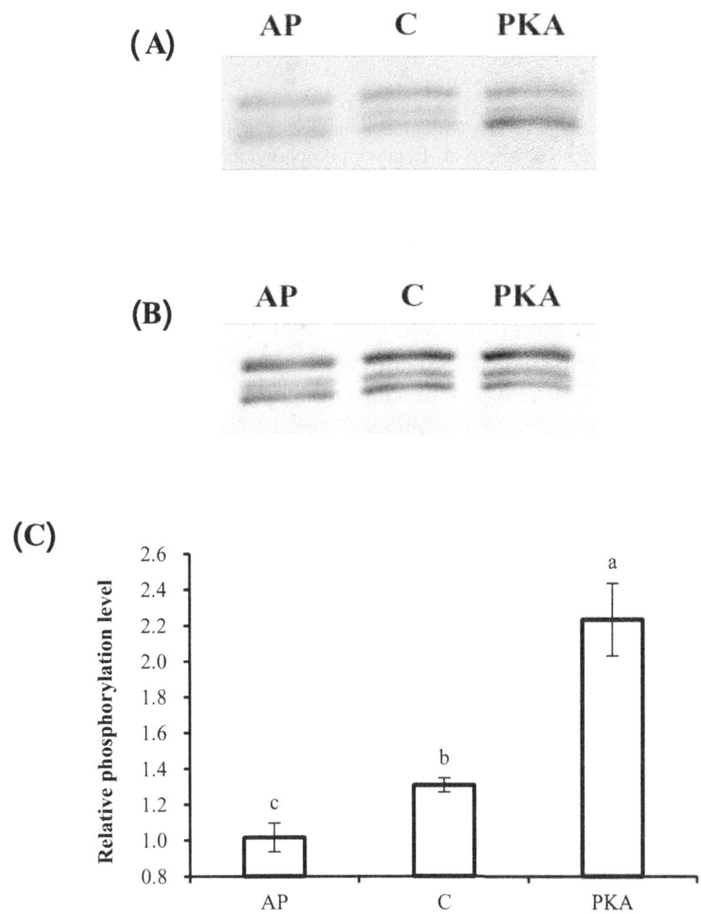

Fig. 8.9 Phosphorylation level of μ-calpain after incubation with AP and PKA at 30 °C for 30 min. (**a–b**), SDS-PAGE gels stained with Pro-Q Diamond (**a**) and SYPRO Ruby (**b**). (**c**) Relative phosphorylation level of μ-calpain. Values with different letters are significantly different ($P < 0.05$). *AP* Alkaline phosphatase, *PKA* Protein kinase A (Du et al. 2017b)

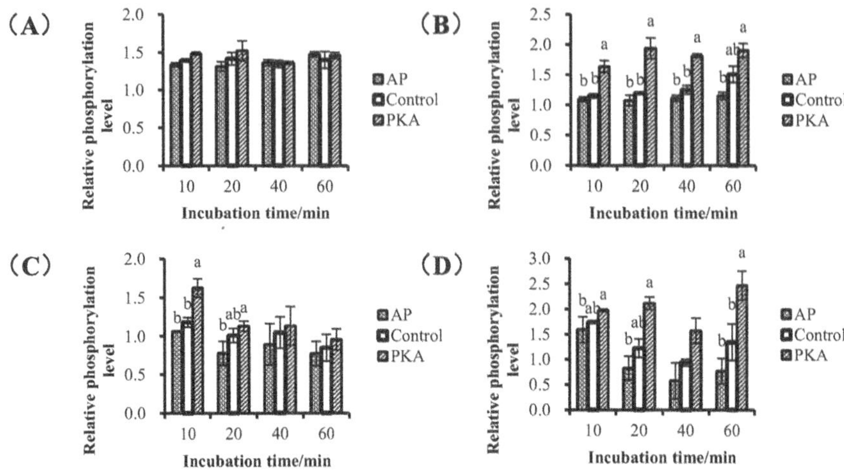

Fig. 8.10 Phosphorylation level of μ-calpain incubated at different temperatures and in the presence or absence of PKA/AP. (**a**–**d**) Relative phosphorylation level of μ-calpain incubated at controlled freezing point (**a**), 4 °C (**b**), 25 °C (**c**), and 37 °C (**d**). [ab] At the same time point, values lacking a common letter are significantly different ($P < 0.05$). *AP* Alkaline phosphatase, *PKA* Protein kinase A (Du et al. 2017b)

demonstrated that, as expected, PKA and AP had effectively phosphorylated and dephosphorylated μ-calpain after incubation at 30 °C for 30 min, respectively.

The incubation of μ-calpain at controlled freezing point (−1 °C), 4, 25, and 37 °C for 10, 20, 40, and 60 min was to study the impact of temperature on μ-calpain phosphorylation and dephosphorylation. Generally, for incubation temperature of 4, 25, and 37 °C, the phosphorylation level of μ-calpain in control group was higher than in AP group, but lower than in PKA group (Fig. 8.10b, c, d), which indicated that phosphorylation and dephosphorylation process effectively kept three treatments maintain established phosphorylation state. However, there was no significant difference in phosphorylation level of μ-calpain among the three treatments when incubated at controlled freezing point ($P > 0.05$) (Fig. 8.10a). Food products stored at controlled freezing point (also called superchilling storage), between freezing and refrigeration where ice crystals could not be generated (Beaufort et al. 2009), have been an increasingly popular way to maintain food freshness and high quality and increase their shelf life (Kaale et al. 2011). Under such storage condition, the microbial growth and lipid oxidation in surimi were reduced, and the proteolytic degradation was limited (Liu et al. 2014). Previous studies of our group about the effects of controlled freezing point storage on the protein phosphorylation level of mutton showed that controlled freezing point reduced protein phosphorylation level at early postmortem by inhibiting the activity of protein kinases (Zhang et al. 2016; Li et al. 2016), which was consistent with this research. Compared with refrigeration, controlled freezing point could delay muscle aging (Li et al. 2016). Therefore, the limited biochemistry activities under superchilling resulted in insignificant

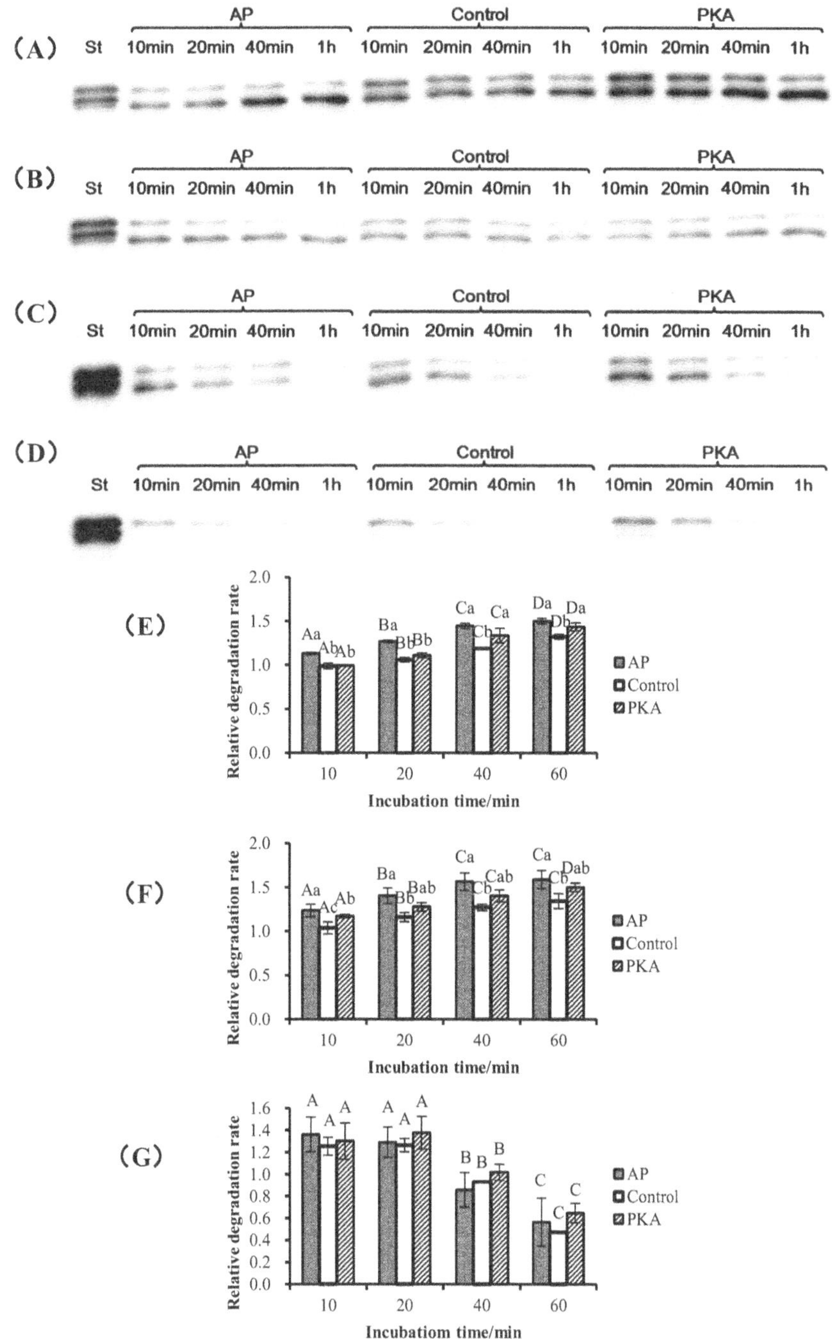

Fig. 8.11 Degradation of 80 kDa μ-calpain subunit during incubation at different temperatures. (**a–d**) The degradation of 80 kDa μ-calpain subunit incubated at controlled freezing point (**a**), 4 (**b**), 25 (**c**), 37 °C (**d**). (**e–g**) Relative degradation rate of 80 kDa μ-calpain subunit incubated at controlled freezing point (**e**), 4 (**f**), 25 °C (**g**). [ab] At the same time point, values lacking a common

difference in phosphorylation level of μ-calpain when incubating at controlled freezing point.

8.4.3 The Degradation of μ-Calpain

The detection of μ-calpain degradation has previously been used to evaluate μ-calpain activity and to analyze the existing state of μ-calpain in real time (Lomiwes et al. 2014b; Melody et al. 2004). The μ-calpain degraded from 80 kDa to 76 kDa peptide in all three groups during incubation (Fig. 8.11). A gradual increase of 76 kDa μ-calpain was observed at controlled freezing point and 4 °C (Fig. 8.11a, b). With the increase of incubation temperature, μ-calpain degraded faster. After incubation at 37 °C for 10 min, it was no longer possible to detect any 78 or 76 kDa subunit of μ-calpain in all three groups (Fig. 8.11d). At controlled freezing point and 4 °C, μ-calpain in AP group had the faster degradation rate than in the other two groups. Meanwhile, the degradation rate of μ-calpain in PKA group was higher than that in control group (Fig. 8.11e, f). However, there was no significant difference in the degradation rate of μ-calpain among AP, control, and PKA groups at 25 °C ($P > 0.05$) (Fig. 8.11g). Generally, the 80 kDa subunit of μ-calpain was degraded to a 76 kDa form through a 78 kDa intermediate by autolysis for its activation. A faster degradation rate of μ-calpain suggested a higher activity of this enzyme (Lomiwes et al. 2014a). Therefore, the higher degradation rate of μ-calpain in AP group at controlled freezing point and 4 °C revealed that dephosphorylation contributed to higher activity of μ-calpain. This was in agreement with our previous study which showed an increase in μ-calpain activity when sarcoplasmic proteins were dephosphorylated by AP (Du et al. 2017a). PKC phosphorylation of both μ- and m-calpain activated these enzymes, leading to enhanced invasion and migration of cancer cells (Xu and Deng 2006a). MAPK (mitogen-activated protein kinase)/ERK1 and MAPK/ERK2 directly phosphorylated μ- and m-calpain in vitro to promote their activation, secretion, and proteolysis ability (Xu and Deng 2004). However, activation of m-calpain was limited by PKA phosphorylation (Shiraha et al. 2002). The opposite effects of PKA and ERK phosphorylation on m-calpain activity demonstrated that different types of protein kinase may have different regulatory roles on calpain. However, the effect of PKA phosphorylation on μ-calpain activity still remains unknown. The present study showed that phosphorylation of μ-calpain by PKA led to increased degradation of μ-calpain compared to control group, indicating that PKA phosphorylation increased μ-calpain activity.

Controlled freezing point storage has been an effective approach to extend meat shelf life and maintain meat quality by reducing microbial growth and limiting

Fig. 8.11 (continued) letter are significantly different ($P < 0.05$). ^{ABCD} Within the same group, values lacking a common letter are significantly different ($P < 0.05$). *AP* Alkaline phosphatase, *PKA* Protein kinase A (Du et al. 2017b)

biochemical reactions in meat (Kaale et al. 2011; Liu et al. 2014). The controlled enzyme activity and limited proteolytic degradation delayed the aging of meat. In the present study, no difference in the degradation rate of 80 kDa μ-calpain subunit was detected between control and PKA groups after incubated at controlled freezing point for 10 and 20 min (Fig. 8.11e). Considering the reduced phosphorylation level of μ-calpain in PKA group at controlled freezing point, the results indicated that controlled freezing point incubation decreased μ-calpain activity by delaying its phosphorylation process. μ-calpain was a temperature sensitive enzyme. Higher incubation temperature rapidly activated μ-calpain followed by an increase in the degradation of μ-calpain (Pomponio and Ertbjerg 2012). In addition, the activity of μ-calpain was optimal at 25 °C (Pomponio and Ertbjerg 2012), resulting in the eliminated difference among the three groups. After incubation at 25 °C for 20 min or at 37 °C for 10 min, the intensity of both intact and autolyzed μ-calpain decreased, suggesting that the 76 kDa peptide of μ-calpain may further degrade to smaller forms.

The further degradation products of μ-calpain catalytic subunit have never been studied in previous research. Two specific bands at about 50 kDa were detected on SDS-PAGE gels by western blotting (Fig. 8.12a, b, c, d), which were probably the further degradation products of 76 kDa μ-calpain peptide. To prove it, proteins in these two bands were identified by LC-MS/MS. As shown in Table 8.2, μ-calpain was identified in both bands, proving that the 76 kDa μ-calpain peptide degraded to about 50 kDa peptide after incubation. The intensity of band 1 and band 2 increased during incubation at controlled freezing point and 4 °C (Fig. 8.12a, b) and was higher in AP and PKA groups than in control group. When incubation temperature increased to 25 °C, band 1 disappeared after incubation for 10 min and the intensity of band 2 decreased after incubation for 40 min (Fig. 8.12c). Despite the degradation rate of μ-calpain from 80 kDa to 76 kDa showed no significant difference among the three groups, the content of about 50 kDa products was lower in control group than in the other two groups when incubated at 25 °C (Fig. 8.12c), indicating that the degradation rate of μ-calpain was faster in AP and PKA groups than in control group. At 37 °C, it was no longer possible to detect band 1 and the intensity of band 2 significantly decreased (Fig. 8.12d, h), which meant that about 50 kDa peptide of μ-calpain may degrade to even smaller molecules. All these results showed that degradation of μ-calpain was phosphorylation and temperature dependent. The AP dephosphorylation and PKA phosphorylation of μ-calpain promoted its degradation and activation.

8.5 Effects of Phosphorylation of Sarcoplasmic Proteins on μ-Calpain Activity at Different Ca^{2+} Concentrations

The indirect effects of phosphorylation on the activity of μ-calpain and the sensitivity of μ-calpain to Ca^{2+} were investigated in this part. Sarcoplasmic proteins were extracted from three Fat Tail Han sheep (8 months) at 30 min postmortem, adjusted

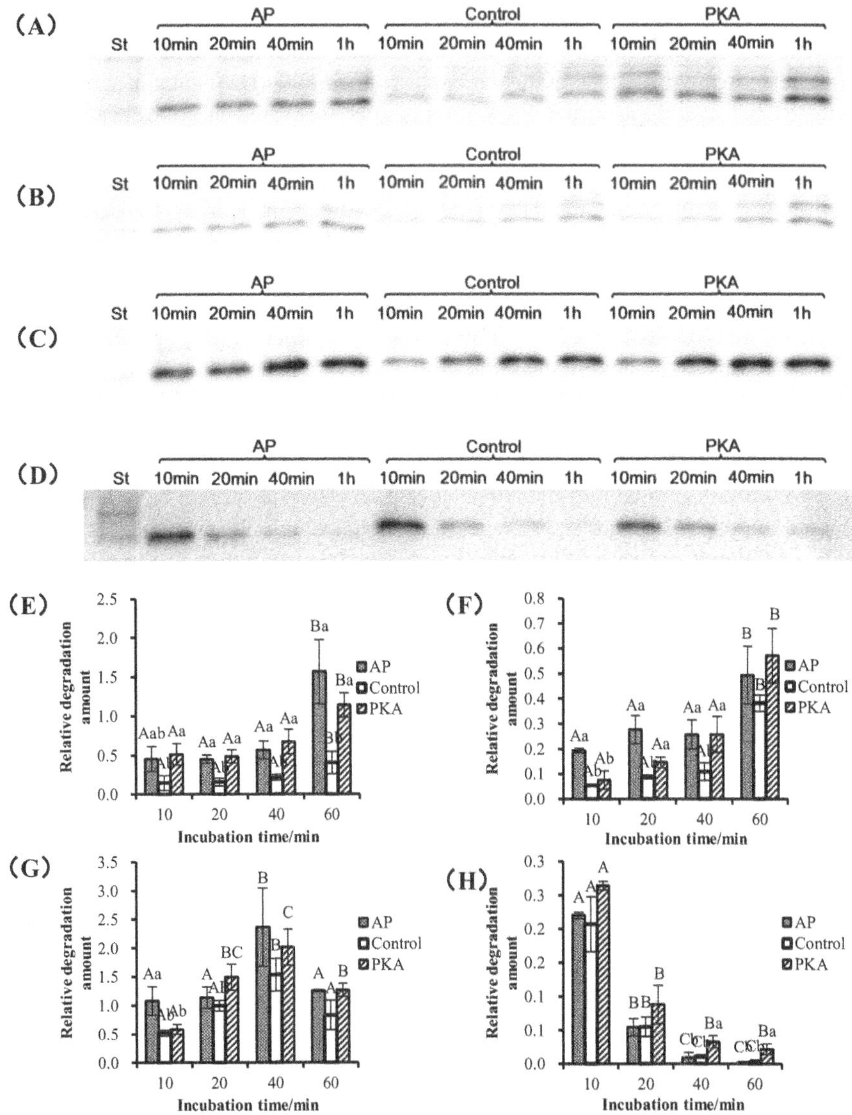

Fig. 8.12 Degradation of 80 kDa μ-calpain subunit to about 50 kDa peptides. (**a–d**) Detection of about 50 kDa products after incubation of μ-calpain at controlled freezing point (**a**), 4 (**b**), 25 (**c**), 37 °C (**d**). (**e–h**) Relative intensity of μ-calpain smaller peptides in three groups incubated at controlled freezing point (**e**), 4 (**f**), 25 (**g**), 37 °C (**h**). [ab] At the same time point, values lacking a common letter are significantly different ($P < 0.05$). [ABC] Within the same group, values lacking a common letter are significantly different ($P < 0.05$). *AP* Alkaline phosphatase, *PKA* Protein kinase A (Du et al. 2017b)

Table 8.2 LC-MS/MS identified proteins in about 50 kDa protein bands on SDS-PAGE gels (Du et al. 2017b)

Band	Accession	Protein name	Score	Mass	Matches	Sequences
1	Q27970	Calpain-1 catalytic subunit	2411	82,782	137 (91)	25 (24)
2	Q27970	Calpain-1 catalytic subunit	6291	82,782	325 (235)	26 (24)

to 12 μg/μL, and subjected to three treatments: (1) 25 U/100 μg protein AP (AP); (2) One tablet/10 mL supernatant phosphatase inhibitor (PI); (3) Control. For calcium sensitivity analysis, samples treated with AP and PI were incubated at 0.01, 0.05, 0.1, and 1 mM Ca^{2+}, respectively.

8.5.1 *Phosphorylation Level Analysis of Sarcoplasmic Proteins*

The phosphorylation level of prepared sarcoplasmic proteins was measured to verify whether AP and PI worked effectively in regulating protein phosphorylation. Results showed that the phosphorylation level of PI group was significantly higher than control, which was significantly higher than AP group ($P < 0.05$) (Fig. 8.13).

8.5.2 *μ-Calpain Activity*

The degradation of μ-calpain and the corresponding degradation rate in the presence of Ca^{2+} were shown in Fig. 8.14. In the presence of 0.01, 0.05, and 0.1 mM Ca^{2+}, the degradation of 80 kDa μ-calpain subunit to 76 kDa form was fastest in AP samples, of which the rate was significantly higher than that of the other two groups ($P < 0.05$). PI samples had the lowest degradation rate (Fig. 8.14a, b, c). The 80 kDa subunit of μ-calpain remarkably degraded when the Ca^{2+} concentration was higher than 0.05 mM, but the degradation level remained unchanged when $[Ca^{2+}] = 0.1$ mM. The differences in degradation rate among the three groups disappeared when Ca^{2+} concentration was 1 mM. μ-calpain was a calcium dependent protease and required micro-molar of calcium for its activation. The optimal Ca^{2+} concentration for its activity has been intensively studied (Biswas et al. 2016). Injection of $CaCl_2$ increased calpain activity and improved meat tenderness (Jaturasitha et al. 2004). And injection of $CaCl_2$ also activated μ-calpain, which eventually resulted in activity loss due to autolysis (Koohmaraie et al. 1989; Koohmaraie et al. 1990). The increased Ca^{2+} activated μ-calpain and promoted the degradation of μ-calpain at the same time. The AP group had a greater degradation degree of μ-calpain at 0.01, 0.05, and 0.1 mM Ca^{2+} concentrations, which indicated that μ-calpain with lower phosphorylation had higher activity, which was consistent with experiment 1. As the level of phosphorylation increased, Ca^{2+}

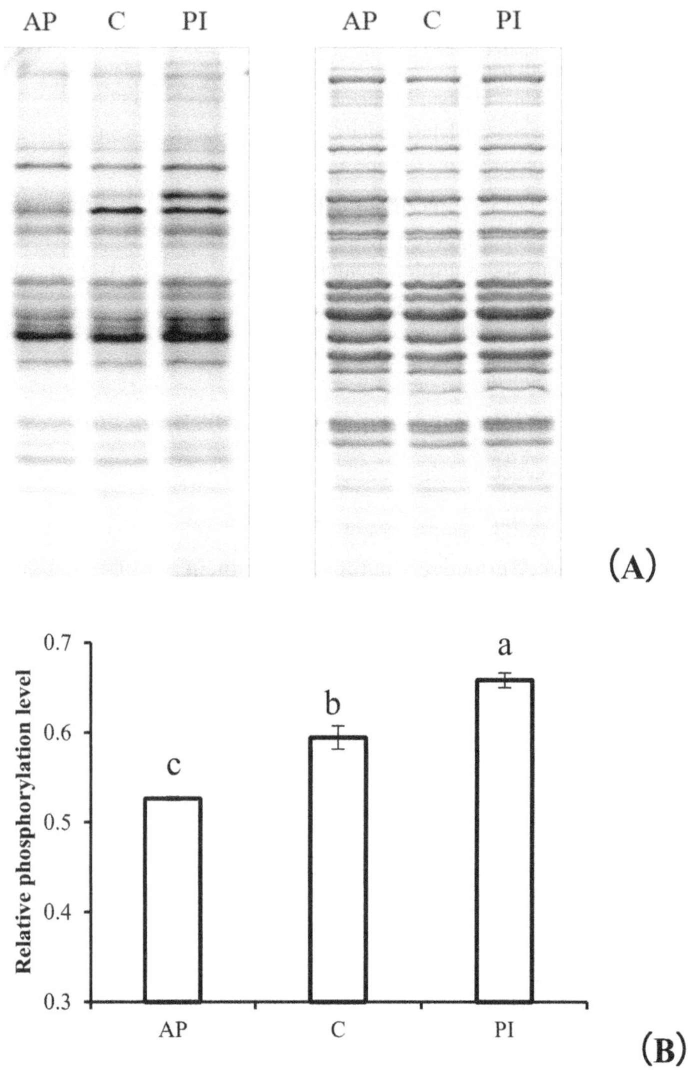

Fig. 8.13 Phosphorylation of sarcoplasmic proteins before incubating at different Ca^{2+} concentration. (**a**) Left: SDS-PAGE gel stained by Pro-Q Diamond stains; right: SDS-PAGE gel stained by Ruby stains. (**b**) Relative phosphorylation level of three treatments (Du et al. 2017a)

concentration required for μ-calpain activation increased. Furthermore, as the concentration of Ca^{2+} increased, the difference in μ-calpain degradation among AP, control, and PI groups reduced. The effect of phosphorylation on μ-calpain activity reduced and even disappeared as Ca^{2+} concentration increased. This could be explained by the increased activity of μ-calpain in the presence of calcium.

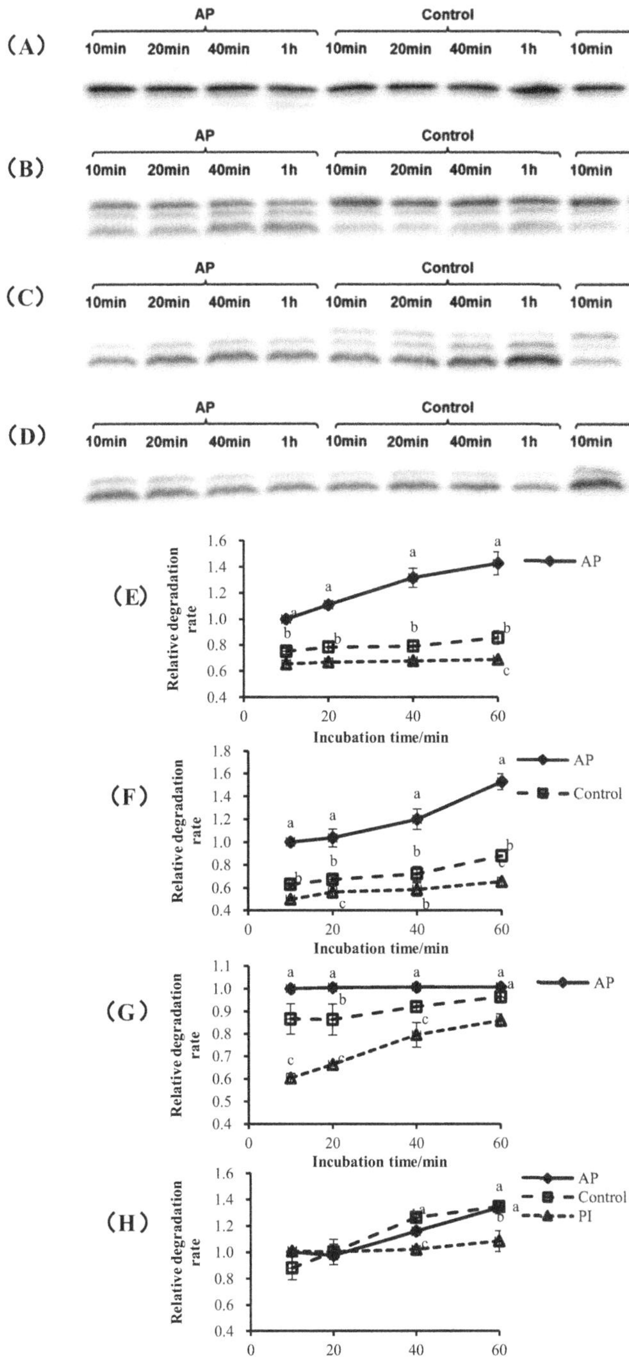

Fig. 8.14 Western blotting analysis during incubation at different Ca^{2+} concentration. (**a**), (**b**), (**c**), (**d**) The degradation of μ-calpain 80 kDa subunit in three groups incubated at 0.01, 0.05, 0.1, and 1 mM Ca^{2+}, respectively. (**e**), (**f**), (**g**), (**h**) Relative degradation rate of μ-calpain 80 kDa subunit in

8.6 Effects of Phosphorylation on μ-Calpain Activity at Different Ca^{2+} Concentrations In Vitro

The effect of phosphorylation/dephosphorylation regulated by PKA and AP on μ-calpain activity at different Ca^{2+} concentrations was investigated in this part. The purified μ-calpain was treated with AP or PKA at 0.01, 0.05, 0.1, and 1 mM Ca^{2+}. The pH value decreased in the AP group but remained stable in the control and PKA groups during incubation. Except for samples incubated at 0.01 and 0.1 mM Ca^{2+} for more than 20 min, μ-calpain incubated with PKA showed a higher level of autolysis than control, but lower than the AP group. The content of α-helix structure of μ-calpain increased as phosphorylation level rose. Phosphorylation of μ-calpain at serines 255, 256, 476, 417, and 420 was identified. PKA catalyzed μ-calpain phosphorylation at serines 255, 256, and 476, located at domains II and III, positively regulated μ-calpain activity. These data demonstrated that dephosphorylation and PKA phosphorylation positively regulated μ-calpain activity, which was limited by increased Ca^{2+} concentration.

8.6.1 *The Changes in pH Values During Incubation*

The pH values of μ-calpain solutions treated with or without AP/PKA were presented in Fig. 8.15. Before incubation with AP and PKA, there was no significant difference in the pH values of μ-calpain solutions. After 30 min incubation, the pH of AP group decreased from 6.91 to 6.70, which was significantly lower than that of the other two groups ($P < 0.05$). There was no difference in pH values between the PKA group and the control ($P > 0.05$). AP was a hydrolase that catalyzed the protein dephosphorylation. The phosphate group removed from μ-calpain by AP was the acidic ingredient in AP group that reduced the pH value (Longo et al. 2015).

8.6.2 *The Phosphorylation of μ-Calpain*

The phosphorylation level of μ-calpain after incubation with AP and PKA is shown in Fig. 8.16a, b, c. On the Pro-Q Diamond stain gels, the intensity of phosphorylated μ-calpain band was the highest in PKA group and the lowest in AP group (Fig. 8.16a). Correspondingly, μ-calpain in PKA group had the highest phosphorylation level, but the AP group had the lowest phosphorylation level ($P < 0.05$) (Fig. 8.16c). The process of removing phosphate group from proteins induced by AP

Fig. 8.14 (continued) three groups incubated at 0.01, 0.05, 0.1, and 1 mM Ca^{2+}, respectively. AP: Alkaline phosphatase; PI: Phosphatase inhibitor (Du et al. 2017a)

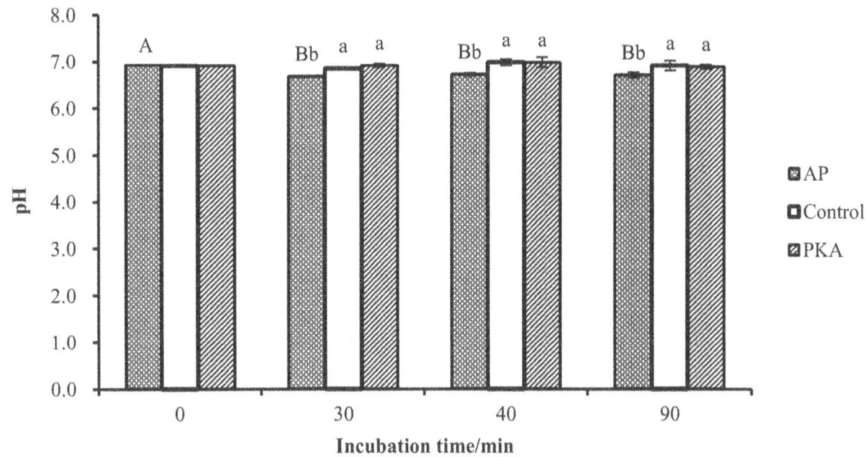

Fig. 8.15 The changes in pH values during incubation at 0, 30, 40, 90 min. Values with different lowercase letters show significant difference in the results of different groups at the same incubation time ($P < 0.05$). Values with different capital letters show significant difference in the results in same groups at different incubation times ($P < 0.05$). *AP* Alkaline phosphatase; *PKA* Protein kinase A (Du et al. 2018)

was called dephosphorylation. Conversely, PKA had the ability to catalyze proteins phosphorylation at serine/threonine residues (Shiraha et al. 2002). These data showed that AP and protein kinase A worked as expected in our experiment to dephosphorylate and phosphorylate μ-calpain, respectively. μ-calpain with different phosphorylation level was obtained after incubation with or without AP/PKA.

The phosphorylation level of μ-calpain in incubation with Ca^{2+} was also quantified. Generally, the phosphorylation of μ-calpain treated with PKA was greater than that of control, which was greater than that of μ-calpain incubated with AP (Fig. 8.16d, e, f, g). However, samples incubated in the presence of 0.05 and 0.1 mM Ca^{2+} showed no significant difference in phosphorylation level after incubation for 20 min (Fig. 8.16f, g). As the Ca^{2+} concentration increased, an increase in μ-calpain degradation level in all treatments was observed (Fig. 8.16a, b, c, d). The content of intact and autolyzed μ-calpain differed between three groups and was significantly different from the initial content of μ-calpain, leading to a variable and undifferentiated phosphorylation level of μ-calpain in three treatments at later stage of incubation. Overall, incubation with PKA or AP in the presence of ATP effectively changed the phosphorylation state of μ-calpain.

8.6.3 The Degradation of μ-Calpain

To investigate the changes in μ-calpain activity induced by phosphorylation, the degradation/autolysis of μ-calpain was evaluated by western blotting. μ-calpain

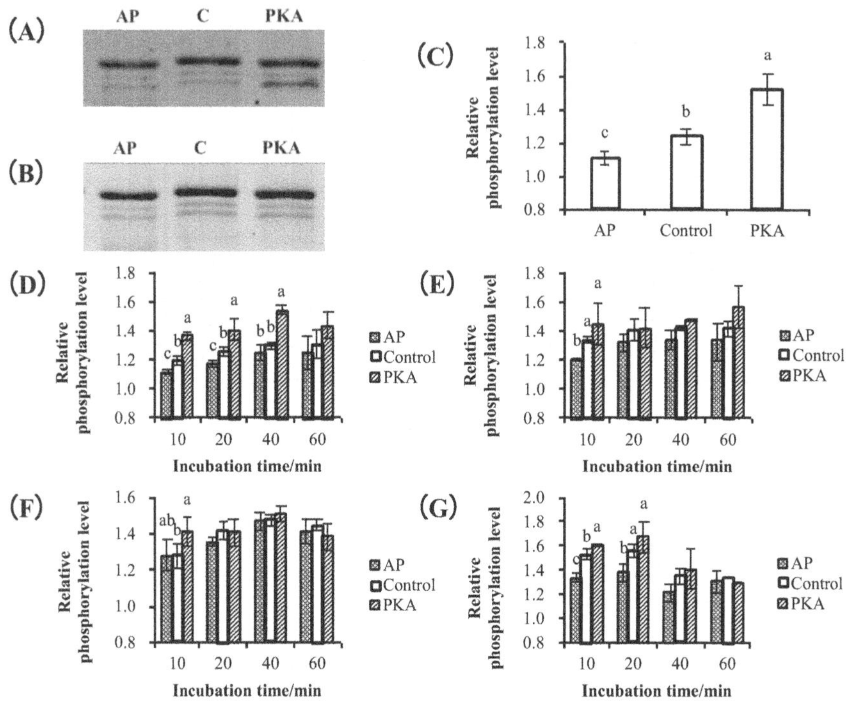

Fig. 8.16 Phosphorylation of μ-calpain in three groups after phosphorylation and dephosphorylation treatments and during incubation at different Ca^{2+} concentrations. (**a**), (**b**) SDS-PAGE gels stained by Pro-Q Diamond stains and Ruby stains, respectively. (**c**) Relative phosphorylation level of μ-calpain after phosphorylation and dephosphorylation treatments. (**d**), (**e**), (**f**), (**g**) Relative phosphorylation level of μ-calpain in three treatments incubated at 0.01, 0.03, 0.05, 0.1 mM Ca^{2+}, respectively. Values with different letters show significant difference in the results in different groups at the same incubation time ($P < 0.05$). *AP* Alkaline phosphatase, *PKA* Protein kinase A (Du et al. 2018)

degraded gradually at all tested Ca^{2+} concentrations during incubation (Fig. 8.17a, b, c, d). Increased degradation of μ-calpain was detected as Ca^{2+} concentration increased. The degradation degree of AP treated μ-calpain was significantly higher than that of control ($P < 0.05$), except for incubation at 0.1 mM Ca^{2+} after 20 min. μ-calpain of PKA treatment showed reduced degradation when compared to AP treated μ-calpain, but a higher degradation level than control except for samples incubated at 0.01 and 0.1 mM Ca^{2+} for more than 20 min (Fig. 8.17e, f, g, h). The activation of μ-calpain was paralleled by its autolysis of the intact 80 kDa large subunit to the 78 kDa than 76 kDa form and was associated with proteolytic activity of μ-calpain in postmortem muscles (Goll et al. 2003; Lomiwes et al. 2014a). A previous study by our team discovered that protein phosphorylation level of tender meat was lower than that of tough meat (Chen et al. 2016). Degradation of AP treated myofibrillar proteins by μ-calpain was greater than untreated proteins (Li et al. 2017). These studies indicated that dephosphorylation promotes proteolysis

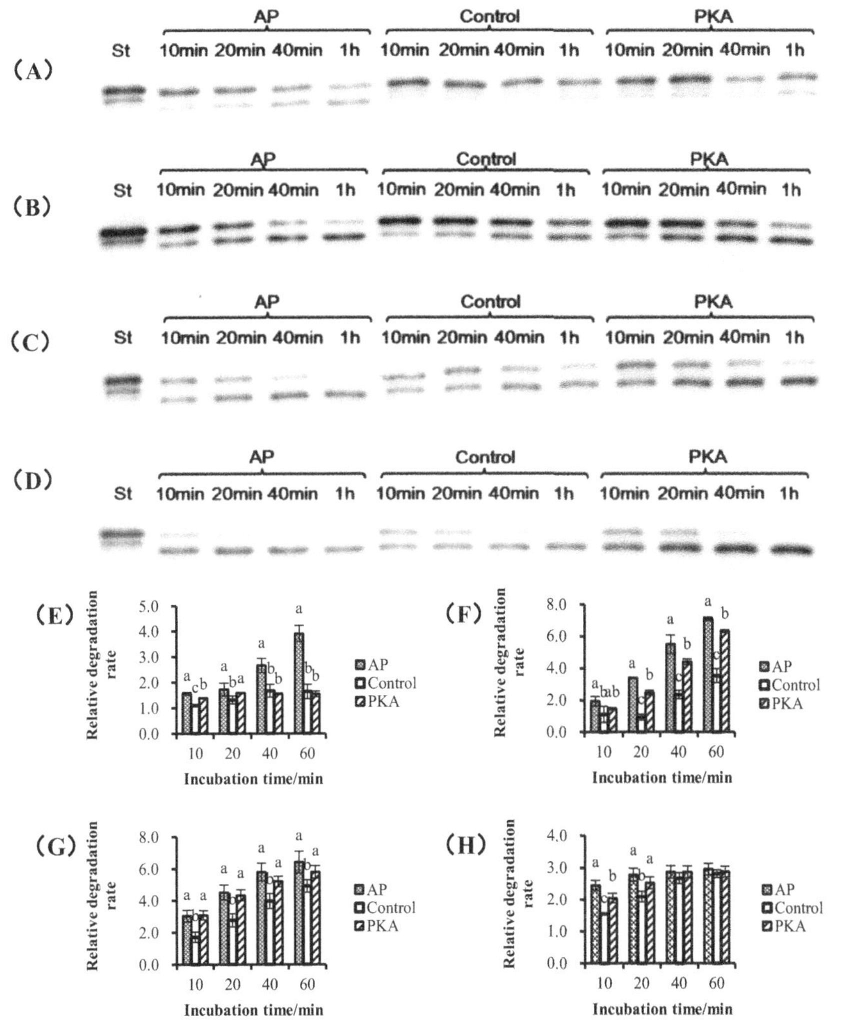

Fig. 8.17 Western blotting analysis during incubation at 0.01, 0.03, 0.05, and 0.1 mM Ca^{2+}. (**a**), (**b**), (**c**), (**d**) The degradation of μ-calpain 80 kDa subunit in three groups incubated at 0.01, 0.03, 0.05, and 0.1 mM Ca^{2+}, respectively. (**e**), (**f**), (**g**), (**h**) Relative degradation rate of μ-calpain 80 kDa subunit in three groups incubated at 0.01, 0.03, 0.05, and 0.1 mM Ca^{2+}, respectively. Values with different letters show significant difference in the results in different groups at the same incubation time ($P < 0.05$). *AP* Alkaline phosphatase, *PKA* Protein kinase A (Du et al. 2018)

and plays a positive role in regulating μ-calpain activity. Further study revealed that higher μ-calpain activity of sarcoplasmic proteins could be dephosphorylated by AP (Du et al. 2017a). The activity of μ-calpain was maximum at pH near neutral (Goll et al. 2003; Koohmaraie and Geesink 2006). Based on the determined pH values presented in Fig. 8.15, PKA and control groups with a higher pH value should have a higher μ-calpain activity than AP group. In other words, the degradation degree of μ-calpain in PKA and control group was supposed to be higher than AP group under such the pH condition without considering other factors. However, the degradation degree of μ-calpain in AP group was significantly higher than that of PKA and control groups, which indicated that dephosphorylation of μ-calpain promoted its degradation and enhanced the activity of μ-calpain. The effects of dephosphorylation on μ-calpain activity were greater than that of pH.

μ-calpain of PKA treatment degraded faster compared to control, which meant that incubation of μ-calpain with PKA promoted its autolysis and activation. μ-calpain can be phosphorylated by protein kinase like PKA, PKC, and extracellular signal-regulated kinase (ERK) (Storr et al. 2011). Phosphorylation of μ-calpain and m-calpain by MAPK (mitogen-activated protein kinase)/ERK (extracellular signal-regulated kinase) could lead to its activation as well as secretion, increasing migration and invasion of lung cancer cells (Xu and Deng 2004). The same result was found by investigating the phosphorylation of μ-calpain and m-calpain catalyzed by PKC (Xu and Deng 2006a). ERK directly phosphorylated m-calpain at serine 50 in domain I, which was autolyzed when calpain was activated by calcium (Glading et al. 2004). However, activation of calpain by epidermal growth factor (EGF) was inhibited by PKA. PKA phosphorylated m-calpain at serine 369 in domain III, which contained the characteristic C2 domains and was involved in structural changes during calcium binding (Hanna et al. 2008; Moldoveanu et al. 2008). Phosphorylation of serine 369 could restrict domain movement and keep m-calpain in an inactive state (Shiraha et al. 2002). All these studies demonstrated that phosphorylation of calpain at different sites by different kinases had different effects on its activity. Different from m-calpain, the present study showed that phosphorylation of μ-calpain by PKA improved its degradation and led to activation. AP catalyzed the removal of phosphate group from proteins to decrease protein phosphorylation level. Its broad-spectrum dephosphorylation ability could limit phosphorylation of protein by almost all kinds of protein kinases. In the present study, the increased autolysis of μ-calpain after incubation with AP demonstrated that dephosphorylation of μ-calpain by AP promoted its activation. In addition, the disparity of degradation degree among the three treatments decreased when Ca^{2+} concentration increased. After incubation at 0.1 mM Ca^{2+} for 20 min, the difference in degradation rate of μ-calpain disappeared among three groups ($P > 0.05$), demonstrating that high concentrations of Ca^{2+} eliminated the impact of phosphorylation/dephosphorylation on μ-calpain activity. Further work is required to determine the influences of phosphorylation and dephosphorylation on μ-calpain catalyzed by other protein kinase and phosphatase.

8.6.4 Changes in μ-Calpain Secondary Structure

To determine the changes in secondary structures that responded to phosphorylation or dephosphorylation of μ-calpain, circular dichroism spectroscopy was used to characterize the secondary structures of μ-calpain treated with or without AP/PKA.

Table 8.3 Estimated secondary-structure contents of μ-calpain after phosphorylation and dephosphorylation treatments (Du et al. 2018)

Treatment	α-helix	β-sheet	β-turn	Random coil	Total sum
AP	0.188	0.205	0.312	0.295	1
Control	0.591	0.409	0	0	1
PKA	0.799	0.001	0	0.199	1

The content of α-helix, β-sheet, β-turn, and random coil structures was presented in Table 8.3. For all the three treatments, the sum of structure fractions of μ-calpain was 1, showing that the analysis met the requirements. Studies about μ-calpain structure concentrated on the crystal structure, while circular dichroism has been rarely used. α-helix and β-sheet were the essential second structures of μ-calpain (Edmunds et al. 1991). The circular dichroism spectra of μ-calpain showed a notable negative peak at around 223 nm and a weaker negative peak at around 236 nm (Fig. 8.18). The negative peak at around 223 nm, which represented α-helix structure (Edmunds et al. 1991), became larger as phosphorylation level of μ-calpain increased, indicating the increased content of α-helix. This suggested that dephosphorylation of μ-calpain by AP caused α-helical structure to break down, while PKA induced the formation of α-helix. All the four domains of 80 kDa μ-calpain subunit contained α-helix, however, α-helix mainly concentrated on domain II and domain IV, which also contained five EF hands (Goll et al. 2003; Strobl et al. 2000). Any changes in α-helix in μ-calpain may influence the combining of calcium to EF hands, the proteolysis, and activity of μ-calpain. In addition, the content of β-sheet structure in AP and PKA treated μ-calpain was lower, but random coil structure was higher than in control. The lower content of β-sheet structure and the higher content of random coil structure may be the reason why the activity of μ-calpain in AP and PKA groups was significantly higher than that of control (Fig. 8.17). β-sheet principally existed in catalytic domain (II) and regulatory domain (III) which plays important roles in substrate binding and catalysis (Reverter et al. 2001; Strobl et al. 2000). The difference in β-sheet, β-turn, and random coil content of μ-calpain among the three treatments indicated that phosphorylation and dephosphorylation may be involved in regulating the activity of μ-calpain by changing its secondary structure, sheltering or exposing the binding sites of substrate proteins or calcium on μ-calpain.

8.6.5 Detection of Phosphorylated Peptides and Phosphorylated Sites of μ-Calpain

A total of three unique phosphopeptides with five phosphorylated sites of μ-calpain was identified in the present study. For PKA treated μ-calpain, the three phosphopeptides and the five phosphorylation sites (serine 476, 417, 420, 255, and 256) were all determined. For the control and AP treated μ-calpain, the three unique phosphopeptides and only two phosphorylation sites (serine 417 and serine 420)

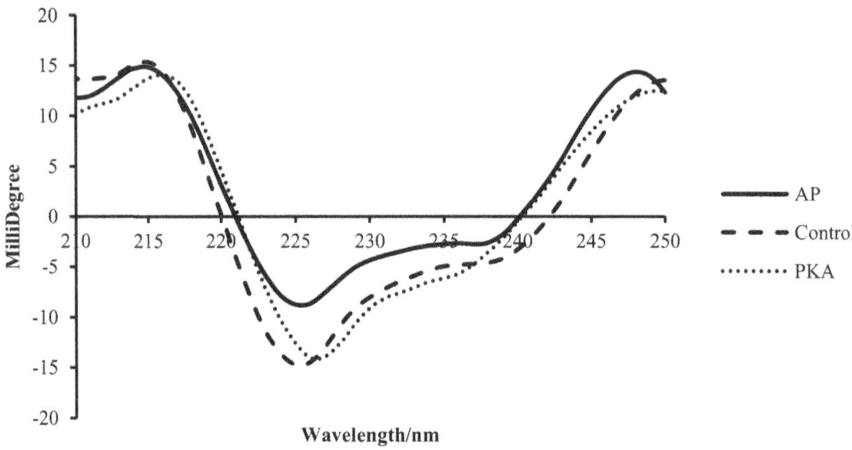

Fig. 8.18 The influence of phosphorylation on the circular dichroism spectra of μ-calpain after phosphorylation and dephosphoryaltion treatments. *AP* Alkaline phosphatase, *PKA* Protein kinase A (Du et al. 2018)

Table 8.4 Q Exactive LC-MS/MS for the detection of phosphorylated peptides of μ-calpain after phosphorylation and dephosphorylation treatments (Du et al. 2018)

Treatment	Phosphorylated peptides	Phosphorylation sites	IonScore[a]
Standard	–	–	–
AP	EsGcsFLLALMQK	Ser 417, Ser 420	19
Control	EsGcsFLLALMQK	Ser 417, Ser 420	26
PKA	ARsEQFINLR	Ser 476	44
	EsGcsFLLALMQK	Ser 417, Ser 420	65
	GSLLGcSIDIssILDmEAVTFK	Ser 255, Ser 256	76

[a]The Mascot score of the phosphorylated peptides

were figured out (Table 8.4). At the same time, no phosphopeptides and phosphorylation sites were determined in untreated standard sample, which meant that the phosphorylation of μ-calpain occurred during incubation. In the present study, only phosphoserine sites were determined. This was consistent with literature that phosphorylation of μ-calpain appeared mostly at serine residues (Vazquez et al. 2008). The results showed that the number of phosphopeptides and phosphorylation sites of μ-calpain identified in PKA group was much higher than that of the other two groups. Although serine 417 and 420 were identified in all the three treatments, the IonScore of the peptides was different. IonScore represented the Mascot score of the identified phosphorylated peptides. Higher score meant a higher content of the peptide and a higher accuracy of the matched result. Therefore, the higher IonScore of phosphorylated peptides in PKA group than in control, which was higher than in AP group, indicated that PKA effectively phosphorylated and AP dephosphorylated μ-calpain at serine residues. This was in agreement with the measured μ-calpain phosphorylation levels.

The serines 255 and 256 were in domain II and serines 417, 420, and 476 in domain III of μ-calpain 80 kDa subunit. Phosphorylation of serines 255 and 256 may change the ability of μ-calpain to catalyze substrate degradation. Regulatory domain III was also involved in structural changes in μ-calpain. Thus, phosphorylation of serine 476 may influence the self-regulatory activity of μ-calpain and facilitate domain movement to activate μ-calpain (Shiraha et al. 2002). In addition, the increased α-helix in domain II and IV of PKA treated μ-calpain suggested that phosphorylation of μ-calpain at serines 255 and 256 regulated μ-calpain activity by changing its structure (Shiraha et al. 2002; Storr et al. 2011). Phosphorylation of serines in domain II and III could be the reason why μ-calpain incubated with PKA had significantly higher activity.

8.6.6 The Ser-Phosphorylation Level of μ-Calpain

To confirm that PKA phosphorylated μ-calpain primarily at serine residues, phosphoserine of μ-calpain was quantified by western blotting using antibodies (Fig. 8.19). The band intensity of phosphorylated serine residues in PKA group was obviously higher compared to control group, while AP group showed the lowest content of phosphoserine at all Ca^{2+} concentrations. μ-calpain from bovine skeletal muscle or human placenta has been reported to be phosphorylated at four serine sites, three threonine sites, and two tyrosine sites (Vazquez et al. 2008). In the present study, only phosphoserine was detected in μ-calpain. As PKA was a serine/threonine kinase, it was reasonable to detect no phosphotyrosine in the present study. Nevertheless, it was unclear why the previously reported phosphothreonine was not determined. The highest content of phosphoserine in PKA samples, but the lowest in AP samples was consistent with the results of μ-calpain phosphorylation level evaluated by SDS-PAGE and imaging analysis, indicating that PKA effectively phosphorylated μ-calpain at serine residues.

8.7 The Inhibition of Calpastatin to the Activity of Phosphorylated μ-Calpain

Calpastatin, as one of the well-characterized components in the calpain system, is the only known protein inhibitor specific for μ-calpain and m-calpain. Each calpastatin can inhibit four calpain molecules to limit postmortem muscle proteolysis and meat tenderization. The effect of phosphorylation on the sensitivity of μ-calpain to the inhibition induced by calpastatin was investigated in this part. Purified μ-calpain was incubated with AP or PKA to regulate the phosphorylation level of μ-calpain.

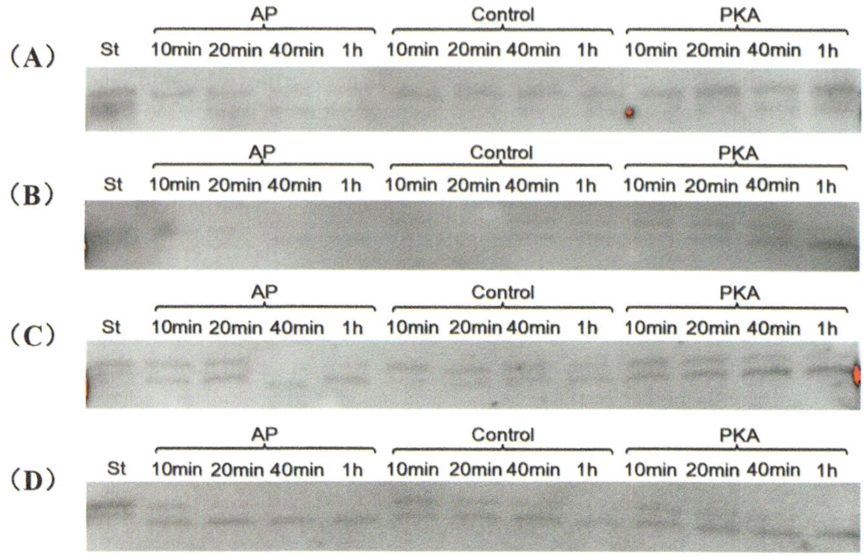

Fig. 8.19 Detection of μ-calpain phosphorylated on serine residues during incubation at 0.01, 0.03, 0.05, and 0.1 mM Ca^{2+}. (**a**), (**b**), (**c**), (**d**) Phosphorylation of μ-calpain at serine residues in three groups incubated at 0.01, 0.03, 0.05, and 0.1 mM Ca^{2+}, respectively (Du et al. 2018)

Accurately 25, 50, 100, and 150 units of AP/PKA treated μ-calpain were mixed with the same amount of heat stable proteins and incubated at 4 °C. In calpastatin-free system, AP and PKA treated μ-calpain had higher proteolytic activity compared to control. Intact AP treated μ-calpain degraded fastest in the 50, 100, and 150 units - μ-calpain incubation systems. However, the degradation rate of μ-calpain in control and PKA groups was non-significant in 100 and 150 units μ-calpain systems. Our results demonstrated that, compared to dephosphorylated and control μ-calpain, calpastatin presented greater inhibiting effect on PKA phosphorylated μ-calpain. This study further enhanced the understanding of mechanism of μ-calpain activity regulated by phosphorylation.

8.7.1 The pH Values of μ-Calpain Solution Measured Before and After Treated with AP/PKA

The pH values of μ-calpain solution before and after incubation with AP/PKA were showed in Table 8.5. No significant difference was detected between AP, control, and PKA treated μ-calpain at 0 and 30 min ($P > 0.05$). Biochemical changes in postmortem muscles were pH dependent, especially the activity of enzymes. μ-Calpain was a neutral protease with an optimum pH 7.5 (Pulford et al. 2009). Muscles with higher pH values have been shown to have greater degradation degree

Table 8.5 The pH values of μ-calpain solution measured before and after incubation at 30 °C for 30 min (Du et al. 2019)

Treatment	0 min	30 min
AP	6.65 ± 0.02	6.57 ± 0.02
Control	6.68 ± 0.03	6.61 ± 0.03
PKA	6.65 ± 0.02	6.60 ± 0.02

AP Alkaline phosphatase, *PKA* Protein kinase A

and activation rate of μ-calpain than muscles with lower pH values (Bee et al. 2007). Thus, the results indicated that the activity of μ-calpain was not affected by pH in three treatments.

8.7.2 Phosphorylation Level of Heat Stable Proteins and μ-Calpain

As shown in Fig. 8.20a, μ-calpain treated with AP had lower band intensity than that of control after stained with Pro-Q Diamond gel. PKA group had the most abundant phosphorylated μ-calpain. Non-significant difference in total proteins was determined among the three treatments (Fig. 8.20b). Correspondingly, the phosphorylation level of μ-calpain was the highest in PKA group and the lowest in AP group ($P < 0.05$) (Fig. 8.20c). These data showed that phosphorylation level of μ-calpain incubated with AP and PKA was significantly regulated as expected.

To determine whether AP/PKA addition, which altered the phosphorylation level of μ-calpain, changed the phosphorylation status of heat stable proteins during incubation at 4 °C, the phosphorylation level of heat stable proteins after incubation with μ-calpain for 1, 2, 12, and 24 h was detected. Heat stable proteins incubated with 25, 50, 100, and 150 units of μ-calpain presented non-significant difference in phosphorylation level among three treatments ($P > 0.05$) (Fig. 8.21i, j, k, l). The intensity of protein bands on both Pro-Q Diamond and SYPRO Ruby stain gel showed no significant difference among AP, control, and PKA treatments (Fig. 8.21a–h). The amount of ATP added to regulate the phosphorylation level of μ-calpain can only be used to phosphorylate less than 200 μg protein. When the AP/PKA treated μ-calpain was mixed with heat stable proteins, the protein content of each mixture was more than 7000 μg. Under such circumstances, the added ATP was not adequate to influence the phosphorylation status of heat stable proteins and the amount of AP/PKA was also limited in each treatment, resulting in the lack of significant difference in phosphorylation level of proteins among three treatments in the four incubation systems, which achieved the effect we desired.

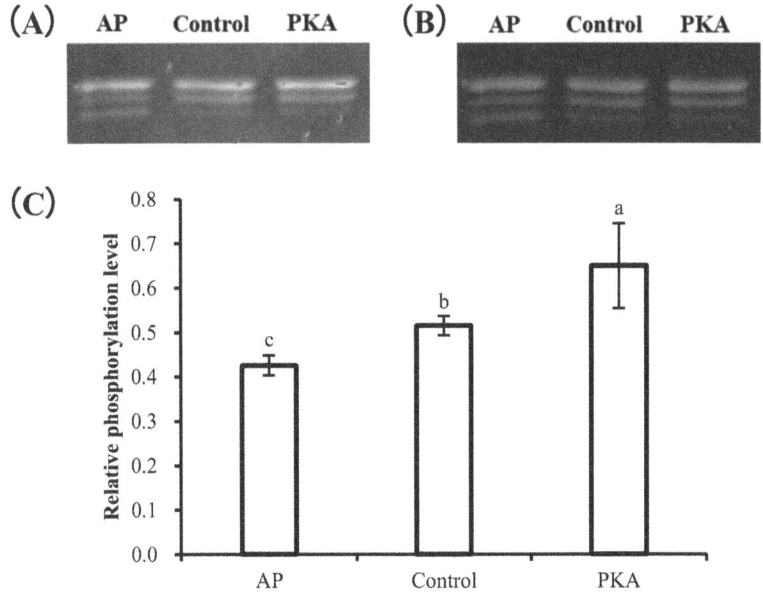

Fig. 8.20 Phosphorylation level of μ-calpain after incubation with AP and PKA at 30 °C for 30 min. (**a**) SDS-PAGE gels stained with Pro-Q Diamond. (**b**) SDS-PAGE gels stained with SYPRO Ruby. (**c**) Relative phosphorylation level of μ-calpain. Values with different letters are significantly different (*P* < *0.05*). *AP* Alkaline phosphatase, *PKA* Protein kinase A (Du et al. 2019)

8.7.3 The Activity of μ-Calpain

The 80 kDa subunit of intact μ-calpain progressively degrades to 78 kDa and then 76 kDa forms postmortem, which is considered as the activation process of μ-calpain (Goll et al. 2003). In the present study, the degradation degree of μ-calpain increased as the amount of μ-calpain added to the system increased (Fig. 8.22a, b, c, d). When heat stable proteins were incubated with 25 units of μ-calpain, there was no significant degradation rate of μ-calpain among AP, control, and PKA groups (*P* > *0.05*) (Fig. 8.22e). Overexpression of calpastatin reduced the degradation extent and degradation rate of μ-calpain, indicating a decline of postmortem proteolysis (Kent et al. 2004). Therefore, the non-significant difference in degradation rate of μ-calpain among the three treatments was caused by a relative excess of calpastatin to 25 units - μ-calpain. When the amount of μ-calpain in incubation was more than 50 units, the degree of degradation of μ-calpain in AP group was the greatest and the corresponding degradation rate of μ-calpain was significantly higher than the other two groups (*P* < *0.05*) (Fig. 8.22b–d, f–h). μ-Calpain in AP and PKA groups had significantly higher degradation rate than control group in calpastatin-free control system during incubation (*P* < *0.05*) (Fig. 8.23a, b). Dephosphorylation and phosphorylation induced by PKA of μ-calpain accelerated its degradation and activation. Studies have shown that dephosphorylation improved the activity of μ-calpain

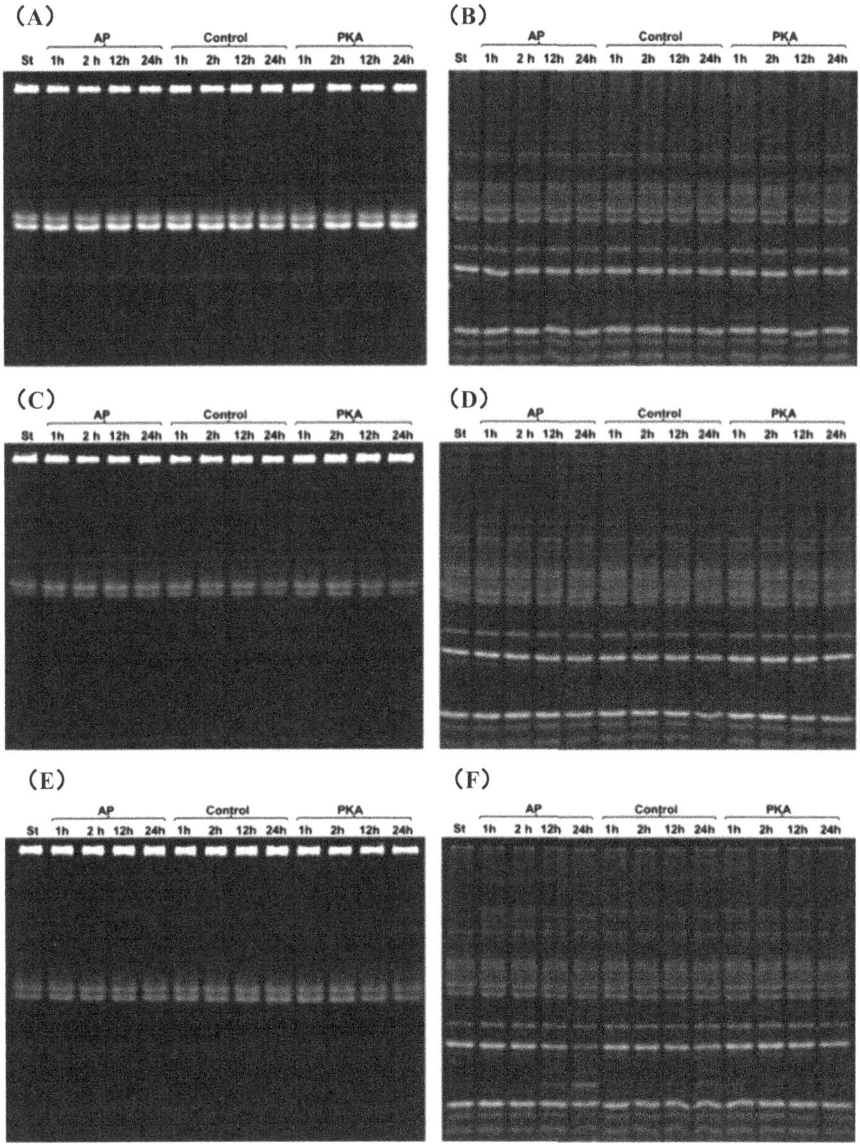

Fig. 8.21 Phosphorylation of heat stable proteins during incubation with μ-calpain. (**a**), (**b**) Proteins incubated with 25 units μ-calpain. (**c**), (**d**) Proteins incubated with 50 units μ-calpain. (**e**), (**f**) Proteins incubated with 100 units μ-calpain. (**g**), (**h**) Proteins incubated with 150 units μ-calpain. (**a**), (**c**), (**e**), (**g**) SDS-PAGE gels stained with Pro-Q Diamond. (**b**), (**d**), (**f**), (**h**) SDS-PAGE gels stained with SYPRO Ruby. (**i**), (**j**), (**k**), (**l**) Relative phosphorylation level of proteins incubated with 25, 50, 100, 150 units μ-calpain, respectively. *AP* Alkaline phosphatase, *PKA* Protein kinase A (Du et al. 2019)

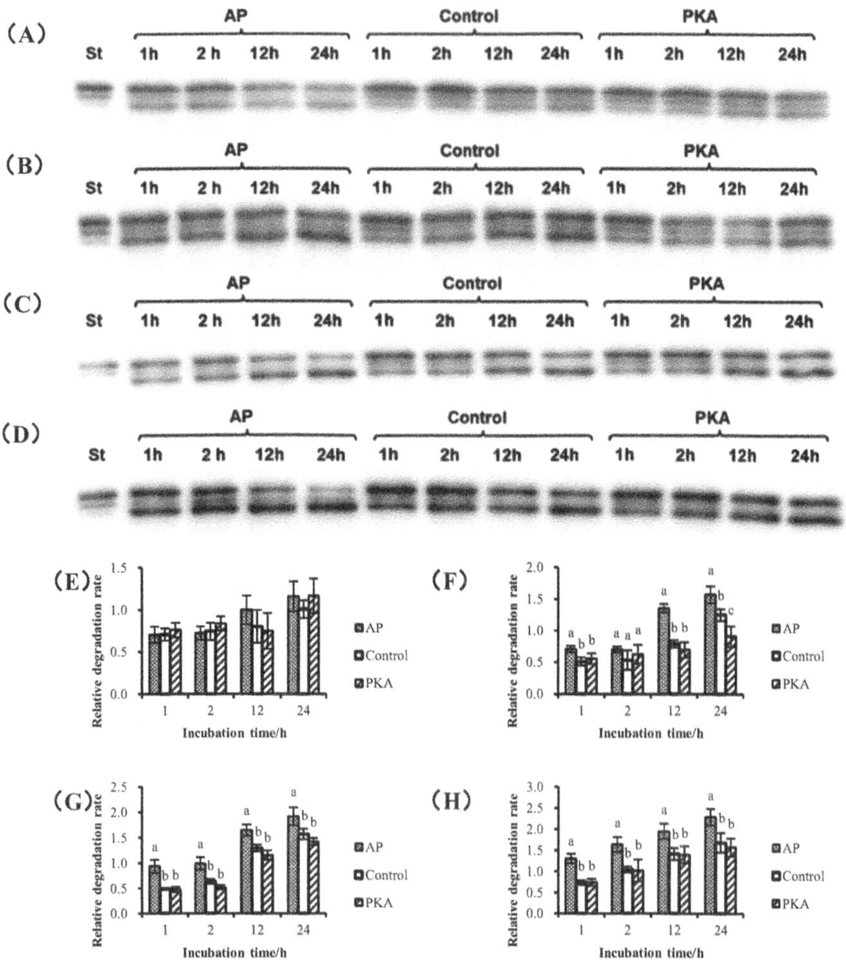

Fig. 8.22 Degradation of 80 kDa μ-calpain subunit during incubation. (**a–d**) The degradation of 80 kDa μ-calpain subunit incubated with 25 (**a**), 50 (**b**), 100 (**c**), 150 units (**d**) μ-calpain. (**e–h**) Relative degradation rate of 80 kDa μ-calpain subunit incubated with 25 (**e**), 50 (**f**), 100 (**g**), 150 units (**h**) μ-calpain. Values with different letters differ at the same incubation time ($P<0.05$). *AP* Alkaline phosphatase, *PI* Phosphatase inhibitor (Du et al. 2019)

(Du et al. 2017a). But different protein kinases functioned differently in regulating μ-calpain activity (Storr et al. 2011; Leloup et al. 2010). In the calpastatin-free control system in present study, PKA phosphorylated μ-calpain also enhanced the activity of μ-calpain, which was consistent with the results of our recent research (Du et al. 2017b, 2018). However, μ-calpain in PKA treatment showed significantly lower degradation rate than control at 24 h in 50 units μ-calpain incubation system ($P < 0.05$) (Fig. 8.22f). Besides, in 100 and 150 units μ-calpain incubation system, there was no significant difference in the degradation rate of μ-calpain between PKA

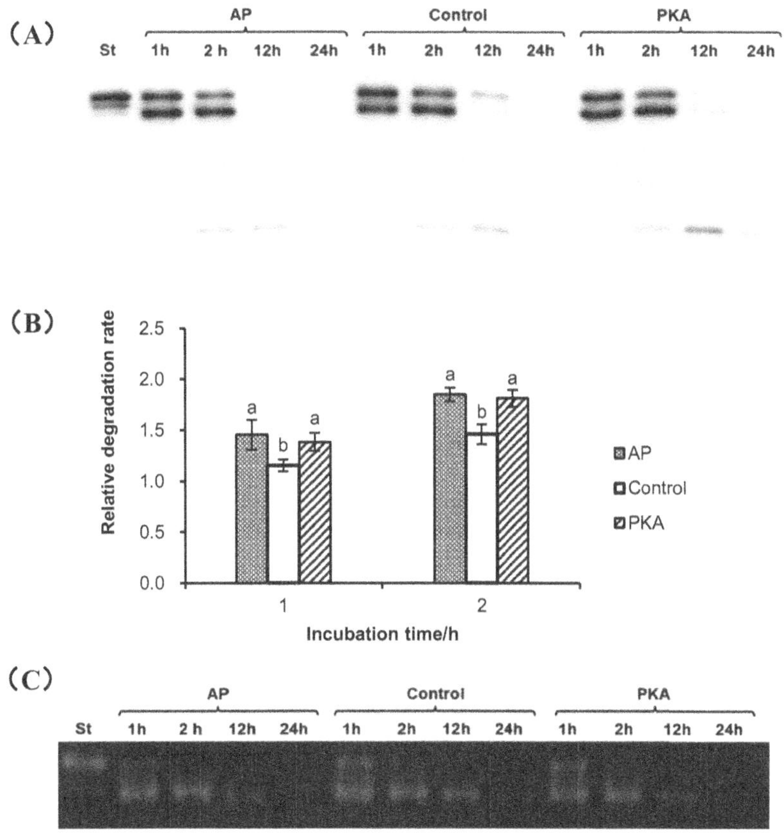

Fig. 8.23 Western blotting analysis and casein zymography analysis of μ-calpain incubated in absence of calpastatin. (**a**) Detection of degradation of 80 kDa μ-calpain subunit by western blot. (**b**) Relative degradation rate of 80 kDa μ-calpain subunit. (**c**) Casein zymography analysis of μ-calpain. Values with different letters show significant difference in the results in different groups at the same incubation time ($P < 0.05$). *AP* Alkaline phosphatase, *PI* Phosphatase inhibitor (Du et al. 2019)

and control treatments during incubation ($P > 0.05$) (Fig. 8.22g, h). These results suggested that the degradation of PKA phosphorylated μ-calpain could probably be easily inhibited by calpastatin.

Casein zymography detects the native and autolyzed μ-calpain which has the proteolytic activity to degrade casein in gels. Casein zymography analysis of μ-calpain incubated without calpastatin is shown in Fig. 8.23c. The top row of bands was native μ-calpain, the bottom row of bands was autolyzed μ-calpain. The results showed that μ-calpain in AP and PKA groups autolyzed faster than that of control. After 24 h incubation, μ-calpain was still detected in control group but not in AP and PKA groups, indicating that without calpastatin, dephosphorylation and PKA phosphorylation positively regulated μ-calpain degradation/autolyzation. When 25 units of μ-calpain were added to heat stable proteins, the autolysis of

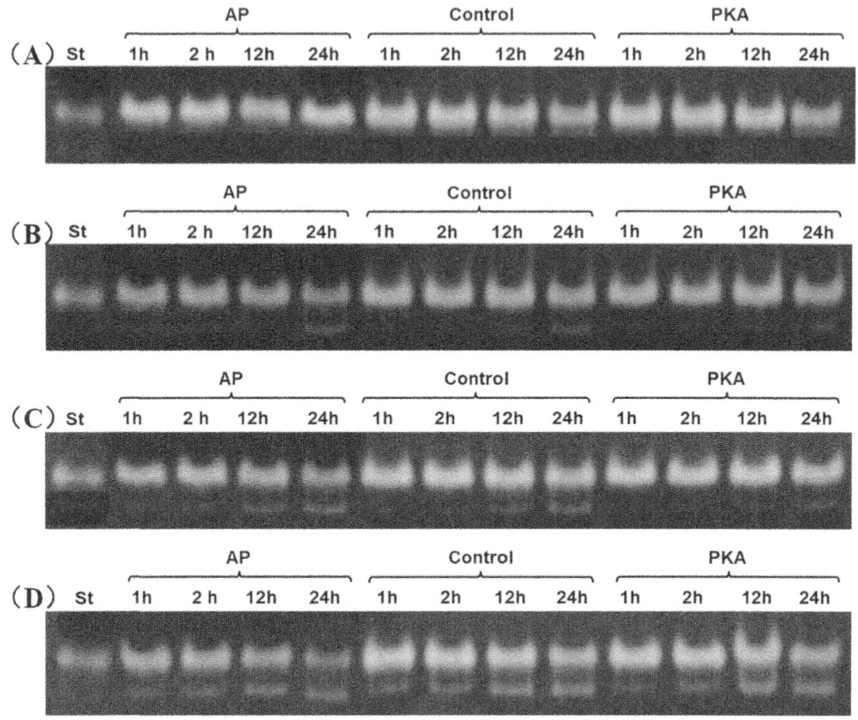

Fig. 8.24 Casein zymography analysis of μ-calpain incubated with calpastatin. (**a**), (**b**), (**c**), (**d**) 25, 50, 100, 150 unit μ-calpain was incubated with heat stable proteins, respectively. *AP* Alkaline phosphatase, *PKA* Protein kinase A (Du et al. 2019)

μ-calpain was prevented and the intensity of intact μ-calpain was not different among the three treatments (Fig. 8.24a). As the content of μ-calpain added to heat stable proteins increased, the autolyzed μ-calpain was detected in casein gels (Fig. 8.24b, c, d). In 50 and 100 units μ-calpain incubation systems, the intensity of intact μ-calpain in AP group decreased faster than in the other two groups. Besides, the amount of autolyzed μ-calpain was higher in control group than that of PKA group at 24 h incubation. But when the incubation system contained 150 units of μ-calpain, the autolysis status of μ-calpain in control and PKA groups showed no difference. These results were consistent with the degradation rate of μ-calpain measured by western blotting. Increased degradation and activity of PKA phosphorylated μ-calpain were limited by calpastatin.

Previous studies have reported that dephosphorylation of μ-calpain enhanced its activity (Du et al. 2017a, b). However, phosphorylation of μ-calpain induced by different kinases may have opposite effects. Protein kinase Cι (PKCι) induced μ-calpain and m-calpain phosphorylation enhanced their activity and was related to increased cell migration (Xu and Deng 2006a). In the meantime, m-calpain phosphorylated by PKA reduced its activity by blocking the binding sites of m-calpain to its activator (Leloup et al. 2010; Shao et al. 2006). Calpain phosphorylated by different protein kinases usually occurred at different amino acid residues, accounting for the different roles they played in regulating calpain activity (Storr et al. 2011). Our recent study showed that increased phosphorylation level of sarcoplasmic proteins induced by phosphatase inhibitor (PI) inhibited μ-calpain activity (Du et al. 2017a). PKA was also reported to positively regulate μ-calpain activity (Du et al. 2017b). Both AP and phosphatase inhibitor had broad-spectrum ability to decrease or increase the phosphorylation level of proteins. Phosphatase inhibitor improved the phosphorylation of proteins by inhibiting dephosphorylation process rather than working directly on protein kinases, indicating that the overall effects of phosphorylation induced by all protein kinases on μ-calpain activity might be negative. Besides, in the present study, calpastatin had higher inhibitory ability on the activity of PKA phosphorylated μ-calpain, which might be another reason why PI treated sarcoplasmic proteins showed lower μ-calpain activity.

8.7.4 The Degradation of Calpastatin

The undegraded calpastatin in lamb *longissimus* muscles is about 130 kDa in size and degrades gradually to small fragments during postmortem storage (Doumit and Koohmaraie 1999). The discrete degradation products of calpastatin include immunoreactive peptides about 100, 80, and 50 kDa. In 25 units μ-calpain system, there were no distinct changes in calpastatin in AP, control, and PKA groups during incubation. As the amount of μ-calpain in incubation system increased, calpastatin degradation increased. The 130, 100, and 80 kDa calpastatins were no longer detectable in all three treatments when 150 units of μ-calpain were added (Fig. 8.25d). Apparently, calpastatin in AP group degraded much faster than control and PKA groups in 100 and 150 units μ-calpain incubation systems (Fig. 8.25c, d). In addition, PKA group presented the lowest degradation rate of calpastatin when samples were incubated with 50 units of μ-calpain (Fig. 8.25b). Proteolytic enzymes, such as μ-calpain, m-calpain, cathepsin, and proteasome, degraded calpastatin, but μ-calpain was the major contributor to this process in postmortem muscles (Doumit and Koohmaraie 1999; Huang et al. 2014). Thus, the μ-calpain added to the systems was not only inhibited by calpastatin but also degraded calpastatin in turn. Therefore, the relatively stable state of calpastatin during incubation with 25 units of μ-calpain could be illustrated by the greatly suppressed and indifferent μ-calpain activity in the three treatments. Meanwhile, the results of calpastatin degradation in 50, 100, and 150 units μ-calpain incubation systems agreed well with the corresponding activity of μ-calpain. Calpastatin in skeletal muscle contained four repeated calpain-inhibiting domains (Lee et al. 1992). The ability of μ-calpain to degrade calpastatin reduced activity of calpastatin, but cannot completely remove the inhibition of μ-calpain by calpastatin (Demartino et al. 1988; Doumit and Koohmaraie 1999; Koohmaraie et al. 1990). According to the results of μ-calpain activity in calpastatin-free system, PKA treated μ-calpain degraded more calpastatin than control. Considering μ-calpain activity in PKA treatment was extensively inhibited by calpastatin,

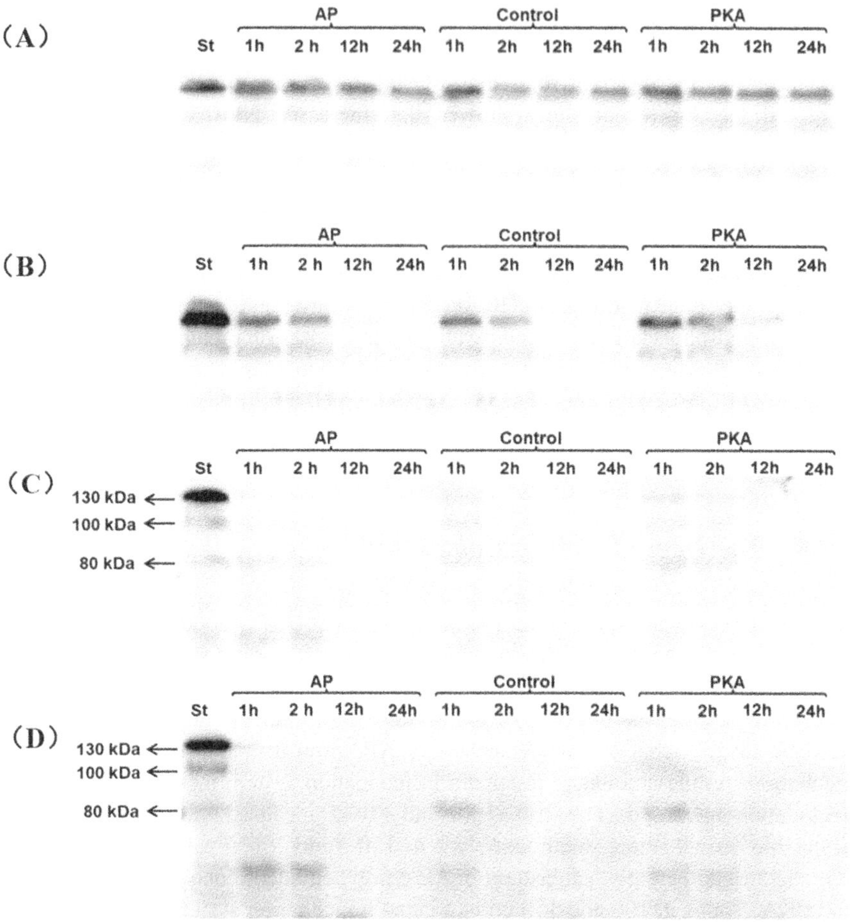

Fig. 8.25 Degradation of calpastatin during incubation. (**a**), (**b**), (**c**), (**d**) 25, 50, 100, 150 unit μ-calpain was incubated with heat stable proteins, respectively. *AP* Alkaline phosphatase, *PKA* Protein kinase A (Du et al. 2019)

the weakened proteolytic ability of μ-calpain contributed to the less degradation of calpastatin than in the other two treatments in 50 units μ-calpain incubation system. Consequently, PKA phosphorylated μ-calpain had stronger inhibitory effects on calpastatin.

8.8 Conclusions

Broad-spectrum catalyzing phosphorylation and dephosphorylation played negative or positive roles in regulating μ-calpain activity, respectively. However, phosphorylation induced by PKA enhanced the activity of μ-calpain. Meanwhile, PKA phosphorylated μ-calpain was more sensitive to calpastatin and had stronger inhibitory effects on calpastatin. Therefore, although PKA phosphorylation enhanced μ-calpain activity, phosphorylation negatively regulated the activity of μ-calpain in general. The mechanism of the effects of phosphorylation on μ-calpain activity was investigated in this study, which helped us to better understand the influencing mechanism of μ-calpain activity and meat tenderness.

Acknowledgments Parts of this chapter are reprinted from Food Chemistry, 228, Du, M., et al. Phosphorylation inhibits the activity of μ-calpain at different incubation temperatures and Ca^{2+} concentrations in vitro, 649–655; Food Research International, 100, Du, M., et al. Effects of phosphorylation on μ-calpain activity at different incubation temperature, 318–324; Food Chemistry, 252, Du, M. et al., Phosphorylation regulated by protein kinase A and alkaline phosphatase play positive roles in μ-calpain activity, 33–39. Food Chemistry, 274, Du, M., et al. Calpastatin inhibits the activity of phosphorylated μ-calpain in vitro, 473–479. Copyright (2020), with permission from Elsevier.

References

Beaufort, A., Cardinal, M., Le-Bail, A., & Midelet-Bourdin, G. (2009). The effects of superchilled storage at −2 °C on the microbiological and organoleptic properties of cold-smoked Salmon before retail display. *International Journal of Refrigeration, 32*, 1850–1857.

Bee, G., Anderson, A. L., Lonergan, S. M., & Huff-Lonergan, E. (2007). Rate and extent of Ph decline affect proteolysis of cytoskeletal proteins and water-holding capacity in pork. *Meat Science, 76*, 359–365.

Biswas, A. K., Tandon, S., & Beura, C. K. (2016). Identification of different domains of Calpain and Calpastatin from chicken blood and their role in post-mortem aging of meat during holding at refrigeration temperatures. *Food Chemistry, 200*, 315–321.

Boehm, M. L., Kendall, T. L., Thompson, V. F., & Goll, D. E. (1998). Changes in the Calpains and Calpastatin during postmortem storage of bovine muscle. *Journal of Animal Science, 76*, 2415–2434.

Chen, L. J., Li, X., Ni, N., Liu, Y., Chen, L., Wang, Z. Y., et al. (2016). Phosphorylation of Myofibrillar proteins in post-mortem ovine muscle with different tenderness. *Journal of Science and Food Agricultural, 96*, 1474–1483.

Claeys, E., De Smet, S., Demeyer, D., Geers, R., & Buys, N. (2001). Effect of rate of Ph decline on muscle enzyme activities in two pig lines. *Meat Science, 57*, 257–263.

Cong, J. Y., Goll, D. E., Peterson, A. M., & Kapprell, H. P. (1989). The role of autolysis in activity of the Ca-2+−dependent proteinase (M-Calpain and M-Calpain). *Journal of Biological Chemistry, 264*, 10096–10103.

Crawford, C. (1990). *Protein and peptide inhibitors of Calpains. Intracellular calcium-dependent proteolysis* (pp. 75–89). Boca Raton: CRC Press.

Cruzen, S. M., Paulino, P. V., Lonergan, S. M., & Huff-Lonergan, E. (2014). Postmortem proteolysis in three muscles from growing and mature beef cattle. *Meat Science, 96*, 854–861.

Demartino, G. N., Wachendorfer, R., Mcguire, M., & Croall, D. E. (1988). Proteolysis of the protein inhibitor of calcium-dependent proteases produces lower molecular weight fragments that retain inhibitory activity. *Archives of Biochemistry and Biophysics, 262*, 189–198.

Doumit, M. E., & Koohmaraie, M. (1999). Immunoblot analysis of Calpastatin degradation: Evidence for cleavage by Calpain in postmortem muscle. *Journal of Animal Science, 77*, 1467–1473.

Du, M. (2018). *Influence mechanism of protein phosphorylation on Calpain activity[D]*. Beijing: Chinese Academy of Agricultural Sciences.

Du, M., Li, X., Li, Z., Li, M., Gao, L., & Zhang, D. (2017a). Phosphorylation inhibits the activity of M-Calpain at different incubation temperatures and Ca2+ concentrations in vitro. *Food Chemistry, 228*, 649–655.

Du, M., Li, X., Li, Z., Shen, Q., Wang, Y., Li, G., et al. (2017b). Effects of phosphorylation on mu-Calpain activity at different incubation temperature. *Food Research International, 100*, 318–324.

Du, M., Li, X., Li, Z., Shen, Q., Wang, Y., Li, G., et al. (2018). Phosphorylation regulated by protein kinase A and alkaline phosphatase play positive roles in mu-Calpain activity. *Food Chemistry, 252*, 33–39.

Du, M., Li, X., Li, Z., Shen, Q., Ren, C., & Zhang, D. (2019). Calpastatin inhibits the activity of phosphorylated μ-calpain in vitro. *Food Chemistry, 274*, 473–479.

Edmunds, T., Nagainis, P. A., Sathe, S. K., Thompson, V. F., & Goll, D. E. (1991). Comparison of the autolyzed and unautolyzed forms of M- and M-Calpain from bovine skeletal muscle. *Biochimica et Biophysica Acta, 1077*, 197–208.

Geesink, G. H., Bekhit, A. D., & Bickerstaffe, R. (2000). Rigor temperature and meat quality characteristics of lamb Longissimus muscle. *Journal of Animal Science, 78*, 2842–2848.

Geesink, G. H., Kuchay, S., Chishti, A. H., & Koohmaraie, M. (2006). M-Calpain is essential for postmortem proteolysis of muscle proteins. *Journal of Animal Science, 84*, 2834–2840.

Geesink, G. H., Van Der Palen, J. G. P., Kent, M. P., Veiseth, E., Hemke, G., & Koohmaraie, M. (2005). Quantification of Calpastatin using an optical surface Plasmon resonance biosensor. *Meat Science, 71*, 537–541.

Glading, A., Bodnar, R. J., Reynolds, I. J., Shiraha, H., Satish, L., Potter, D. A., et al. (2004). Epidermal growth factor activates M-Calpain (Calpain Ii), at least in part, by extracellular signal-regulated kinase-mediated phosphorylation. *Molecular and Cellular Biology, 24*, 2499–2512.

Goll, D. E., Thompson, V. F., Li, H. Q., Wei, W., & Cong, J. Y. (2003). The Calpain system. *Physiological Reviews, 83*, 731–801.

Hanna, R. A., Campbell, R. L., & Davies, P. L. (2008). Calcium-bound structure of Calpain and its mechanism of inhibition by Calpastatin. *Nature, 456*, 409–412.

Huang, F., Huang, M., Zhang, H., Guo, B., Zhang, D., & Zhou, G. (2014). Cleavage of the Calpain inhibitor, Calpastatin, during postmortem ageing of beef skeletal muscle. *Food Chemistry, 148*, 1–6.

Huang, H., Larsen, M. R., & Lametsch, R. (2012). Changes in phosphorylation of Myofibrillar proteins during postmortem development of porcine muscle. *Food Chemistry, 134*, 1999–2006.

Hwang, I. H., Park, B. Y., Cho, S. H., & Lee, J. M. (2004). Effects of muscle shortening and proteolysis on Warner-Bratzler shear force in beef Longissimus and semitendinosus. *Meat Science, 68*, 497–505.

Hwang, I. H., & Thompson, J. M. (2001). The interaction between Ph and temperature decline early postmortem on the Calpain system and objective tenderness in electrically stimulated beef Longissimus Dorsi muscle. *Meat Science, 58*, 167–174.

Jaturasitha, S., Thirawong, P., Leangwunta, V., & Kreuzer, M. (2004). Reducing toughness of beef from Bos Indicus Draught steers by injection of calcium chloride: Effect of concentration and time postmortem. *Meat Science, 68*, 61–69.

Kaale, L. D., Eikevik, T. M., Rustad, T., & Kolsaker, K. (2011). Superchilling of food: A review. *Journal of Food Engineering, 107*, 141–146.

Kemp, C. M., Sensky, P. L., Bardsley, R. G., Buttery, P. J., & Parr, T. (2010). Tenderness--An enzymatic view. *Meat Science, 84*, 248–256.

Kent, M. P., Spencer, M. J., & Koohmaraie, M. (2004). Postmortem proteolysis is reduced in transgenic mice overexpressing Calpastatin. *Journal of Animal Science, 82*, 794–801.

Killefer, J., & Koohmaraie, M. (1994). Bovine skeletal muscle Calpastatin: Cloning, sequence analysis, and steady-state mRNA expression. *Journal of Animal Science, 72*, 606–614.

Kim, Y. H., Luc, G., & Rosenvold, K. (2013). Pre rigor processing, ageing and freezing on tenderness and colour stability of lamb loins. *Meat Science, 95*(2), 412–418.

Koohmaraie, M. (1992). The role of Ca2+−dependent proteases (Calpains) in Postmortem proteolysis and meat tenderness. *Biochimie, 74*, 239–245.

Koohmaraie, M., Crouse, J. D., & Mersmann, H. J. (1989). Acceleration of postmortem tenderization in ovine carcasses through infusion of calcium chloride: Effect of concentration and ionic strength. *Journal of Animal Science, 67*, 934–942.

Koohmaraie, M., & Geesink, G. H. (2006). Contribution of postmortem muscle biochemistry to the delivery of consistent meat quality with particular focus on the Calpain system. *Meat Science, 74*, 34–43.

Koohmaraie, M., Schollmeyer, J. E., & Dutson, T. R. (1986). Effect of low-calcium-requiring calcium activated factor on myofibrils under varying Ph and temperature conditions. *Journal of Food Science, 51*, 28–32.

Koohmaraie, M., Whipple, G., & Crouse, J. D. (1990). Acceleration of postmortem tenderization in lamb and Braham-cross beef carcasses through infusion of calcium chloride. *Journal of Animal Science, 67*, 1278–1283.

Lee, W. J., Ma, H., Takano, E., Yang, H. Q., Hatanaka, M., & Maki, M. (1992). Molecular diversity in amino-terminal domains of human Calpastatin by exon skipping. *Journal of Biological Chemistry, 267*, 8437–8442.

Leloup, L., Shao, H., Bae, Y. H., Deasy, B., Stolz, D., Roy, P., et al. (2010). M-Calpain activation is regulated by its membrane localization and by its binding to phosphatidylinositol 4,5-Bisphosphate. *Journal of Biological Chemistry, 285*, 33549–33566.

Li, C., Zhou, G. H., Xu, X. L., Lundstrom, K., Karlsson, A., & Lametsch, R. (2015). Phosphoproteome analysis of sarcoplasmic and myofibrillar proteins in bovine longissimus muscle in response to Postmortem electrical stimulation. *Food Chemistry, 175*, 197–202.

Li, M., Li, X., Xin, J., Li, Z., Li, G., Zhang, Y., et al. (2017). Effects of protein phosphorylation on color stability of ground meat. *Food Chemistry, 219*, 304–310.

Li, P., Li, X., Li, Z., Chen, L., Li, Z., Chen, L., et al. (2016). Effects of controlled freezing point storage on aging from muscle to meat. *Scientia Agricultura Sinica, 49*, 554–562.

Liu, Q., Kong, B., Han, J., Chen, Q., & He, X. (2014). Effects of superchilling and cryoprotectants on the quality of common carp (*Cyprinus Carpio*) Surimi: Microbial growth, oxidation, and physiochemical properties. *LWT - Food Science and Technology, 57*, 165–171.

Lomiwes, D., Farouk, M. M., Wu, G., & Young, O. A. (2014a). The development of meat tenderness is likely to be compartmentalised by ultimate Ph. *Meat Science, 96*, 646–651.

Lomiwes, D., Hurst, S. M., Dobbie, P., Frost, D. A., Hurst, R. D., Young, O. A., et al. (2014b). The protection of bovine skeletal myofibrils from proteolytic damage post mortem by small heat shock proteins. *Meat Science, 97*, 548–557.

Longo, V., Lana, A., Bottero, M. T., & Zolla, L. (2015). Apoptosis in muscle-to-meat aging process: The Omic witness. *Journal of Proteomics, 125*, 29–40.

Maddock, K. R., Huff-Lonergan, E., Rowe, L. J., & Lonergan, S. M. (2005). Effect of Ph and ionic strength on Calpastatin inhibition of M- and M-Calpain. *Journal of Animal Science, 83*, 1370–1376.

Marsh, B. B., Ringkob, T. P., Russell, R. L., Swartz, D. R., & Pagel, L. A. (1987). Effects of early-postmortem glycolytic rate on beef tenderness. *Meat Science, 21*, 241–248.

Melody, J. L., Lonergan, S. M., Rowe, L. J., Huiatt, T. W., Mayes, M. S., & Huff Lonergan, E. (2004). Early postmortem biochemical factors influence tenderness and water-holding capacity of three porcine muscles. *Journal of Animal Science, 82*, 1195–1205.

Mohrhauser, D. A., Lonergan, S. M., Huff-Lonergan, E., Underwood, K. R., & Weaver, A. D. (2014). Calpain-1 activity in bovine muscle is primarily influenced by temperature, not Ph decline. *Journal Animal Science, 92*, 1261–1270.

Moldoveanu, T., Gehring, K., & Green, D. R. (2008). Concerted multi-pronged attack by Calpastatin to occlude the catalytic cleft of Heterodimeric Calpains. *Nature, 456*, 404–408.

O'Halloran, G. R., Troy, D. J., Buckley, D. J., & Reville, W. J. (1997). The role of endogenous proteases in the tenderisation of fast glycolysing muscle. *Meat Science, 47*(3-4), 187–210.
Ohno, S., Emori, Y., & Suzuki, K. (1986). Nucleotide sequence of a cDNA coding for the small subunit of Human calcium-dependent protease. *Nucleic Acids Research, 14*, 5559.
Pomponio, L., & Ertbjerg, P. (2012). The effect of temperature on the activity of mu- and M-Calpain and Calpastatin during post-mortem storage of porcine Longissimus muscle. *Meat Science, 91*, 50–55.
Pomponio, L., Lametsch, R., Karlsson, A. H., Costa, L. N., Grossi, A., & Ertbjerg, P. (2008). Evidence for post-mortem M-Calpain autolysis in porcine muscle. *Meat Science, 80*, 761–764.
Pulford, D. J., Dobbie, P., Fraga Vazquez, S., Fraser-Smith, E., Frost, D. A., & Morris, C. A. (2009). Variation in bull beef quality due to ultimate muscle Ph is correlated to Endopeptidase and small heat shock protein levels. *Meat Science, 83*, 1–9.
Reverter, D., Sorimachi, H., & Bode, W. (2001). The structure of calcium-free human M-Calpain implications for calcium activation and function. *Trends in Cardiovascular Medicine, 11*, 222–229.
Rhee, M. S., Ryu, Y. C., & Kim, B. C. (2006). Postmortem metabolic rate and Calpain system activities on beef Longissimus tenderness classifications. *Bioscience Biotechnology And Biochemistry, 70*, 1166–1172.
Salamino, F., Tullio, R. D., Michetti, M., et al. (1994). Modulation of calpastatin specificity in rat tissues by reversible phosphorylation and dephosphorylation. *Biochemical and biophysical research communications, 199*(3), 1326–1322.
Shao, H., Chou, J., Baty, C. J., Burke, N. A., Watkins, S. C., Stolz, D. B., et al. (2006). Spatial localization of M-Calpain to the plasma membrane by Phosphoinositide Biphosphate binding during epidermal growth factor Receptor mediated activation. *Molecular & Cellular Biology, 26*, 5481–5496.
Sharma, U., Pal, D., & Prasad, R. (2014). A novel role of alkaline phosphatase in the Erk1/2 Dephosphorylation in renal cell carcinoma cell lines: A new plausible therapeutic target. *Biochimie, 107*, 406–409.
Shiraha, H., Glading, A., Chou, J., Jia, Z., & Wells, A. (2002). Activation of M-Calpain (Calpain Ii) by epidermal growth factor is limited by protein kinase a phosphorylation of M-Calpain. *Molecular and Cellular Biology, 22*, 2716–2727.
Storr, S. J., Carragher, N. O., Frame, M. C., Parr, T., & Martin, S. G. (2011). The Calpain system and cancer. *Nature Reviews Cancer, 11*, 364–374.
Strobl, S., Fernandez-Catalan, C., Braun, M., Huber, R., Masumoto, H., Nakagawa, K., et al. (2000). The crystal structure of calcium-free human M-Calpain suggests an electrostatic switch mechanism for activation by calcium. *Proceedings of the National Academy of Science of the United States of America, 97*, 588–592.
Vazquez, R., Wendt, A. S., Thompson, V. F., Novak, S. M., Ruse, C., Yates, J. R., et al. (2008). Phosphorylation of the Calpains. *The FASEB Journal, 22*, 793.5.
Whipple, G., Koohmaraie, M., Dikeman, M. E., & Crouse, J. D. (1990). Effects of high temperature conditioning on enzymatic-activity and tenderness of Bos-Indicus Longissimus muscle. *Journal of Animal Science, 68*, 3654–3662.
Xu, L., & Deng, X. (2004). Tobacco-specific nitrosamine 4-(Methylnitrosamino)-1-(3-Pyridyl)-1-butanone induces phosphorylation of mu- and M-Calpain in association with increased secretion, cell migration, and invasion. *Journal of Biological Chemistry, 279*, 53683–53690.
Xu, L., & Deng, X. (2006a). Protein kinase Ciota promotes nicotine-induced migration and invasion of cancer cells via phosphorylation of micro- and M-Calpains. *Journal of Biological Chemistry, 281*, 4457–4466.
Xu, L., & Deng, X. (2006b). Suppression of cancer cell migration and invasion by protein phosphatase 2a through dephosphorylation of Mu- and M-Calpains. *Journal of Biological Chemistry, 281*, 35567–35575.
Zhang, Y., Li, X., Li, Z., Li, M., Liu, Y., & Zhang, D. (2016). Effects of controlled freezing point storage on the protein phosphorylation level. *Scientia Agricultura Sinica, 49*, 4429–4440.

Chapter 9
Mechanism of the Effect of Protein Phosphorylation on Myoglobin

Abstract As the most important pigment for meat color, myoglobin mainly exists in sarcoplasm, and meat color is determined by its absolute content as well as the relative content of three types of myoglobin (oxymyoglobin, deoxymyoglobin, and metmyoglobin). Previous studies have shown that protein phosphorylation may negatively regulate the stability of meat color through the regulation of glycolysis pathway and myoglobin. The purpose of this research was to study the influence of phosphorylation on myoglobin stability, in order to provide a theoretical basis for improving the stability of meat color through regulating phosphorylation level. Sodium hydrosulfite was used to reduce myoglobin from skeletal muscle, and later was removed by ultrafiltration. Then alkaline phosphatase (AP) was added to in vitro incubation system to catalyze the dephosphorylation of myoglobin. The changes in myoglobin phosphorylation level were measured by SDS-polyacrylamide gel electrophoresis (SDS-PAGE) and Pro-Q Diamond and Ruby phosphoprotein gel staining. We also kept a record of the changes in pH and used ultraviolet spectrophotometer to measure the relative content of three myoglobin redox forms. The ultraviolet spectrophotometer of myoglobin was determined by circular dichroism (CD) spectroscopy. Myoglobin phosphorylation may lead to changes in its secondary structure, therefore reducing myoglobin stability and increasing its autoxidation rate, which further accelerated the accumulation of metmyoglobin. As a result, meat color was found to have a low stability, and this might be one of the explanations that protein phosphorylation negatively regulated meat color stability.

Keywords Meat color · Myoglobin · Protein phosphorylation · Redox stability · Secondary structure

9.1 Introduction

The most intuitive impression of consumers towards meat and meat products comes from meat color. Meat with poor color gives people the feeling of unsanitary and poor quality; therefore, meat color is one of the most important factors affecting consumers' purchase decisions. Myoglobin is the most important pigment in

D. Zhang et al., *Protein Phosphorylation and Meat Quality*,
https://doi.org/10.1007/978-981-15-9441-0_9

postmortem meat. Studying the influencing factors of myoglobin stability can provide a theoretical basis to improve meat color stability.

Canto et al. (2015) found out that an acid transfer happened to myoglobin isoelectric point, which indicated that there was a post-translation modification in myoglobin (PI = 6.78), and the content of the modified point of myoglobin was high in the unstable meat color group. Thus, we could say that this kind of post-translation modification in myoglobin would affect meat color stability. It has been proved that the phosphorylation modification existed in human and mouse myoglobin (www. photosite. ORG). Therefore, the speculation was that phosphorylation modification in myoglobin might affect meat color stability. Previous research from our team showed that protein phosphorylation could negatively regulate meat color stability, and the possible pathway might be through regulating the glycolysis process and myoglobin stability.

At present, there are few studies about the effect of protein phosphorylation on meat color stability, and neither much on the effect of phosphorylation on myoglobin stability. Based on the pure myoglobin from skeletal muscles, we used sodium dithionite as reductant to obtain deoxymyoglobin. And the incubation of myoglobin and alkaline phosphatase (AP) was conducted in vitro at 37 °C to regulate myoglobin phosphorylation level and obtain myoglobin samples with different phosphorylation level. We studied the relationship between myoglobin phosphorylation and its redox stability, and clarified the effect of phosphorylation on the secondary structure of myoglobin.

9.2 Changes of Myoglobin Relative Contents in Meat After Regulation of Protein Phosphorylation

Myoglobin is the most important pigment affecting meat color. It mainly exists in the sarcoplasm. The absolute content of myoglobin and the relative content of the three kinds of myoglobin (oxymyoglobin, deoxymyoglobin, and metmyoglobin) determine the meat color. The influence of protein phosphorylation on meat color stability was investigated in Chap. 2. Phosphatase and protein kinase inhibitors were added to minced ovine *Longissimus thoracis et lumborum* (LTL) muscle to adjust the global phosphorylation of sarcoplasmic proteins. (1) Phosphatase inhibitor group (high phosphorylation level), muscle pieces were added with phosphatase inhibitor PhosStop (Roche, Mannheim, Germany, one tablet per 5 g muscle pieces), which was dissolved in saline to inhibit the dephosphorylation of proteins; (2) Saline control group, muscle was added with the vehicle of saline; (3) Kinase inhibitor group (low phosphorylation level), muscle was added with kinase inhibitors (M2066, Abmole, Houston, TX, USA, 0.144 lmole per 5 g muscle pieces), which were dissolved in dimethyl sulfoxide (DMSO) and saline to inhibit the activity of protein kinases; (4) DMSO control group, muscle was added with DMSO and saline. After adding the reagents, muscles were minced and stored at 4 °C for 7 days. All the operations were done within 40 min after exsanguination. Minced muscle samples were collected from all four groups at 2 h, 24 h, 2 days, 3 days, 5 days, and 7 days postmortem, snap frozen in liquid nitrogen, and stored at −80 °C until analysis. The data obtained showed that the rate and extent of pH decline, along with lactate accumulation in postmortem muscle, were related to protein phosphorylation. Analysis of meat color stability and the relative content of myoglobin redox forms revealed that meat color stability was inversely related to the phosphorylation of sarcoplasmic proteins. Thus, this study suggested that protein phosphorylation may be involved in meat color development by regulating glycolysis and the redox stability of myoglobin.

Meat color is affected by the proportions of three myoglobin forms, which are influenced by the oxygen availability, the autoxidation rate of myoglobin (Fe^{2+}), and the reducing capacity of metmyoglobin (MetMb) (Mancini and Hunt 2005). Oxidation of deoxymyoglobin (DeoxyMb) and oxymyoglobin (OxyMb) to MetMb leads to brown discoloration. The relative content of DeoxyMb and MetMb in this study was shown in Table 9.1. The relative content of OxyMb was higher ($P < 0.05$) in the kinase inhibition group than in the EMSO control samples at all times, except for 120 h. The relative content of OxyMb showed no difference between Saline control and DMSO control groups. Compared with the other three groups, the relative content of OxyMb was the lowest ($P < 0.05$) in the phosphatase inhibition group during the whole period except for 2 h, indicating a higher relative content of OxyMb in meat with lower protein phosphorylation. The observed changes in the relative content of OxyMb for treatment groups matched with changes of a* values. No difference was found in the relative content of DeoxyMb between the phosphatase inhibition group and the Saline control group for the first 24 h, which then differed afterwards. DeoxyMb in the meat with phosphatase inhibitor was significantly lower than in the kinase inhibitor added meat at all times except for 2 h ($P < 0.05$), indicating a higher relative content of DeoxyMb in meat of lower protein phosphorylation. The highest ($P < 0.05$) relative content of MetMb was determined in the phosphatase inhibition group all the time in the four treatments. Although no difference in MetMb content existed between the kinase inhibition group and the DMSO control group at 48, 120, and 168 h, higher MetMb was detected at 2, 24, and 72 h in DMSO control group ($P < 0.05$).

Among the four treatments, the relative content of DeoxyMb and OxyMb decreased the fastest ($P < 0.05$) in the phosphatase inhibited meat, while MetMb content increased the fastest ($P < 0.05$) in this group during the first 48 h and then kept constant afterwards. MetMb in the kinase inhibited meat increased much slower ($P < 0.05$) when compared with the other three groups during the first 48 h. On the contrary, the relative content of the OxyMb in the kinase inhibition group increased more rapidly ($P < 0.05$) than in the other three groups during the first 48 h and then declined till the end of the experimental period. The changes of relative content of different myoglobin forms were similar between Saline control and DMSO control groups, which were intermediate among the four treatments all the time. In summary, the relative content of different myoglobin forms differed significantly ($P < 0.05$) between the kinase inhibition group and phosphatase inhibition group,

Table 9.1 Effects of inhibitors and storage time (hour) on relative pigment content (Li et al. 2017a)

Attribute	Group	Display time (hours)					
		2	24	48	72	120	168
Deoxymyoglobin (1.5-K/S ratio)	Phosphatase inhibition	0.68 ± 0.04^{aX}	0.54 ± 0.02^{bX}	0.47 ± 0.08^{cX}	0.46 ± 0.02^{cX}	$0.44 \pm 0.02^{c\ X}$	0.43 ± 0.02^{cX}
	Saline control	0.68 ± 0.04^{aX}	0.56 ± 0.01^{bXY}	0.69 ± 0.02^{aY}	0.57 ± 0.01^{bY}	0.56 ± 0.02^{bY}	0.56 ± 0.02^{bY}
	DMSO control	$0.68 \pm 0.05^{a\ X}$	0.57 ± 0.01^{bY}	0.68 ± 0.03^{aY}	0.57 ± 0.02^{bY}	0.57 ± 0.02^{bY}	0.57 ± 0.01^{bY}
	Kinase inhibition	0.68 ± 0.04^{aX}	0.58 ± 0.01^{bY}	0.75 ± 0.02^{cZ}	0.57 ± 0.01^{bY}	0.57 ± 0.01^{bY}	0.57 ± 0.01^{bY}
Oxymyoglobin (1-K/S ratio)	Phosphatase inhibition	0.59 ± 0.04^{aX}	0.49 ± 0.04^{bX}	0.43 ± 0.05^{cX}	0.44 ± 0.03^{cX}	0.40 ± 0.04^{cX}	0.39 ± 0.01^{cX}
	Saline control	0.61 ± 0.02^{aX}	0.61 ± 0.02^{aYZ}	0.61 ± 0.03^{aY}	0.57 ± 0.02^{bY}	0.56 ± 0.02^{bY}	0.55 ± 0.02^{bYZ}
	DMSO control	0.59 ± 0.02^{aX}	0.59 ± 0.03^{abY}	0.60 ± 0.03^{aY}	0.56 ± 0.02^{bcY}	0.55 ± 0.02^{cY}	0.53 ± 0.03^{cY}
	Kinase inhibition	0.61 ± 0.02^{aX}	0.64 ± 0.03^{bZ}	0.68 ± 0.02^{cZ}	0.61 ± 0.01^{aZ}	0.58 ± 0.01^{dY}	0.56 ± 0.02^{dZ}
Metmyoglobin (2-K/S ratio)	Phosphatase inhibition	0.80 ± 0.02^{aX}	1.20 ± 0.06^{bX}	1.28 ± 0.04^{cX}	1.31 ± 0.03^{cX}	1.29 ± 0.03^{cX}	1.28 ± 0.01^{cX}
	Saline control	0.77 ± 0.02^{aYZ}	1.00 ± 0.04^{bY}	1.16 ± 0.02^{cY}	1.07 ± 0.05^{dY}	1.09 ± 0.05^{dY}	1.11 ± 0.05^{dY}
	DMSO control	0.78 ± 0.01^{aY}	0.97 ± 0.05^{bY}	1.16 ± 0.04^{cY}	1.08 ± 0.05^{dY}	1.09 ± 0.05^{dY}	1.10 ± 0.04^{dY}
	Kinase inhibition	0.75 ± 0.01^{aZ}	0.87 ± 0.04^{bZ}	1.11 ± 0.03^{cY}	1.01 ± 0.04^{dZ}	1.07 ± 0.03^{eY}	1.12 ± 0.04^{cY}

DeoxyMyoglobin, OxyMyoglobin, and MetMyoglobin content are evaluated by the K/S ratios and expressed as (1.5-K/S ratio), (1-K/S ratio), and (2-K/S ratio), respectively. Data are presented as mean ± standard deviation ($n = 5$). Data with different lowercase letters in a row are significantly different ($P < 0.05$). Data with different capital letters in the same column are significantly different ($P < 0.05$)

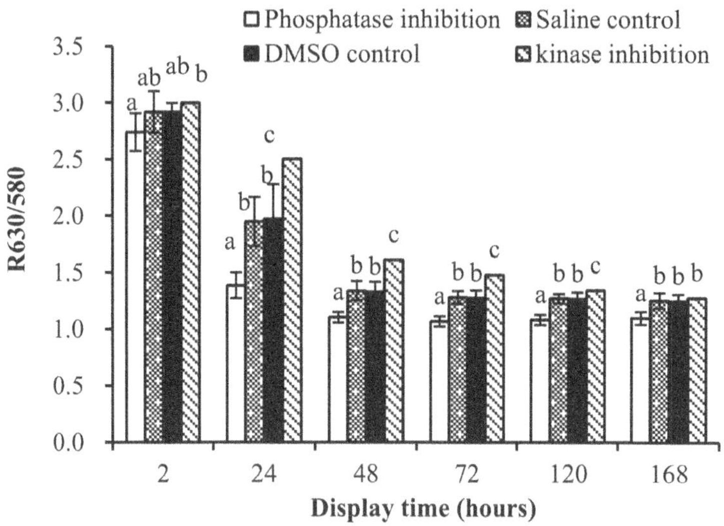

Fig. 9.1 Effects of protein phosphorylation on R630/580 values. Data with different letters at the same time point are significantly different (*P* < *0.05*). Data are presented as mean ± standard deviation, *n* = 5 (Li et al. 2017a)

indicating that the redox stability of myoglobin was influenced by the protein phosphorylation.

The R630/580 values (Fig. 9.1) were not different between Saline control and DMSO control groups at any time. However, the R630/580 values of the phosphatase inhibition group were lower than those of all other groups (*P* < *0.05*) except for 2 h. Kinase inhibition group demonstrated greater (*P* < *0.05*) R630/580 values than those of the rest groups except for 2 h and 168 h. R630/580 was an indirect indicator of meat color stability, a larger ratio indicated more redness due to either OxyMb or DeoxyMb and greater color stability (AMSA 2012). The observed changes in the R630/580 values for treatment groups matched well with the relative content of DeoxyMb, OxyMb, and MetMb. This suggested that kinase inhibition group (low phosphorylation level) was the most color-stable group, and phosphatase inhibition group (high phosphorylation level) was most color-labile group.

All data suggested that meat color stability was significantly affected by the protein phosphorylation/dephosphorylation. According to Canto (2015), nine proteins were differentially abundant in color-stable and color-labile steaks, which included pyruvate kinase M2, glyceraldehyde-3-phosphate dehydrogenase, Myosin regulatory light chain 2, and myoglobin. Joseph et al. (2012) reported that heat shock protein-27 kDa, pyruvate dehydrogenase, and pyruvate kinase isozymes M1/M2 isoform were correlated with meat color. Those color-related proteins were identified to be phosphorylated (Chen et al. 2016; Huang et al. 2011, 2012). In the present study, difference of the phosphorylation level of those color-related proteins between groups may cause the difference in meat color stability.

9.3 Changes of Myoglobin Redox Stability After Regulation of Protein Phosphorylation In Vitro

Studies have shown that protein phosphorylation might have a negative effect on meat color development by regulating glycolysis and the redox stability of myoglobin. The effects of phosphorylation on myoglobin stability were investigated in this study, which would provide theoretical basis for improving meat color stability through regulating protein phosphorylation level. The pure MetMb from skeletal muscle was used in this experiment. MetMb was reduced by sodium dithionite, which was then removed by ultrafiltration. After that, AP was added to catalyze the dephosphorylation of myoglobin in vitro. According to the results, the phosphorylation level of myoglobin was significantly lower ($P < 0.05$) in AP group than that in control group at 6 h, which indicates that AP can catalyze the dephosphorylation of myoglobin in vitro, thus decreasing the phosphorylation level of myoglobin. The relative content of oxymyoglobin in AP group was significantly higher than that in control group, and the relative content of MetMb was significantly lower than that in control group after 2 h. In brief, the automatic oxidation rate of myoglobin was lower and the redox stability of myoglobin was higher in AP group than that in control group. However, no significant difference ($P > 0.05$) was observed between AP group and control group, which meant that pH value of the incubation system was not changed by adding AP. The results showed that the secondary structure of myoglobin was mostly α-helix. From 0 min to 6 h incubation, the content of α-helix and β-sheet of myoglobin was almost unchanged in AP group, while the α-helix content of myoglobin increased and β-sheet content of myoglobin decreased in control group, indicating that the secondary structural stability of myoglobin was increased after dephosphorylation. It was speculated that the secondary structure of myoglobin might be changed after phosphorylation. The secondary structure stability was decreased and automatic oxidation rate of myoglobin was increased after phosphorylation, thus leading to the accumulation of MetMb and color deterioration. This might be one of the reasons by which protein phosphorylation played a negative role in regulating meat color stability.

9.3.1 Changes of Myoglobin Phosphorylation Level

Proteins are phosphorylated by kinase catalysis and dephosphorylated by phosphatase catalysis. AP can catalyze the dephosphorylation of proteins and reduce the phosphorylation level of proteins (Wang et al. 2014). The results of myoglobin phosphorylation levels were shown in Fig. 9.2. At 6 h of incubation, there was a significant difference in the phosphorylation level of myoglobin between the two groups. The phosphorylation level of myoglobin in AP treatment (AP group) was significantly lower than that of the control group ($P < 0.05$), indicating that AP

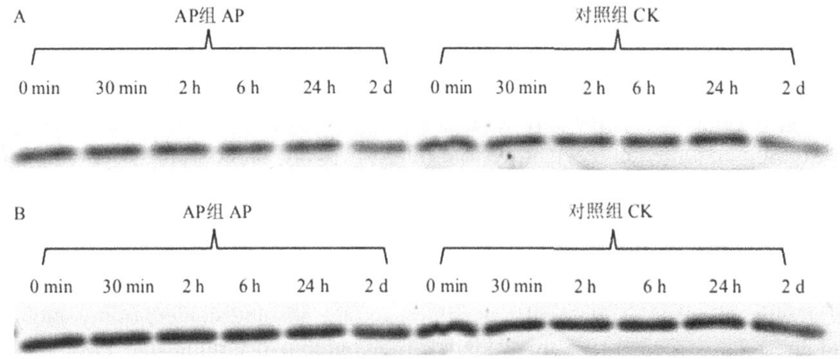

Fig. 9.2 Gel images of phosphorylated and total myoglobin by 1-DE. (**a**): Image of phosphorylated myoglobin stain with Pro-Diamond; (**b**): Image of total myoglobin stain with SYPRO Ruby. (Li et al. 2017b)

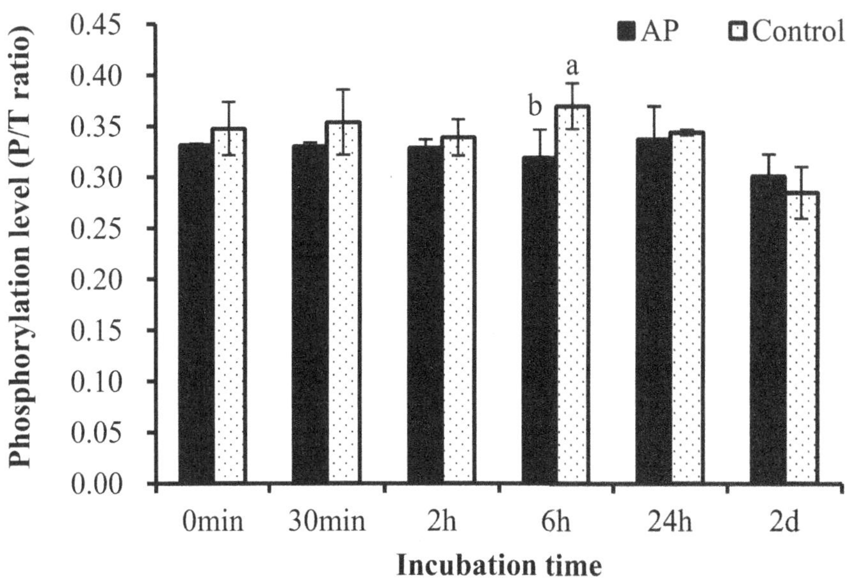

Fig. 9.3 Analysis of the phosphorylation level of myoglobin between AP group and control group. Data with different letters at the same time points are significantly different between two groups ($P < 0.05$). (Li et al. 2017b)

catalyzed the dephosphorylation of myoglobin and reduced the phosphorylation level of myoglobin (Fig. 9.3).

9.3.2 Changes of Myoglobin Relative Contents

Myoglobin consists of a molecule of globin (153 amino acids) and a heme cofactor (Kendrew 1963). The heme cofactor is composed of an iron porphyrin consisting of an iron ion and a tetramolecular pyrrole ring. There are six coordination bonds for iron ions, of which the sixth one can bind to some small molecules, such as oxygen, nitrogen, water, carbon monoxide, etc. Due to the different ligands bound by the sixth ligand of myoglobin, myoglobin exists in many states, including DMb, OxyMb, and MetMb. Myoglobin is the most important pigmentation substance in meat with sufficient bleeding after slaughter, and the absolute content of myoglobin and the relative content of three kinds of myoglobin determine the meat color (Bekhit and Faustman 2005). The relative contents of the three kinds of myoglobin were shown in Fig. 9.4.

At 0–30 min, DeoxyMb was the main component of the incubation system, and then DeoxyMb combined with oxygen to produce oxymyoglobin. At the same time, myoglobin was oxidized to produce MetMb. When incubated for 2 h, OxyMb and MetMb were dominant in the system. The content of OxyMb in the system decreased gradually and the content of MetMb increased gradually from 2 h. The relative content of DeoxyMb, OxyMb, and MetMb in the incubation system was significantly different between AP group and control group from 2 h after incubation ($P < 0.05$). Among them, the content of OxyMb in control group was significantly lower than that in AP group ($P < 0.05$), the content of DeoxyMb was significantly lower than that in AP group ($P < 0.05$), and the content of MetMb was significantly higher than that in AP group ($P < 0.05$). This indicated that compared with the control group, the content of reduced myoglobin in AP group was higher ($P < 0.05$), and the content of MetMb was lower ($P < 0.05$).

9.3.3 Changes of pH in Myoglobin Incubation System

pH is an important factor affecting the stability of myoglobin. In order to determine whether the pH of incubation system has an effect on the stability of myoglobin, the changes of the pH of incubation system during incubation were measured. The results showed that there was no significant difference in the pH between AP group and control group during the whole incubation process ($P > 0.05$) (Fig. 9.5). This indicated that the addition of AP did not change the pH of incubation system. Therefore, it was speculated that the difference of myoglobin stability between AP group and control group during incubation was not caused by pH.

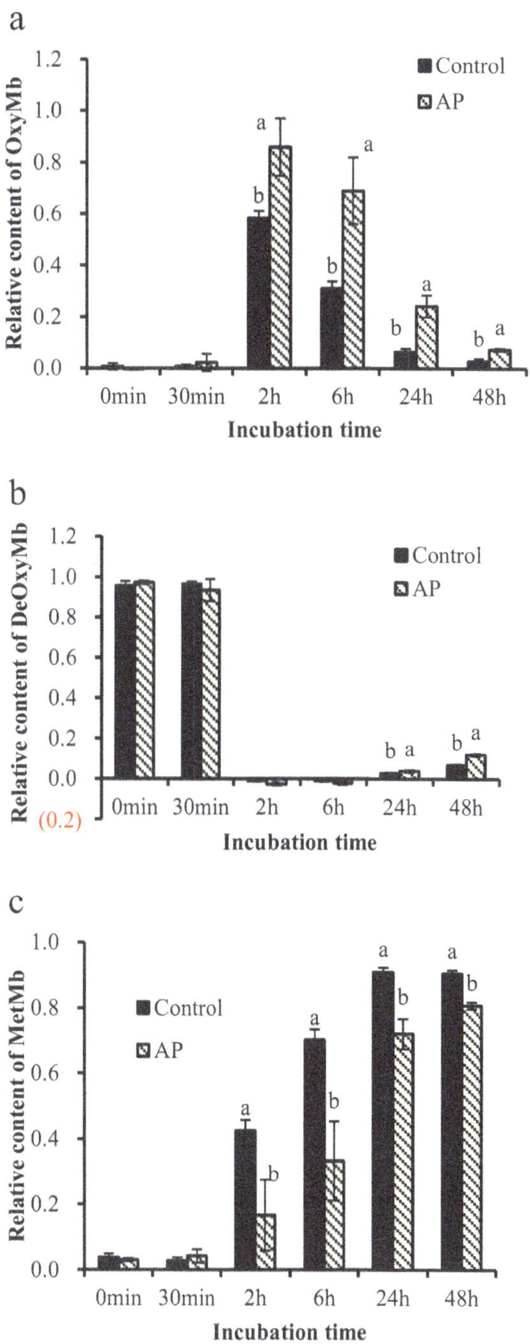

Fig. 9.4 Changes of the relative proportions of deoxymyoglobin, oxymyoglobin, and metmyoglobin over time. (**a**): Changes of the relative proportions of oxymyoglobin over time; (**b**): Changes of the relative proportions of deoxymyoglobin over time; (**c**): Changes of the relative proportions of metmyoglobin over time. Data with different letters at the same time points are significantly different between two groups ($P < 0.05$) (Li et al. 2017b)

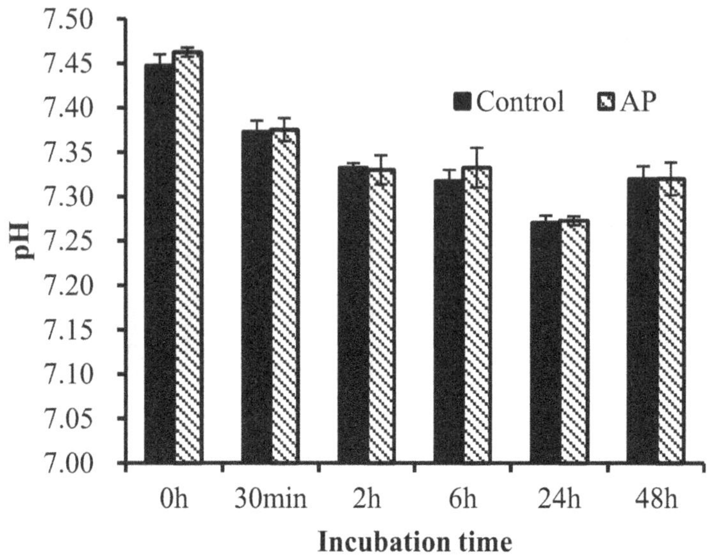

Fig. 9.5 Changes of the pH values in the incubation system over time. Data with different letters at the same time points are significantly different between two groups ($P < 0.05$) (Li et al. 2017b)

Table 9.2 Changes of myoglobin secondary structure (Li et al. 2017b)

		Time				
		0 min	30 min	6 h	24 h	48 h
α-helix (%)	AP	95.2	95	95.2	100	100
	Control	95.2	96.3	98.5	100	100
β-fold (%)	AP	4.9	5.03	4.56	0	0
	Control	4.9	3.75	1.3	0	0

9.3.4 Changes of Myoglobin Secondary Structure

The secondary structure of myoglobin is mainly α-helix (Bisig et al. 1995). Changes of secondary structure of myoglobin before and after incubation with AP were shown in Table 9.2. The secondary structure of myoglobin was mainly α-helix. When incubated for 30 min and 6 h, the content of α-helix in AP group was lower than that in control group, and the content of β-fold was higher than that in control group. At 24 and 48 h of incubation, the secondary structure of myoglobin in AP group and control group was mainly α-helix, and there was no difference in helix content between the two groups. From 0 to 6 h after incubation, the content of α-helix and β-fold of myoglobin in AP group changed slightly. The content of α-helix of myoglobin in control group increased and the content of β-fold decreased. This indicated that the secondary structure stability of myoglobin in AP group was higher than that in control group. It means that the secondary structure of myoglobin was affected by phosphorylation of myoglobin, and the secondary structure stability of myoglobin at low phosphorylation level was higher.

When DeoxyMb was dominant on the surface of meat, the meat was purple red or purple pink. When the surface of meat was mainly oxymyoglobin, the meat appeared to be bright red. When the surface of meat was predominantly MetMb, the meat was brown, so the production of MetMb was accompanied by the deterioration of meat color (Suman et al. 2007). The content of MetMb in AP group was significantly lower than that in control group, indicating that the automatic oxidation rate of myoglobin in AP group was lower than that in control group, and the redox stability was higher than that in control group. In addition, the phosphorylation level of myoglobin in AP group was lower than that in control group. It was speculated that the auto-oxidation rate of myoglobin in low phosphorylation level group was lower than that in high phosphorylation level group. It means that the redox stability of myoglobin at low phosphorylation level was higher than that of myoglobin at high phosphorylation level, which might negatively regulate the redox stability of myoglobin.

The declined rate and degree of pH postmortem is one of the most important factors affecting meat color (Scheffler and Gerrard 2007), and the redox rate of myoglobin is significantly affected by pH. The stability of globin and hemoglobin covalent binding bond in myoglobin molecule could be reduced at lower pH, which would increase the automatic oxidation rate of myoglobin and reduce the oxidation-reduction stability and make it easy to be oxidized. Gutzke and Trout (2002) found out that the oxidation rate of myoglobin was significantly affected by pH at different species and temperature. Shikama and Sugawara (1978) studied the effect of pH (4.8–2.6) on myoglobin oxidation. The results showed that the oxidation of myoglobin was directly affected by H^+ concentration. In this study, the pH of incubation system between AP group and control group had no difference, so the effect of pH on myoglobin stability could be excluded. That is, the difference of myoglobin stability between AP group and control group was not caused by pH. In addition, since incubation tests were conducted with the same materials and at the same ambient temperature, the effect of other external factors on the rate of myoglobin autoxidation could be excluded. It was speculated that the difference in stability between AP group and control group was due to the difference in phosphorylation level of myoglobin between the two groups.

Myoglobin is a monomer globulin that stores oxygen and transmits oxygen to mitochondria. Studies have shown that covalent or non-covalent modification was a common phenomenon to regulate the activity of monomer globulin (Ascenzi et al. 2013). For example, human neuroglobulin (monomer globulin) can be phosphorylated (covalent modification) to protect nerve cells under hypoxic conditions both in vivo and in vitro. It is speculated that the phosphorylation of myoglobin may be related to oxygen content, and alters the binding ability of myoglobin to oxygen, which in turn affects the existence of myoglobin. Studies have shown that phosphorylation can regulate the structure, activity, and stability of proteins. For example, PK can be transformed into an acid-stable isomer after phosphorylation, and then maintain high activity in PSE meat (Schwägele et al. 1996). The serine at the 14th position of GP can be phosphorylated, causing structural changes and activation (Sprang et al. 1988; Johnson 1992). After phosphorylation, fructokinase phosphate forms a complex with actin, which regulates the activity of fructokinase phosphate and provides energy for cells (Cai et al. 1997; Kuo et al. 1986). In addition, study by Diaz-Moreno et al. (2009) has shown that phosphorylation of KH-type splicing regulatory protein (KSRP) can extend the irregular structure in the unstable KH1 region, thereby altering the structure of the protein. Zhang and Liu (2017) showed that serine of (*Gossypium hirsutum* pyruvate kinase 6) GhPK6 was phosphorylated at positions 215 and 402. Phosphorylation of serine at positions 215 inhibited the activity of the enzyme, and phosphorylation at positions 215 and 402 promoted the degradation of the protein. In this study, the secondary structure stability of myoglobin with different phosphorylation levels was also different. Therefore, it is speculated that the difference of phosphorylation level of myoglobin between AP group and control group may lead to the difference of secondary structure of myoglobin, and the difference of structure may lead to the difference of automatic oxidation rate and redox stability of myoglobin. After phosphorylation or dephosphorylation of myoglobin, it may regulate its secondary structure, and then affect the binding state of myoglobin with oxygen, and ultimately affect the automatic oxidation rate and redox stability of myoglobin.

9.4 Conclusions

AP could catalyze dephosphorylation of myoglobin in vitro which decreased the phosphorylation level of myoglobin. The autoxidation rate of myoglobin decreased in AP treated group, i.e. the redox stability of myoglobin increased. The pH of the incubation system was not influenced by adding AP which excluded the possible effects of pH on myoglobin stability. The secondary structure of myoglobin increased after dephosphorylation compared with control group. According to the results from this study, it can be concluded that phosphorylation of myoglobin could decrease its stability of secondary structure and result in decreased myoglobin redox stability and meat color, and this could be a possible mechanism that protein phosphorylation negatively regulated meat color stability.

Acknowledgments Parts of this chapter are reprinted from Food Chemistry, 219, Li, M., et al., Effects of protein phosphorylation on color stability of ground meat, 304-310, Copyright (2020), with permission from Elsevier. Parts of this chapter are translated from Scientia Agricultura Sinica, 50(22), Li, M., et al., Effect of phosphorylation level on myoglobin stability (Chinese), 4382-4388. Copyright (2020), with permission from Journal of Scientia Agricultura Sinica.

References

AMSA. (2012). *Meat color measurement guidelines (M. Hunt & D. A. King, writing committee co-chairs)*. Savoy, IL: American Meat Science Association.

Ascenzi, P., Marino, M., Polticelli, F., Coletta, M., Gioia, M., Marini, S., et al. (2013). Non-covalent and covalent modifications modulate the reactivity of monomeric mammalian globins. *Biochimica et Biophysica Acta, 1834*(9), 1750–1756.

Bekhit, A. E., & Faustman, C. (2005). Metmyoglobin reducing activity. *Meat Science, 71*(3), 407–439.

Bisig, D. A., Di Iorio, E. E., Diederichs, K., Winterhalter, K. H., & Piontek, K. (1995). Crystal structure of Asian elephant (Elephas maximus) cyano-metmyoglobin at 1.78-A resolution. Phe29 (B10) accounts for its unusual ligand binding properties. *The Journal of Biological Chemistry, 270*(35), 20754–20762.

Cai, G. Z., Callaci, T. P., Luther, M. A., & Lee, J. C. (1997). Regulation of rabbit muscle phosphofructokinase by phosphorylation. *Biophysical Chemistry, 64*(1–3), 199–209.

Canto, A. C., Suman, S. P., Nair, M. N., Li, S., Rentfrow, G., Beach, C. M., et al. (2015). Differential abundance of sarcoplasmic proteome explains animal effect on beef Longissimus lumborum color stability. *Meat Science, 102*, 90–98.

Chen, L. J., Li, X., Ni, N., Liu, Y., Chen, L., Wang, Z. Y., et al. (2016). Phosphorylation of myofibrillar proteins in post-mortem ovine muscle with different tenderness. *Journal of the Science of Food and Agriculture, 96*(5), 1474–1483.

Diaz-Moreno, I., Hollingworth, D., Frenkiel, T. A., Kelly, G., Martin, S., Howell, S., et al. (2009). Phosphorylation-mediated unfolding of a KH domain regulates KSRP localization via 14-3-3 binding. *Nature Structural & Molecular Biology, 16*(3), 238–246.

Gutzke, D., & Trout, G. R. (2002). Temperature and pH dependence of the autoxidation rate of bovine, ovine, porcine, and cervine oxymyoglobin isolated from three different muscles longissimus dorsi, gluteus medius, and biceps femoris. *Journal of Agricultural and Food Chemistry, 50*(9), 2673–2678.

Huang, H. G., Larsen, M. R., Karlsson, A. H., Pomponio, L., Costa, L. N., & Lametsch, R. (2011). Gel-based phosphoproteomics analysis of sarcoplasmic proteins in postmortem porcine muscle with pH decline rate and time differences. *Proteomics, 11*(20), 4063–4076.

Huang, H. G., Larsen, M. R., & Lametsch, R. (2012). Changes in phosphorylation of myofibrillar proteins during postmortem development of porcine muscle. *Food Chemistry, 134*(4), 1999–2006.

Johnson, L. N. (1992). Glycogen phosphorylase: Control by phosphorylation and allosteric effectors. *FASEB Journal, 6*(6), 2274–2282.

Joseph, P., Suman, S. P., Rentfrow, G., Li, S. T., & Beach, C. M. (2012). Proteomics of muscle-specific beef color stability. *Journal of Agricultural and Food Chemistry, 60*(12), 3196–3203.

Kendrew, J. C. (1963). Myoglobin and the structure of proteins. *Science, 139*, 1259–1266.

Kuo, H. J., Malencik, D. A., Liou, R. S., & Anderson, S. R. (1986). Factors affecting the activation of rabbit muscle phosphofructokinase by actin. *Biochemistry, 25*(6), 1278–1286.

Li, M., Li, X., Xin, J. Z., Li, Z., Li, G. X., Zhang, Y., et al. (2017a). Effects of protein phosphorylation on color stability of ground meat. *Food Chemistry, 219*, 304–310.

Li, M., Li, Z., Li, X., Du, M. T., Song, X., & Zhang, D. Q. (2017b). Effect of phosphorylation level on myoglobin stability. *Scientia Agricultura Sinica, 50*(22), 4382–4388. (in Chinese).

Mancini, R. A., & Hunt, M. C. (2005). Current research in meat color. *Meat Science, 71*(1), 100–121.

Scheffler, T. L., & Gerrard, D. E. (2007). Mechanisms controlling pork quality development: The biochemistry controlling postmortem energy metabolism. *Meat Science, 77*(1), 7–16.

Schwägele, F., Haschke, C., Honikel, K. O., & Krauss, G. (1996). Enzymological investigations on the causes for the PSE-syndrome, I. Comparative studies on pyruvate kinase from PSE- and normal pig muscles. *Meat Science, 44*(1–2), 27–40.

Shikama, K., & Sugawara, Y. (1978). Autoxidation of native oxymyoglobin. Kinetic analysis of the pH profile. *European Journal of Biochemistry, 91*(2), 407–413.
Sprang, S. R., Acharya, R., Goldsmith, E. J., Stuart, D. I., Varvill, K., Fletterick, R. J., et al. (1988). Structural changes in glycogen phosphorylase induced by phosphorylation. *Nature, 336*(6196), 215–221.
Suman, S. P., Faustman, C., Stamer, S. L., & Liebler, D. C. (2007). Proteomics of lipid oxidation-induced oxidation of porcine and bovine oxymyoglobins. *Proteomics, 7*(4), 628–640.
Wang, X., Ni, M., Niu, C., Zhu, X., Zhao, T., Zhu, Z., et al. (2014). Simple detection of phosphoproteins in SDS-PAGE by quercetin. *EuPA Open Proteomics, 4*, 156–164.
Zhang, B., & Liu, J. Y. (2017). Serine phosphorylation of the cotton cytosolic pyruvate kinase GhPK6 decreases its stability and activity. *FEBS Open Biology, 7*(3), 358–366.

Part III
Improvement of Meat Quality by Regulating Protein Phosphorylation

Chapter 10
Effects of Temperature on Protein Phosphorylation

Abstract Extensive studies have revealed that protein phosphorylation plays an important role in postmortem meat quality, such as tenderness, color, and water holding capacity. During storage of meat, the storage temperature is one of the most important factors that determine meat quality. High temperature promotes meat spoilage by accelerating biochemical reactions and microbial growth. However, whether storage temperature affects the development of meat quality through phosphorylation is not well understood. In order to investigate it clearly, the postmortem ovine muscle samples were stored at 25 °C, 15 °C, 4 °C, and −1.5 °C from 0.75 h to 21 days. The correlation between temperature, pH, adenosine triphosphate (ATP) content, and phosphorylation levels was analyzed, and then differential phosphorylated proteins were identified by using liquid chromatography-tandem mass spectrometry (LC-MS/MS) and gene ontology analysis. Myosin binding protein C, troponin T3, actinin, myosin light chain, heat shock protein 90, glucose-6-phosphate, pyruvate kinase, enolase, fructose-bisphosphate aldolase, L-lactate dehydrogenase, etc. were phosphorylated in postmortem ovine muscles. Moreover, ATP content was a vital factor which had the strongest correlation with the phosphorylation level of individual protein band. In order to prove whether ATP content mainly affected phosphorylation level, ATP was added to ground ovine muscle in vitro, which showed that ATP improved the phosphorylation level of myofibrillar proteins. The phosphorylation reaction was reversible. However, ATP was unnecessary for dephosphorylation. Whether the particular temperature and pH (5.2, 5.8, 6.4) had effect on the dephosphorylation of myofibrillar protein through alkaline phosphatase are need to be studied. The result showed that −1.5 °C and pH 5.2 restrained the activity of alkaline phosphatase and decreased phosphorylation level. Phosphorylation was an essential post-translational modification and more insights were needed for the complicated formation in postmortem muscle.

Keywords Temperature · pH · ATP content · Phosphorylation · Dephosphorylation · Myofibrillar proteins · Sarcoplasmic proteins · Alkaline phosphatase · Postmortem muscle

D. Zhang et al., *Protein Phosphorylation and Meat Quality*,
https://doi.org/10.1007/978-981-15-9441-0_10

10.1 Introduction

Protein phosphorylation is one of the most important post-translational modifications in organisms. Recently, some researches have found out that phosphorylation or dephosphorylation regulates the traits of meat quality, such as tenderness, color, and water holding capacity (D'Alessandro and Zolla 2013; D'Alessandro et al. 2012a; Gao et al. 2017; Li et al. 2017; Li et al. 2018). Besides, the storage temperature is the direct factor that contributes to the shortening, rigor, and aging of postmortem muscle (Fernandez and Tornberg 1994; Kim et al. 2014). Thus, the mechanism of how storage temperature regulates meat quality by protein phosphorylation needs further study.

The reactions of phosphorylation and dephosphorylation are catalyzed by protein kinase and phosphatase, respectively. With the presence of phosphate group provided by ATP, proteins are phosphorylated under the catalysis of protein kinases. And the reversible reaction is catalyzed by phosphatase. Besides, the temperature and pH both can affect the activity of protein kinases and phosphatases. But in postmortem muscle, which one of temperature, pH, and/or ATP mainly affects protein phosphorylation level is still unclear. According to the result of Ren et al. (2020), ATP might be more related to individual protein band, although the global phosphorylation level of sarcoplasmic proteins was stable. With ATP depletion, the phosphorylation will be restrained, but how dephosphorylation changes with different temperatures and pH remains unknown. In the complex system of postmortem muscle, a more comprehensive understanding of phosphorylation or dephosphorylation can help us better understand the development of muscle to meat.

10.2 Effects of Temperature on Protein Phosphorylation in Postmortem Muscle

Postmortem ovine muscle was stored at 25 °C (hot fresh meat is commonly sold at this temperature in China), 15 °C, 4 °C, and −1.5 °C (super chilling temperature). From the correlation analysis of temperature, pH, ATP content, and phosphorylation level, eight myofibrillar protein bands and five sarcoplasmic protein bands were selected for LC-MS/MS.

10.2.1 Effect of Temperature on Glycolysis in Postmortem Muscle

The rate and extent of pH decline, along with metabolism in postmortem, were vital for meat quality (Gratacos-Cubarsi and Lametsch 2008). The pH values of muscles stored at different temperatures are presented in Fig. 10.1a. Generally, pH declined

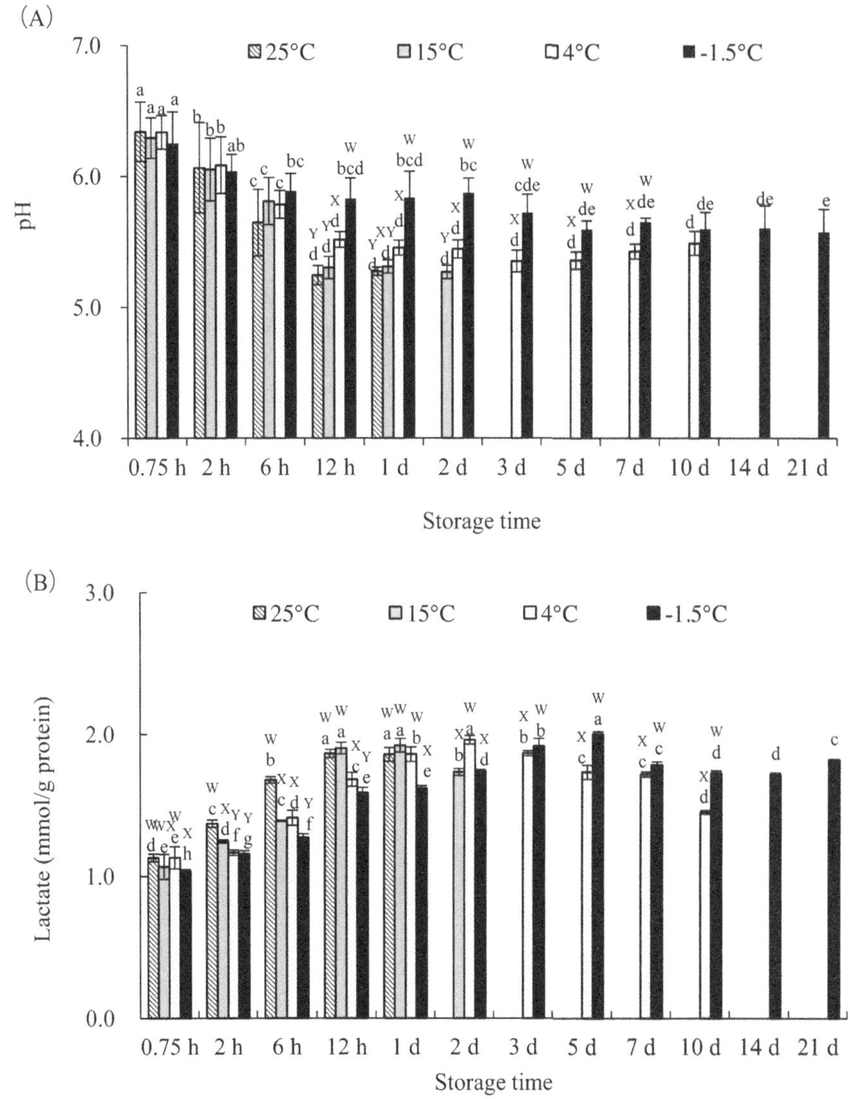

Fig. 10.1 The pH values (**a**) and lactate content (**b**) in ovine muscles stored at different temperatures. Data are means ± standard deviation ($n = 4$). Different lowercase letters represent significant difference between different time points of the same storage temperature ($P < 0.05$). Different capital letters represent significant difference between different storage temperatures at the same time point ($P < 0.05$) (Ren et al. 2020)

quickly at early stage and later kept relatively constant postmortem. For muscles stored at 4 °C and −1.5 °C, pH was higher than those stored at 25 °C and 15 °C at 12 h, 1 day, and 2 days ($P < 0.05$). This result suggested higher temperature promoted glycolysis, contributing to fast pH decline (Scheffler and Gerrard 2007).

The ultimate pH of −1.5 °C muscle was significantly higher than that of the other three treatments ($P < 0.05$) and reached ultimate pH later than others. These data demonstrated that −1.5 °C restrained glycolysis compared with other storage temperatures.

In postmortem muscle, anaerobic glycolysis was the major reason to decrease pH by generating lactate and H^+ (Huang et al. 2011). Lowering pH value could enhance the effect of lactate in inhibiting glycolysis (Liu et al. 2015). The lactate content first increased and then decreased slightly, which was consistent with the change of pH value. Compared with other temperatures, the lactate content of −1.5 °C was significantly lower within 2 days and higher after 3 days ($P < 0.05$, Fig. 10.1b). The lactate content increased and reached the maximum value in 12 h at 25 °C and 15 °C. However, at 4 °C and −1.5 °C, the highest lactate content appeared in 2 days and 5 days, respectively, which was later than 25 °C and 15 °C. This result indicated that the rate of lactate accumulation was faster in high temperature and low temperature inhibited lactate accumulation by limiting glycolysis.

10.2.2 Effect of Temperature on ATP Content in Postmortem Muscle

High ATP consumption caused more metabolic flux, resulting in shortened rigor mortis time (Savell et al. 2005). As shown in Fig. 10.2, ATP content was reduced

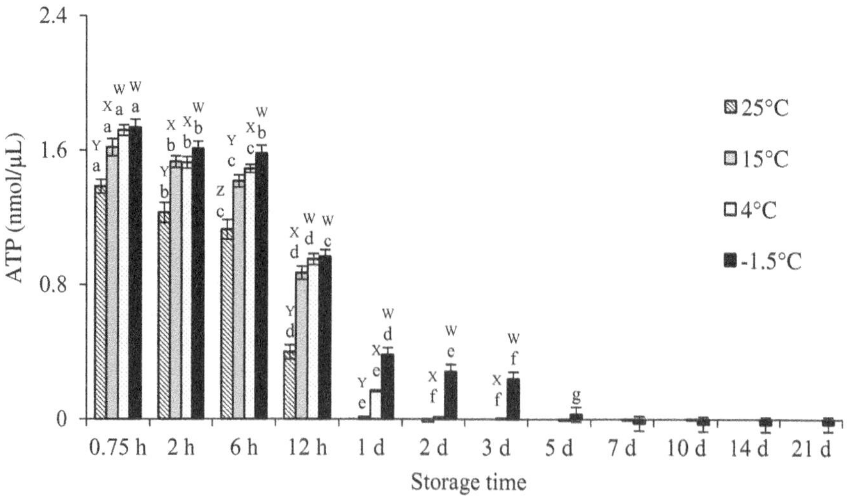

Fig. 10.2 The ATP content in ovine muscle during storage. Data are means ± standard deviation ($n = 4$). Different lowercase letters represent significant difference between different time points of the same storage temperature ($P < 0.05$). Different capital letters represent significant difference between different storage temperatures at the same time point ($P < 0.05$). (Ren et al. 2020)

gradually, which was in agreement with previous studies (Shen et al. 2006; Wang et al. 2013). ATP was exhausted in 1, 2, 5, and 7 days when muscles were stored at 25 °C, 15 °C, 4 °C, and −1.5 °C, respectively. At 0.75 h, the ATP content of muscles stored at 25 °C and 15 °C was lower than that of muscles stored at 4 °C and −1.5 °C ($P < 0.05$), which showed that ATP has been used up quickly in higher temperature during early postmortem time. Some studies reported that ATP could be produced only by anaerobic glycolysis in postmortem muscle, nevertheless, there were multiple processes to consume ATP, such as muscle contraction, transportation of ions, and even protein phosphorylation (Hudson, 2012; Lametsch et al. 2002). Higher temperature could promote anaerobic glycolysis, however, at the same time it would accelerate pathways that increase ATP content. As a result, the ATP content decreased with storage time.

10.2.3 Effect of Temperature on Global Phosphorylation Levels in Postmortem Muscle

Dodecyl sulfate sodium salt and polyacrylamide gel electrophoresis (SDS-PAGE) and fluorescence staining were used to detect phosphoproteins. Pro-Q Diamond and SYPRO Ruby staining images of myofibrillar proteins and sarcoplasmic proteins are shown in Fig. 10.3a, b. Seventeen individual myofibrillar protein bands and 20 individual sarcoplasmic protein bands were selected to calculate global phosphorylation level (Fig. 10.3c).

As shown in Fig. 10.3C-1, the protein phosphorylation level of myofibrillar proteins decreased at 25 °C and 15 °C, which increased early stage and then decreased when muscles were kept at 4 °C and −1.5 °C. The phosphorylation level in the four groups showed significant difference at 6 h, 12 h, and 1 day ($P < 0.05$). After 1 day storage at 4 °C, the phosphorylation level increased to the highest value and then decreased. Same changes happened when muscles were stored at −1.5 °C, and it kept the high level of protein phosphorylation for a longer time. The main reason might relate to ATP content. At high temperature (25 °C, 15 °C), the consumption of ATP was faster, which would inhibit the phosphorylation reaction. However, the consumption of ATP would not affect the dephosphorylation reaction. As a result, the total phosphorylation level decreased gradually at 25 °C and 15 °C. Conversely, lower temperature (4 °C, −1.5 °C) restrained ATP consumption, making the phosphorylation last longer, and the changes in myofibrillar proteins phosphorylation level were different. Thus, the most important impact factor for phosphorylation level of myofibrillar proteins might be the consumption of ATP postmortem.

The phosphorylation level of sarcoplasmic proteins is shown in Fig. 10.3C-2. Generally, the phosphorylation level increased at early storage time and then kept stable. The phosphorylation level was significantly different ($P < 0.05$) at 1 and 2 days, with no difference ($P > 0.05$) at other storage time. It was interesting that the

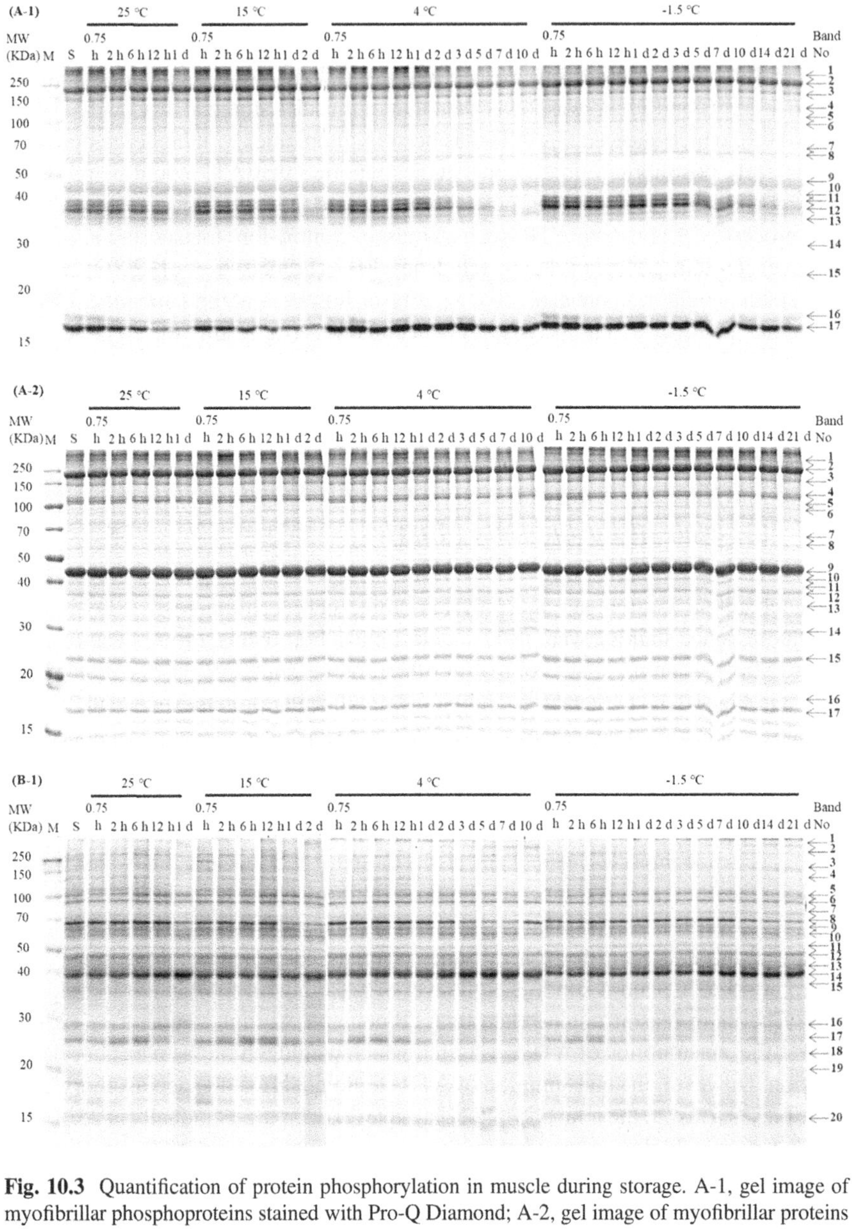

Fig. 10.3 Quantification of protein phosphorylation in muscle during storage. A-1, gel image of myofibrillar phosphoproteins stained with Pro-Q Diamond; A-2, gel image of myofibrillar proteins stained with SYPRO Ruby; B-1, gel image of sarcoplasmic phosphoproteins stained with Pro-Q Diamond; B-2, gel image of sarcoplasmic proteins stained with SYPRO Ruby. The global phosphorylation level of myofibrillar proteins (C-1) and sarcoplasmic proteins (C-2) in muscles stored at different temperatures. Data are mean ± standard deviation (n = 4). Different lowercase letters represent significant difference between different time points of the same storage temperature ($P < 0.05$). Different capital letters represent significant difference between different storage temperatures at the same time point ($P < 0.05$) (Ren et al. 2020)

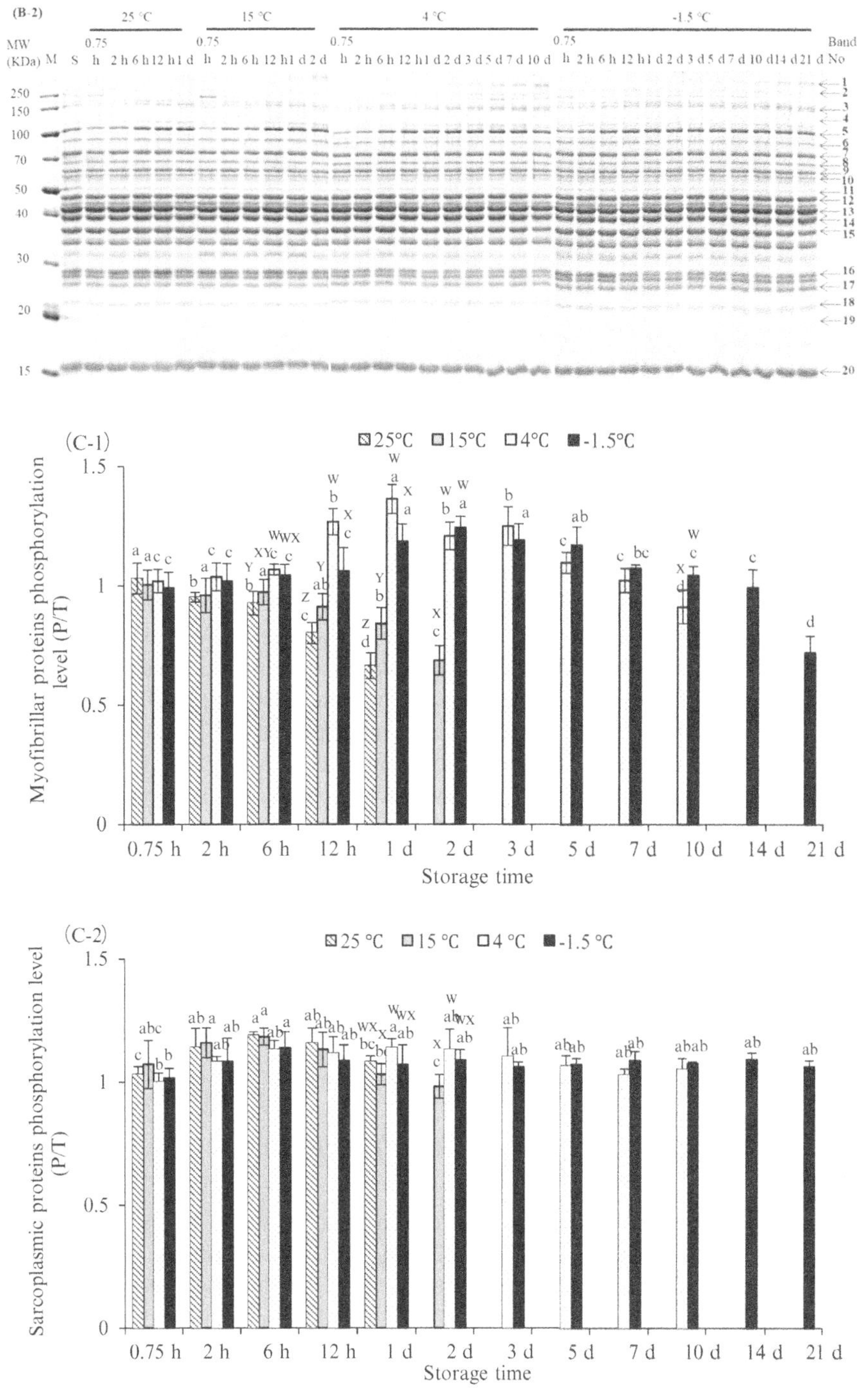

Fig. 10.3 (continued)

changes in phosphorylation level between myofibrillar proteins and sarcoplasmic proteins were diverse. The system in postmortem muscle was complicated. As for myofibrillar proteins and sarcoplasmic proteins of postmortem muscle, phosphorylation and dephosphorylation might happen at different time. Sarcoplasmic proteins might have phosphorylation reaction earlier than myofibrillar proteins. It meant that sarcoplasmic proteins would preferentially utilize ATP. Therefore, sarcoplasmic proteins could keep total phosphorylation level stable rather than create drastic changes like myofibrillar proteins.

10.2.4 Association between Phosphorylation Levels and Temperature, pH, ATP Content

The association between global phosphorylation level or phosphorylation level of individual protein bands and temperature, pH, ATP content was analyzed (Table 10.1). The global phosphorylation level of myofibrillar proteins was negatively correlated ($P < 0.01$) with temperature ($r = -0.548$), while the global phosphorylation level of sarcoplasmic proteins did not correlate ($P > 0.05$) with temperature ($r = 0.277$). There was no significant relation ($P > 0.05$) between global phosphorylation and pH, ATP content for either myofibrillar proteins or sarcoplasmic proteins. As for myofibrillar proteins, the phosphorylation level of bands 13, 14 was positively correlated ($P < 0.05$) with temperature and ATP content, but not ($P > 0.05$) with pH value. The phosphorylation level of bands 5, 7, 10, 12, 13 and 14 was positively correlated ($P < 0.05$) with ATP content. In terms of sarcoplasmic proteins, the phosphorylation level of bands 3, 5, 7, 8, 10, 11, 12, 13, 14, 18, and 19 was only related to pH and ATP content ($P < 0.05$). The association analysis of individual bands suggested that temperature, pH, and ATP content had different influences on protein phosphorylation.

According to association analysis, bands 3, 4, 5, 7, 12, 13, 14, and 15 of myofibrillar proteins and bands 10, 11, 12, 14, and 18 of sarcoplasmic proteins were selected and subjected to protein identification.

10.2.5 Identification of Individual Protein Bands

The 8 myofibrillar protein bands and five sarcoplasmic protein bands were chosen and identified by LC-MS/MS (Table 10.2). The proteins with unique peptide count ≥ 2 were selected for analysis. Gene ontology (GO) analysis was used to classify these proteins into biological process (BP), molecular function (MF), and cellular component (CC) groups.

As for myofibrillar proteins, myosin binding protein C (slow type), myosin binding protein C (fast type), and myomesin 2 were identified in band 3. TPM1,

Table 10.1 Pearson correlation coefficients of the temperature, pH, and ATP content with global phosphorylation levels

P/T ratio	Temperature	pH	ATP content
(A)			
Global	−0.548**	0.162	0.004
Band 1	−0.01	−0.157	−0.098
Band 2	0.414	−0.338	−0.323
Band 3	0.608**	−0.396*	−0.023
Band 4	−0.166	−0.409*	−0.485**
Band 5	0.318	0.436*	0.468**
Band 6	−0.429*	−0.338	−0.6**
Band 7	−0.479**	−0.134	0.345*
Band 8	−0.068	−0.361*	−0.253
Band 9	−0.231	0.155	0.092
Band 10	0.297	0.339	0.57**
Band 11	0.328	−0.264	−0.236
Band 12	0.203	0.466**	0.709**
Band 13	0.523**	0.219	0.503**
Band 14	0.417*	−0.077	0.356*
Band 15	0.175	−0.793**	−0.658**
Band 16	−0.422*	−0.08	−0.131
Band 17	−0.735**	0.157	−0.074
(B)			
Global	0.277	−0.113	0.211
Band 1	0.568**	0.207	0.642**
Band 2	0.252	0.29	0.515**
Band 3	−0.102	0.364*	0.564**
Band 4	0.255	−0.002	0.388*
Band 5	0.107	0.538**	0.434*
Band 6	0.305	−0.164	−0.198
Band 7	−0.22	−0.366*	−0.532**
Band 8	−0.07	0.452**	0.422*
Band 9	0.316	0.142	0.467**
Band 10	0.148	−0.647**	−0.699**
Band 11	−0.202	−0.678**	−0.771**
Band 12	0.226	0.688**	0.867**
Band 13	0.088	−0.355*	−0.516**
Band 14	−0.062	−0.503**	−0.682**
Band 15	−0.582**	0.002	−0.299
Band 16	0.347*	−0.036	0.012
Band 17	0.406*	0.247	0.693**
Band 18	−0.244	−0.523**	−0.731**
Band 19	0.402*	0.401*	0.718**
Band 20	0.066	0.309	0.49**

Note: (A) shows correlation coefficients of the temperature, pH, and ATP content with myofibrillar proteins phosphorylation; (B) shows correlation coefficients of the temperature, pH, and ATP content with sarcoplasmic proteins phosphorylation; Significance levels: *, $0.01 < P < 0.05$; **, $P \leq 0.01$. (Ren et al. 2020)

Table 10.2 Unique proteins identified from SDS-PAGE (Ren et al. 2020)

Protein number	Protein name	Band	UniProt ID	Peptide Count	Unique PepCount	Cover Percent	MW (Da)	PI
1	Myosin binding protein C, slow type	3	W5Q0I1	62	42	33.99%	126,221	6
2	Myosin binding protein C, fast type	3	W5PX04	56	38	26.82%	127,412	6.28
3	Myomesin 2	3	W5P6W4	38	35	22.66%	167,003	5.92
4	Actinin alpha 3	4	W5P9P1	101	46	42.25%	102,521	5.31
5	Actinin alpha 2	4	W5P0L8	93	42	42.52%	104,062	5.28
6	Actinin alpha 1	4	W5QJ62	27	11	10.72%	101,428	5.41
7	Actinin alpha 4	4	W5P707	24	11	11.52%	101,401	5.46
8	Heat shock protein 90 beta family member 1	5	W5Q4E1	13	13	12.66	92,678	4.77
9	Myotilin	7	W5Q1Q5	15	12	22.24%	55,667	9.11
10	Annexin A2	12	A2SW69	15	14	38.05%	38,612	6.92
11	Uncharacterized protein (gene name = CAPZA2)	12	W5NPQ4	6	6	12.78%	36,331	6.77
12	TPM1	13	B2LU28	10	10	28.52%	32,694	4.69
13	Four and a half LIM domains 1 protein	13	C8BKC8	11	9	22.30%	33,609	8.83
14	Four and a half LIM domains 3	13	W5QFC3	4	4	10.63%	33,341	5.93
15	Prohibitin	14	W5P7Q3	9	9	26.47%	29,804	5.57
16	14-3-3 protein gamma	14	W5PWD6	8	8	27.11%	25,736	4.76
17	Troponin T3, fast skeletal type	14	W5NRC7	7	7	19.73%	25,648	10.14
18	Troponin I type 1 variant X1	15	A0A1X9H6E3	32	16	54.55%	21,663	9.68
19	Myosin light chain 1	15	W5QD16	42	15	68.02%	21,487	5.04
20	Myosin light chain 3	15	W5PAN7	5	4	17.59%	21,931	4.99
21	Myosin light chain 6	15	W5PK27	4	3	10.75%	23,442	5.14
22	Heat shock protein family B (small) member 8	15	W5NUW9	3	3	13.61%	21,089	5
23	Glucose-6-phosphate	10	W5P323	23	14	17.95%	62,890	7.35

24	Pyruvate kinase	10	W5QC41	11	9	12.87%	61,591	8.2
25	Eukaryotic translation elongation factor 1 alpha 2	11	W5PN24	17	12	17.58%	52,215	9.28
26	Elongation factor 1-alpha	11	W5PD15	8	7	14.07%	50,140	9.1
27	Enolase 3	12	W5P663	99	25	43.09%	47,068	7.6
28	Enolase 2	12	W5P5C0	34	9	17.05%	47,240	4.94
39	Enolase 1	12	W5PIG7	17	6	12.42%	48,195	6.44
30	Fructose-bisphosphate aldolase	14	W5P1X9	87	28	60.44%	39,436	8.45
31	L-lactate dehydrogenase	14	W5PIN4	8	8	19.11%	39,742	8.58
32	Adenylate kinase isoenzyme 1	18	C5IJA8	23	12	38.14%	21,650	8.4
33	Peroxiredoxin 2	18	C8BKC5	10	6	24.24%	21,845	5.23
34	Lactoylglutathione lyase	18	W5Q0T	3	3	11.17%	21,095	5.22

Note: All protein bands were detected by LC-MS/MS spectra and analyzed by MASCOT engine against the UniProt database. Peptide Count is the number of peptides identified by LC-MS/MS. Unique PepCount is the number of unique peptide count. Cover Percent means the number of amino acids spanned by the assigned peptides divided by the sequence length. Molecular weight (MW) and isoelectric point (pI) are determined using Compute PI/MW tool (http://web.expasy.org/compute_pi/). Protein number 1–22 were for about myofibrillar proteins and 23–34 were for about sarcoplasmic proteins four and a half LIM domains 1 protein, four and a half LIM domains 3 were identified in band 13. Band 14 contained prohibitin, 14-3-3 protein gamma, and troponin T3 (fast skeletal type). The individual phosphorylation level of bands 3, 13 and 14 increased first and then decreased in all four groups. And those proteins in band 3 were relevant with muscle contraction [GO: 0006936]. It meant high phosphorylation level of those proteins in band 3 was related to muscle rigor mortis. Meanwhile, actinin alpha 3 (band 4), actinin alpha 2 (band 4), troponin T3 (band 14), myosin light chain 1 (band15), and myosin light chain 3 (band15) also regulated muscle contraction (D'Alessandro et al. 2012b; Muroya et al. 2007). Bands 13 and 14, functioning in metal ion binding, were positively related to temperature and ATP content. Actinin alpha 3, actinin alpha 2, actinin alpha 1, actinin alpha 4, heat shock protein 90 beta family member 1, myotilin, annexin A2, uncharacterized protein (gene name = CAPZA2), troponin I type 1 variant X1, myosin light chain 1, myosin light chain 3, myosin light chain 6, and heat shock protein family B (small) member 8 were identified in bands 4, 5, 7, 12, and 15, respectively. The changes in phosphorylation level of these five bands were different among four different temperatures. Heat shock protein 90 beta family member 1 is an ATP-dependent chaperone and its phosphorylation level was positively related to ATP content.

As for sarcoplasmic proteins, glucose-6-phosphate and pyruvate kinase were identified in band 10, and enolase 1, enolase 2, enolase 3 were identified in band 12. These proteins were glycolytic enzymes [GO: 0006096]. When muscles were stored at 25 °C and 15 °C, the changes in phosphorylation level of these two bands were inconsistent. The major reason might be that band 10 was negatively related to pH and ATP content, while band 12 was positively related to pH and ATP content. Researchers reported that, when glycolysis proteins were phosphorylated, some of them regulated postmortem glycolysis and bound to muscle filaments (Sola-Penna et al. 2010; Shen and Du 2005). Schwagele et al. (1996) showed that phosphorylated pyruvate kinase transformed to a more stirring and acid-stable isoform. The phosphorylation level of pyruvate kinase was higher at 25 °C and 15 °C, which indicated that phosphorylated pyruvate kinase might be related to accelerated glycolysis. The activity of enolase 3 was associated with the regeneration and development of muscle (Merkulova et al. 2000). Together with this finding, the phosphorylation level of band 10 being negatively related to ATP content, it could be inferred that high phosphorylation level of enolase 3 contributed to ATP consumption and the transformation of muscle to meat. Eukaryotic translation elongation factor 1 alpha 2 and elongation factor 1-alpha, relating to GTPase activity [GO: 0003924], were mainly detected in band 11. The phosphorylation level of band 11 was increased in all muscle in the present study, which was opposite to the decreased ATP content. Besides, fructose-bisphosphate aldolase and L-lactate dehydrogenase were identified in band 14. Adenylate kinase isoenzyme 1, peroxiredoxin 2, and lactoylglutathione lyase were identified in band 18. Lactoylglutathione lyase was related to apoptotic process [GO: 0043066]. During the later stage of storage in the present study, increased phosphorylation level of lactoylglutathione lyase might promote the apoptotic process.

The identified 13 individual bands were intensively related to temperature, pH, and ATP content. And the phosphorylation level changes of individual band were inconsistent. Myosin binding protein C, troponin T3, myosin light chain 1, heat shock protein family B, glucose-6-phosphate, enolase 3, fructose-bisphosphate aldolase, and other proteins were identified, and these proteins were mainly involved in glycolysis and muscle contraction. It could be proved that storage temperature played a vital role in biochemical process and changed phosphorylation level of many proteins in postmortem muscle.

10.3 Effect of ATP on Protein Phosphorylation in Postmortem Muscle with Different Temperatures

The result of Table 10.1 showed that ATP content might be the main factor affecting phosphorylation level. In order to testify this result, the purified ATP and the same volume of solution without ATP were added to ground meat, respectively, which were considered as the ATP treatment group and the control group. Then ground meat was stored at 25 °C, 15 °C, 4 °C, and −1.5 °C. The ATP content, pH, and global phosphorylation level were determined to instigate whether ATP was a key factor of phosphorylation.

10.3.1 Effect of Temperature on ATP Content in Postmortem Muscle

Energy metabolism is an important biochemical process in postmortem muscle. ATP, a substance providing energy, has been widely reported to be correlated with rigor mortis and glycolysis (Watabe et al. 1991; Hamm 1977). Generally, ATP content was decreased with the increase of storage time in all treatment groups (Fig. 10.4). After slaughtering, aerobic glycolysis stopped and anaerobic glycolysis initiation made ATP synthesis reduce largely. The increasing consumption and decreasing synthesis of ATP led to the total ATP content depleted in postmortem muscle. Besides, the decomposition of ATP would affect the biochemical mechanism in postmortem. Adenosine monophosphate (AMP), one of catabolites of ATP, was linked to the activity of glycolytic enzymes. When AMP content was sufficient, glycogen was degraded to glucose-l-phosphate by phosphorylase b. And the activity of phosphor fructokinase (PFK) could be accelerated by AMP (Aberle and Merkel 1968).

Except for 1 day at 25 °C, ATP content in ATP treated group was higher ($P < 0.05$) than the control group at all time of both groups. The reason for no significant difference of 1 d at 25 °C was that ATP content could not meet the need of ATP consumption. However, the ATP content at 1 day of 4 °C was enough to

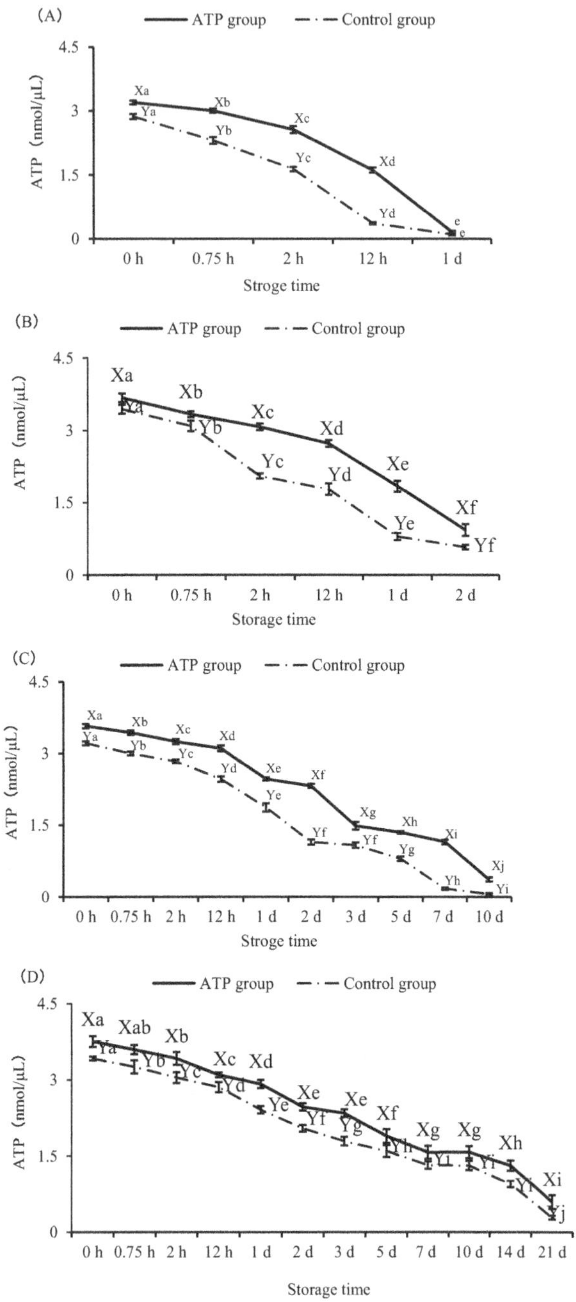

Fig. 10.4 The changes of ATP content in ground muscle at 25 °C (**a**), 15 °C (**b**), 4 °C (**c**), and − 1.5 °C (**d**). Different lowercase letters represent significant difference within the same group ($P < 0.05$). At the same storage time, different capital case letters represent significant difference between ATP treated group and the control group ($P < 0.05$)

maintain ATP consumption. The ATP of 25 °C and 15 °C was almost completely consumed at 1–2 days, but it still existed at 10–21 days of 4 °C and −1.5 °C, which suggested ATP consumption was slower at lower temperature (Li et al. 2015) because of repressed metabolic processes.

10.3.2 *Effect of ATP on pH in Postmortem Muscle*

As shown in Fig. 10.5, pH values decreased in both ATP treated group and the control group, which was the same as Hamm (1977). The ultimate pH (pHu) value at 25 °C was about 5.33 (lower than normal pHu value of intact muscle) suggested that glycolytic rate was faster in ground muscle. Naturally, the lower pHu value appeared at higher temperature, just as $pHu_{(-1.5\ °C)} > pHu_{(4\ °C)} > pHu_{(15\ °C)} > pHu_{(25\ °C)}$ in this study (Kim et al. 2014). The pHu also depended on meat quality and consumer acceptability due to pH-dependent biochemical mechanisms changed sharply in postmortem. Aroma, flavor, juiciness, texture, and tenderness were evaluated correlatively with different pHu (Devine et al. 1993).

At 25 °C and 15 °C, no significant difference ($P > 0.05$) existed between ATP treated group and the control group during later storage time. For 4 °C and − 1.5 °C, the pH value of ATP treated group was higher than the control group within 12 h, which was opposite with later time. In early storage time, there was abundant ATP in ATP treated group which would restrain the activity of phosphofructokinase (England et al. 2014). PFK was a key limited-speed enzyme of glycolysis through allosteric kinetics (Costa Leite et al. 2007). The plentiful ATP would impose negative effect on glycolysis rate and pH decline. With ATP consumption, this kind of inhibition was gradually weakened. More H^+ was produced by ATP hydrolyzation which led to lower pH value in ATP group at later storage time. Thus, the effect of ATP on the glycolytic enzyme activity and enhancement of ATP hydrolyzation resulted in different pH values in ATP treated group and the control group.

10.3.3 *Effect of ATP on Global Phosphorylation Levels in Postmortem Muscle*

To estimate the effect of ATP content on myofibrillar protein phosphorylation of different treatments, the phosphorylation level was measured through fluorescent staining (Fig. 10.6). In the control group at 25 °C and 15 °C, phosphorylated proteins divided by total proteins (P/T ratio) were reduced significantly ($P < 0.05$) after 12 h. While in ATP treated group, P/T ratio reached the highest value at 2 h and later decreased. The result demonstrated that ATP was depleted rapidly and phosphorylation was lower than dephosphorylation. However, the changes in P/T ratio at 4 °C

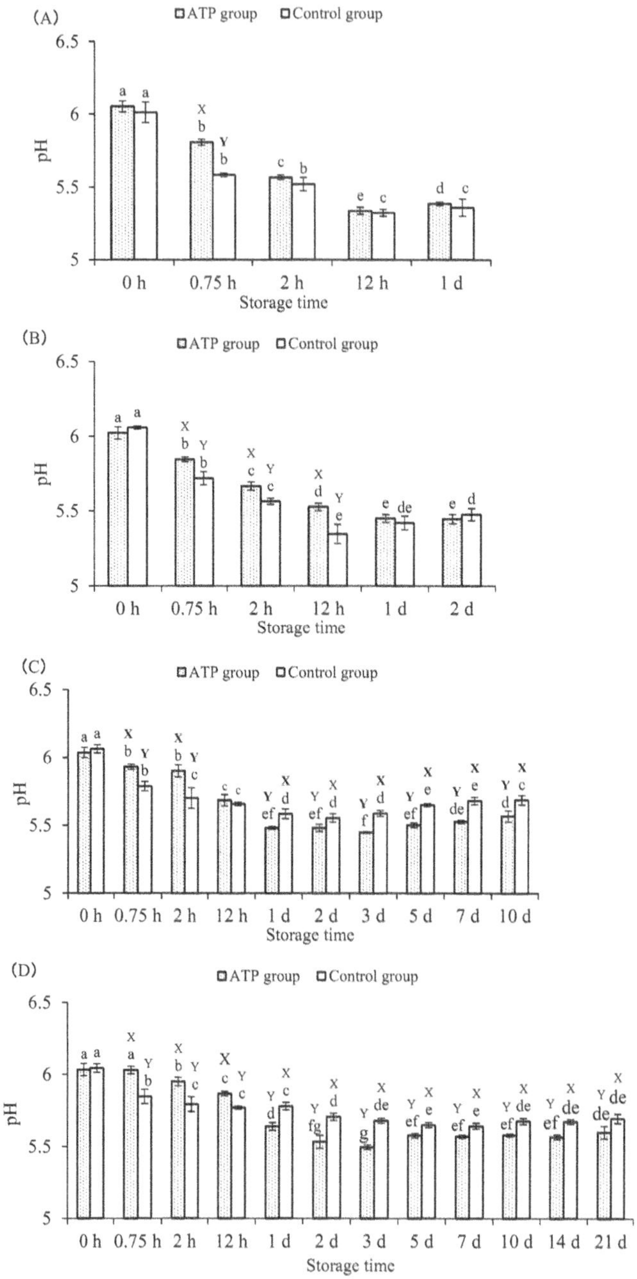

Fig. 10.5 The changes of pH in ground muscle at 25 °C (**a**), 15 °C (**b**), 4 °C (**c**), and − 1.5 °C (**d**). Different lowercase letters represent significant difference within the same group ($P < 0.05$). At the same storage time, different capital case letters represent significant difference between ATP treated group and the control group ($P < 0.05$)

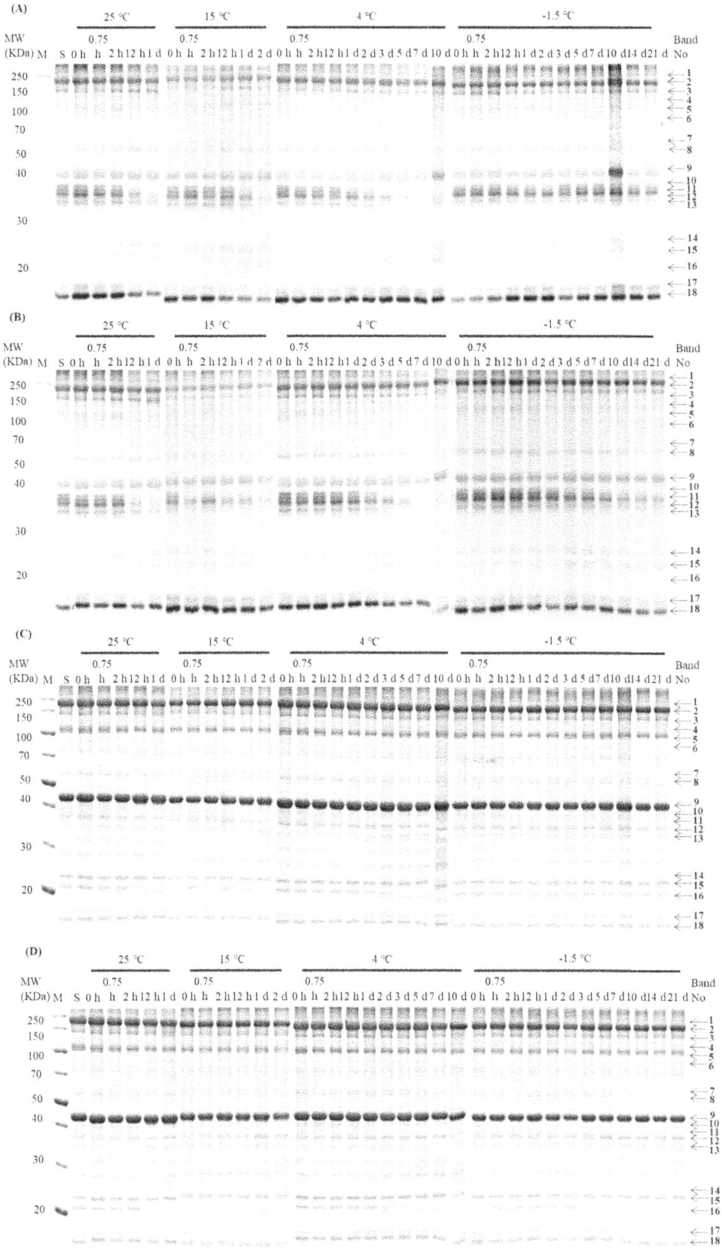

Fig. 10.6 Gel images of and phosphorylation level of myofibrillar proteins. (**a** and **b**) Gel images of myofibrillar phosphoproteins in ATP treated and the control group stained with Pro-Q Diamond; (**c** and **d**), gel images of myofibrillar proteins in ATP treated group and the control group stained with SYPRO Ruby. The phosphorylation level of myofibrillar proteins at 25 °C (**e**), 15 °C (**f**), 4 °C (**g**), and − 1.5 °C (**h**) in ATP treated and the control group. Different lowercase letters represent significant difference within the same group ($P < 0.05$). At the same storage time, different capital case letters represent significant difference between ATP treated group and the control group ($P < 0.05$)

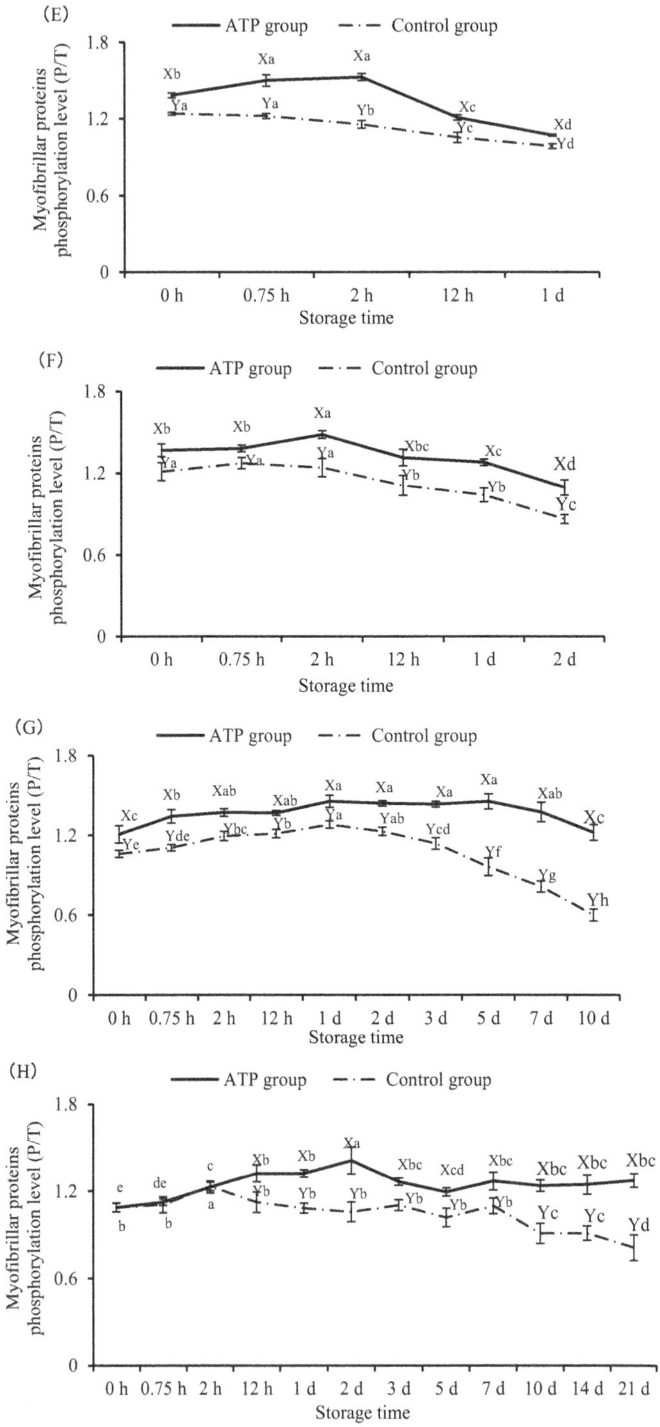

Fig. 10.6 (continued)

and − 1.5 °C were different compared with 25 °C and 15 °C. At 4 °C, the P/T ratio increased first and then decreased significantly ($P < 0.05$) after 3 days in the control group, which indicated that the ATP content could maintain phosphorylation reaction before 2 days and later, due to inadequate ATP content, phosphorylation reaction was weakened and dephosphorylation was the dominated one. While in ATP treated group at 4 °C, the P/T ratio raised within 0.75 h ($P < 0.05$) and then kept slightly changing with no significant difference ($P > 0.05$). The changes in phosphorylation level at −1.5 °C were similar with 4 °C because the phosphate groups offered by ATP were sufficient to promote phosphorylation. The slower consumption rate of ATP at low temperature made more phosphate group could be utilized in postmortem muscle. The phosphorylation level of ATP treated group was higher than the control group ($P < 0.05$) at each temperature, which testified that exogenous ATP stimulated the elevation of phosphorylation level.

Besides the direct effect of phosphate group provided by ATP on phosphorylation level, there was also indirect effect such as pH value impacted by ATP. Because ATP was related to glycolysis, pH value in postmortem muscle mainly relied on lactic acid produced by glycolysis. Huang et al. (2012) reported that the phosphorylation level of myofibrillar protein was influenced by different pH decline rate at early postmortem time, but had little effect on the extension of storage time. As the result in Fig. 10.6, the phosphorylation level at early storage time of 25 °C and 15 °C was higher than 4 °C and −1.5 °C, which was coherent with previous (Huang et al. 2012) conclusion that the phosphorylation level was higher in the group of higher pH decline rate. Combined with the results in this study, the rate of ATP consumption was a decisive factor of phosphorylation level in postmortem.

According to previous studies, the phosphorylated myofibrillar protein is principally involved in muscle contraction (Lametsch et al. 2011; Chen et al. 2016). Muroya et al. (2007) reported that myosin regulatory light chain was phosphorylated during rigor mortis formation. The increasing phosphorylation level of myosin light chain 2, desmin, and actin was measured in aging muscle (Gannon et al. 2008). Conclusively, ATP content might play an important role in muscle contraction and meat tenderness through regulating myofibrillar protein phosphorylation.

10.3.4 Effect of ATP on Protein Degradation in Postmortem Muscle

The degradation level of μ-calpain, desmin, and troponin T was measured by using western blot and the result is shown in Fig. 10.7. With the increase of postmortem time, μ-calpain was the most well-characterized type of calpain and showed more proteolysis through autolysis (Huff-Lonergan et al. 1996). At 25 °C in ATP treated group, the bands of 80 kDa completely disappeared, showing that μ-calpain was degraded to lower molecular weight (78–76 kDa) after 12 h of storage, while at 2 h in the control group. Similarly, at 4 °C, the degradation of 80 kDa was faster in the

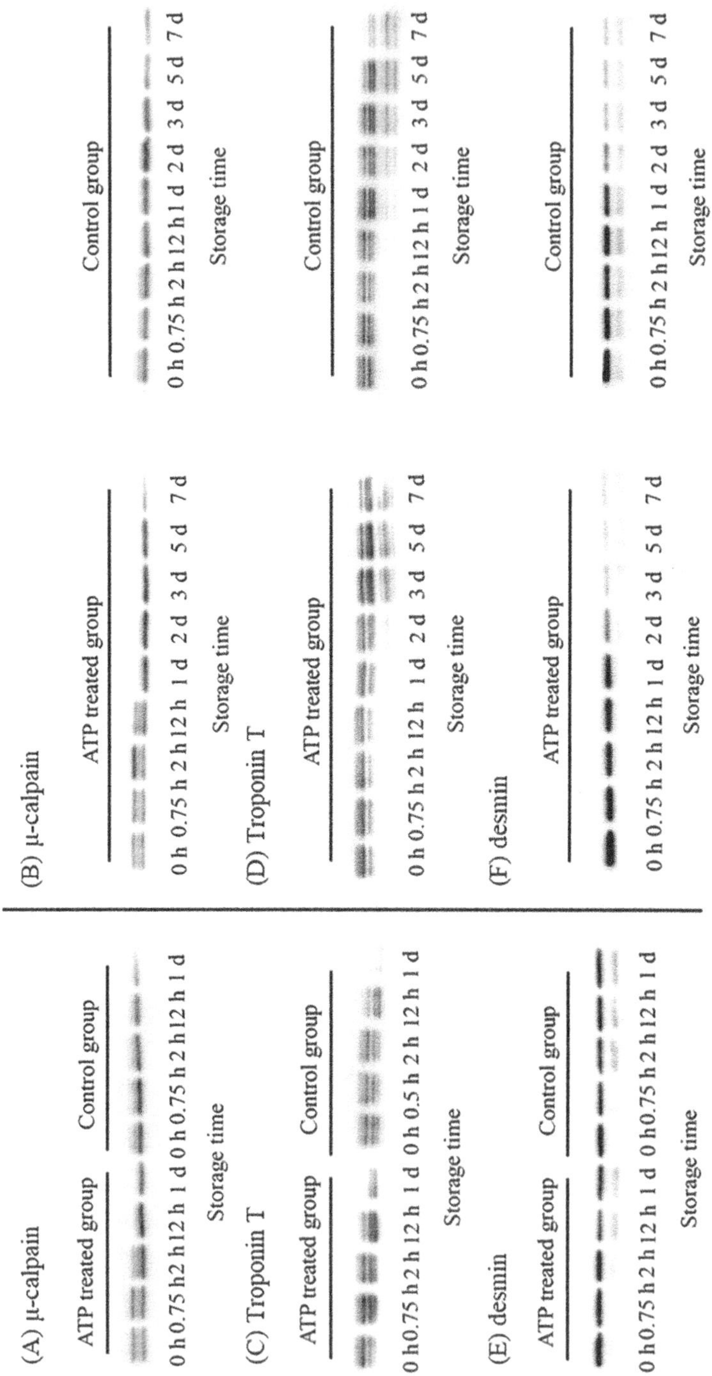

Fig. 10.7 Degradation of μ-calpain (**a**, **b**), troponin T (**c**, **d**), and desmin (**e**, **f**) in ATP treated and the control groups at 25 °C and 4 °C

control group than the ATP treated group, which indicated the earlier autolysis and activity of μ-calpain in control group. Therefore, the ATP content was negatively related to degradation of μ-calpain and through increased phosphorylation level of μ-calpain. However, the faster exhaustion of ATP in the control group led to the early release of Ca^{2+}, which attributed to the earlier autolysis of μ-calpain (Huang et al. 2012). Desmin and troponin T, related to meat tenderness, were structural proteins of the sarcomere and can be degraded by μ-calpain during postmortem (Huff-Lonergan et al. 2010). Desmin was localized at the periphery of the myofibrillar Z-disk in skeletal muscle and degraded during postmortem storage (Huff-Lonergan et al. 1996). Troponin T was strongly related to the shear force and regulated the thin filament during muscle contraction through its interaction with tropomyosin (Lehman et al. 2001). The bands of desmin in ATP treated group (25 °C) were degraded to lower molecular weight (approximately 38 kDa) at 12 h of storage, and more rapid degradation occurred in the control group. The degradation trend of troponin T was similar to that of desmin. At 4 °C, the degradation of desmin and troponin T was quicker in the control group than the ATP treated group. Therefore, the obviously slower degradation of desmin and troponin T was showed in ATP treated group, where the autolysis of μ-calpain was inhibited as well. The reason was that the phosphorylation of myofibrillar proteins reduced the susceptibility of proteolytic degradation by μ-calpain (Li et al. 2017). Thus, ATP might contribute to meat tenderness by affecting the phosphorylation and controlling protein degradation.

10.4 Effect of Temperature and pH on Dephosphorylation of Myofibrillar Protein In Vitro

This part mainly determined the effect of temperature (25°C, 15°C, 4°C, and −1.5°C) and pH (5.2, 5.8, and 6.4) on dephosphorylation of myofibrillar protein, which would provide a theoretical basis in order to improve meat quality. Myofibrillar protein of postmortem muscle was extracted, and alkaline phosphatase (AP) was added to catalyze dephosphorylation. The phosphorylation level was detected by SDS-PAGE. Gels were stained by Pro-Q Diamond and SYPRO Ruby.

10.4.1 Effect of Temperature and pH on the Activity of Alkaline Phosphatase

The optimal pH of AP is about 9–10, and the optimum temperature is about 40 °C (Wang and Yan 2009; Zhang and Guan 2012; Guo et al. 2004; Zhao et al. 2003). The results of AP activity at different temperatures and pH are shown in Fig. 10.8. The activity of AP gradually decreased with the extension of incubation time. At 25 °C

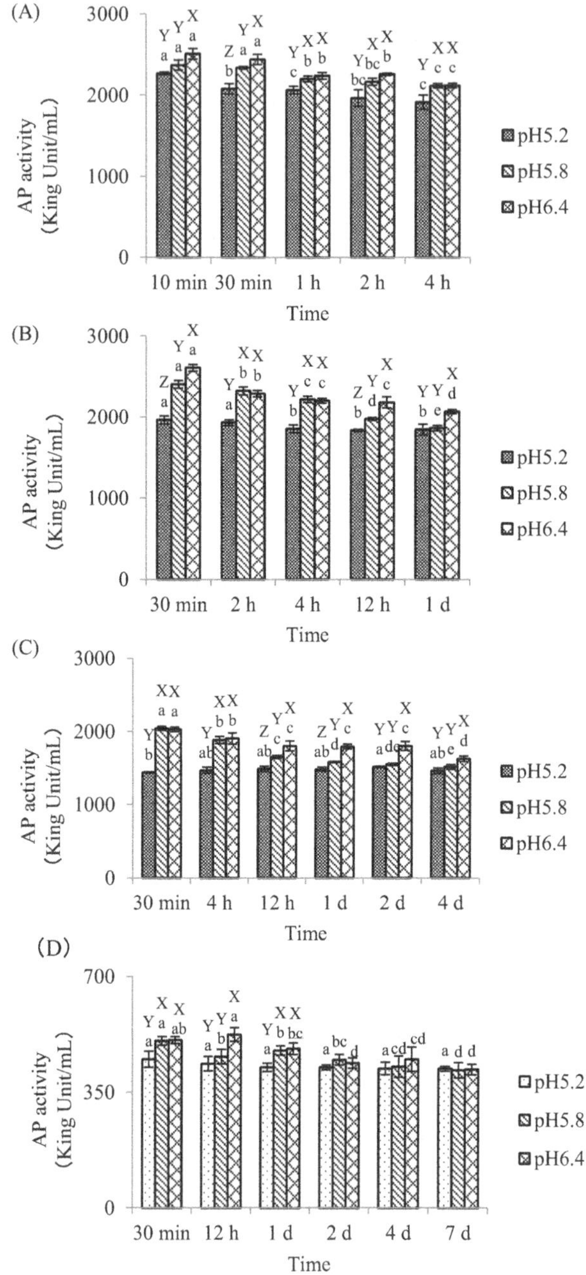

Fig. 10.8 The AP activity of different temperatures and pH. (**a**–**d**) Represent AP activity of 25 °C, 15 °C, 4 °C, and − 1.5 °C. Different lowercase letters represent significant difference between different time points of the same storage temperature ($P < 0.05$). Different capital letters represent significant difference between different storage temperatures at the same time point ($P < 0.05$). Translated from (Ren et al. 2019)

during 1–4 h, AP activity was 2115.27–2258.25 King unit/mL at pH 5.8 and 6.4, which was significantly higher than 1912.53–2063.10 King unit/mL at pH 5.2

($P < 0.05$). The result suggested that pH 5.2 restrained AP activity at 25 °C ($P > 0.05$).

When incubating for 30 min, 12 h, 1 day at 15 °C, the AP activity at pH 6.4 was significantly higher than pH 5.2 and pH 5.8 ($P < 0.05$). This result showed that when temperature decreased, the lower the pH, the stronger the inhibitory effect on AP activity. At 4 °C and pH 5.2, the highest AP activity was 1805.22 King unit/mL at 2 days. It showed that as the incubation time prolonged, the inhibition of lower temperature and pH would decrease.

10.4.2 Effect of Temperature and pH on Phosphorylation Level of Myofibrillar Protein

Myofibrillar protein was stained by Pro-Q Diamond and SYPRO Ruby after SDS-PAGE (Fig. 10.9). Protein dephosphorylation can be catalyzed by phosphatase and reduce protein phosphorylation level (Silverman-Gavrila et al. 2009; Hunter 1995). As shown in Fig. 10.10, with the extension of incubation time, there was no significant difference in phosphorylation level of the control group at different temperatures and pH ($P > 0.05$). The result suggested that myofibrillar protein could not be dephosphorylated without AP. For pH 6.4, when incubating for 4 h at 25 °C, myofibrillar protein phosphorylation level of AP treated group was significantly lower than that of the control group ($P < 0.95$ 0.05); at 15 °C, myofibrillar protein phosphorylation level of AP treated group was significantly lower than that of the control group ($P < 0.05$); at 4 °C and 4 days, myofibrillar protein phosphorylation level of AP treated group was significantly lower than that of the control group ($P < 0.05$). For pH 5.8, when incubating for 4 h, 1 day, 4 days at 25 °C, 15 °C, 4 °C, myofibrillar protein phosphorylation level of AP treated group was significantly lower than that of the control group ($P < 0.05$). These findings suggested high temperature was advantageous for AP catalyzing dephosphorylation of myofibrillar protein. For pH 5.2, myofibrillar protein phosphorylation level of AP treated group and the control group had no significant difference ($P > 0.05$), which showed that AP activity was inhibited at pH 5.2.

In conclusion, AP catalyzed the dephosphorylation of myofibrillar protein, which was regulated by both temperature and pH value. However, when a factor deviated from the optimal condition, it would inhibit AP catalyze dephosphorylation.

10.4.3 Effect of Temperature and pH on Phosphorylation Level of AP

The molecular weight of AP used in this experiment was 140–160 kDa. AP was a homologous dimer phosphomonoesterase (Bi 2015), which would be denatured

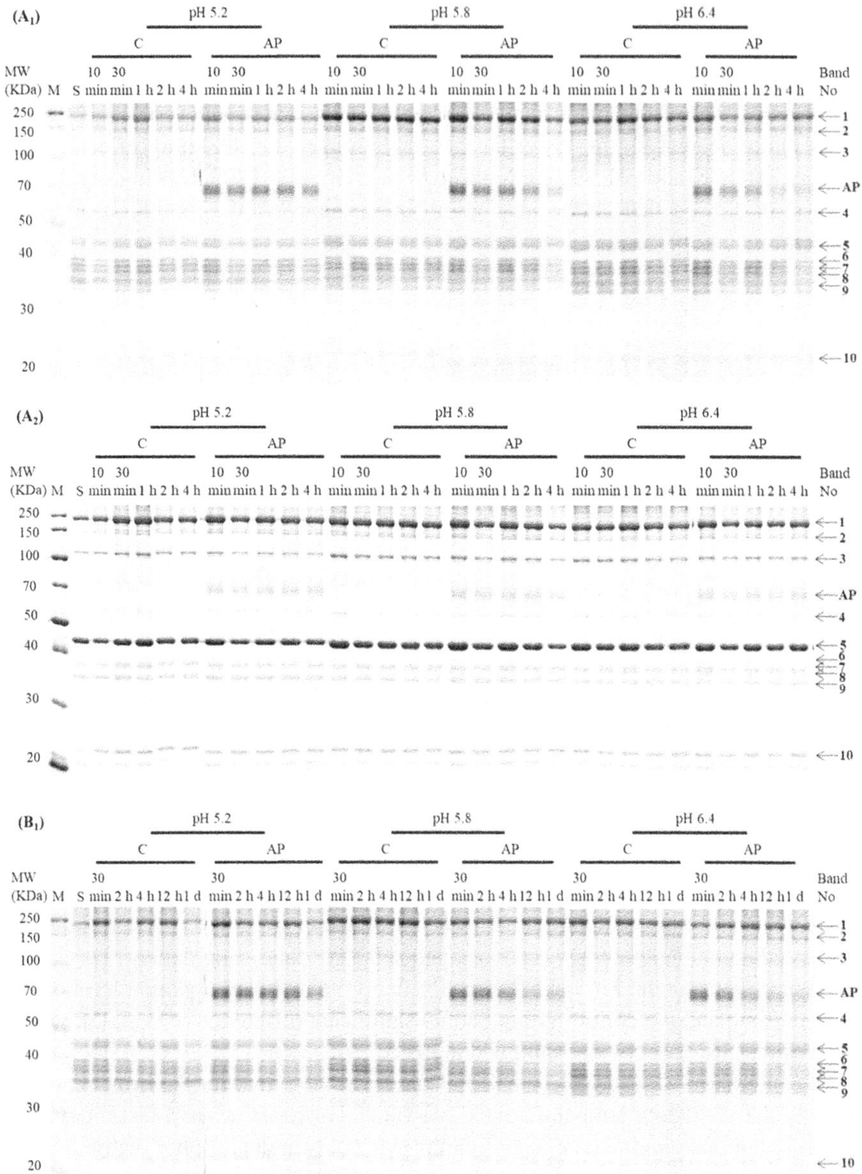

Fig. 10.9 Gel images of Pro-Q and Ruby stained by SDS-PAGE of different temperatures and pH values. AP means AP treated group, C means the control group, M means marker, S means standard. A_1, B_1, C_1, D_1: Pro-Q stained gel images of 25 °C, 15 °C, 4 °C, −1.5 °C; A_2, B_2, C_2, D_2: Ruby stained gel images of 25 °C, 15 °C, 4 °C, −1.5 °C. Translated from (Ren et al. 2019)

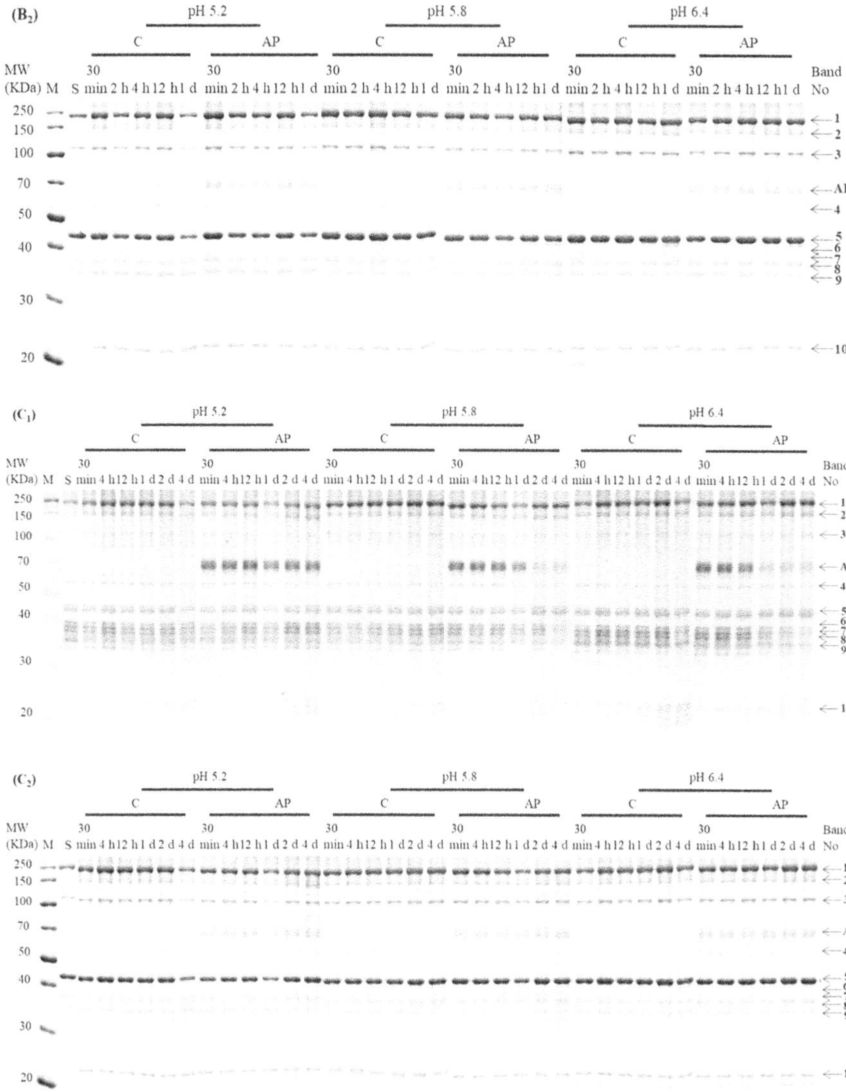

Fig. 10.9 (continued)

when preparing for electrophoresis samples. After the disassembly of the dimer, the molecular weight was about 70 kDa, as shown in Fig. 10.11. The alkaline phosphatase phosphorylation level changed significantly during incubation, as shown in Fig. 10.10. With the extension of incubation time, the phosphorylation level of AP was significantly changed. At different temperatures and pH, the phosphorylation

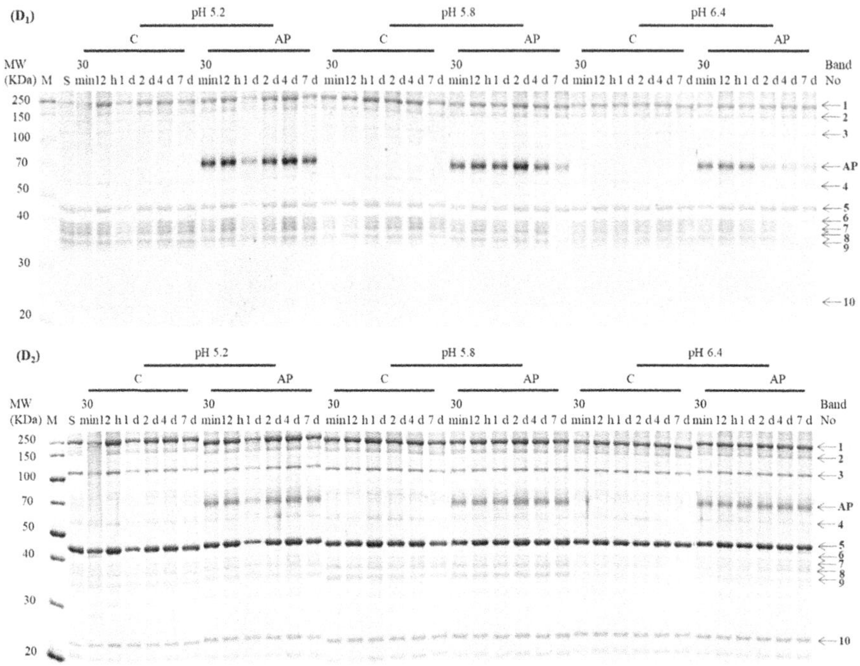

Fig. 10.9 (continued)

level of AP was gradually decreased, indicating that AP could not only catalyze myofibrillar dephosphorylation but also catalyze dephosphorylation of itself. At 25 °C, AP phosphorylation level was significantly different ($P < 0.05$) when incubating for 10 min; at 15 °C, AP phosphorylation level was significantly different ($P < 0.05$) when incubating for 30 min; at 4 °C, AP phosphorylation level was significantly different ($P < 0.05$) after 4 h incubation. This result suggested that rising temperature would move up the time of dephosphorylation. At later stage of incubation, AP phosphorylation level of pH 5.2 was significantly higher than that of other two pH levels ($P < 0.05$), indicating that pH 5.2 had a stronger inhibition effect on AP, even though the temperature increased. When AP was dephosphorylated, its activity would decrease slightly, the main factors affecting AP activity were temperature and pH.

Therefore, postmortem storage temperature affected protein phosphorylation and meat quality development at least by affecting pH decline and ATP consumption in postmortem muscle.

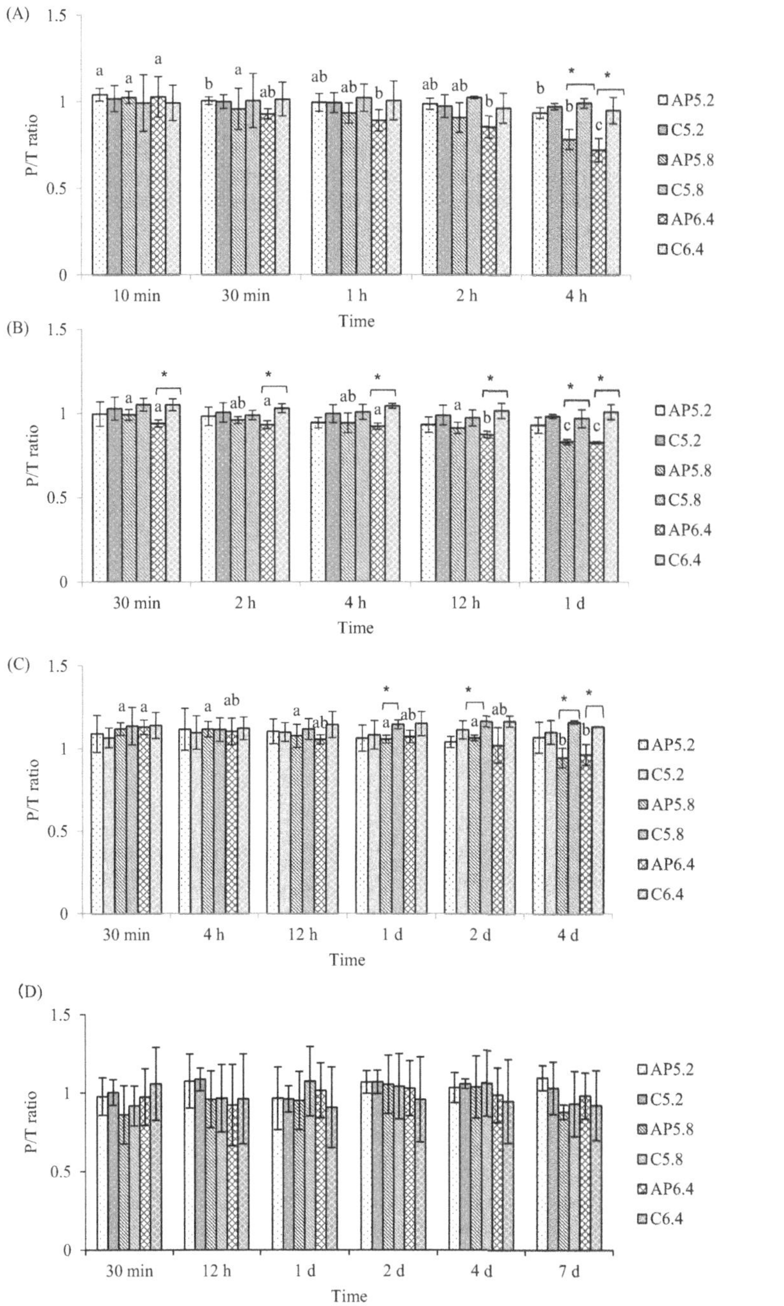

Fig. 10.10 Myofibrillar protein phosphorylation level of different temperatures and pH. (**a–d**) Represent AP activity of 25 °C, 15 °C, 4 °C, and −1.5 °C. AP means AP treated group, (**c**) means the control group. Different lowercase letters represent significant difference between different time points of the same pH ($P < 0.05$). * represents significant difference between different group at the same time point ($P < 0.05$). Translated from (Ren et al. 2019)

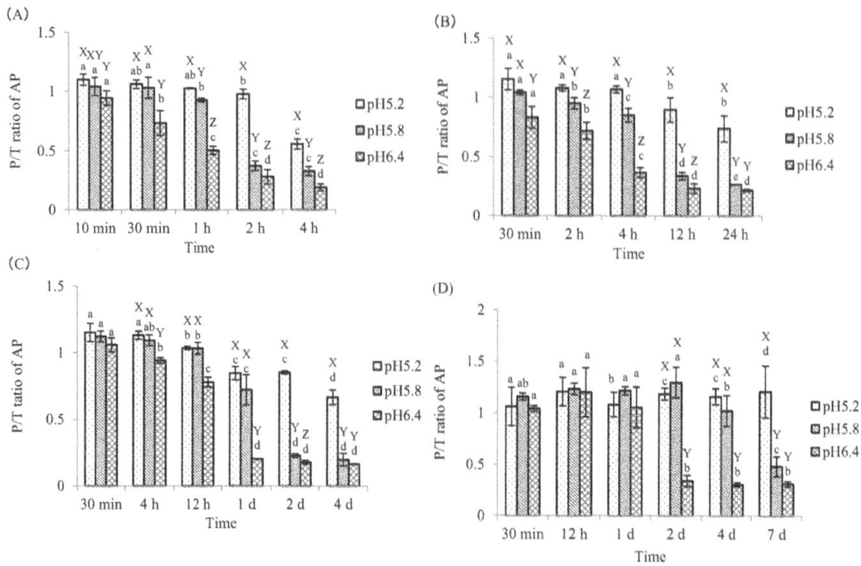

Fig. 10.11 AP phosphorylation level of different temperatures and pH values. A, B, C, and D represent AP phosphorylation level of 25 °C, 15 °C, 4 °C, and −1.5 °C. Different lowercase letters represent significant difference between different time points of the same pH ($P < 0.05$). Different capital letters represent significant difference between different pH at the same time point ($P < 0.05$). Translated from (Ren et al. 2019)

10.5 Conclusions

Temperature affected phosphorylation and dephosphorylation through pH changes and ATP content. The proteins involved in glycolysis and muscle construction were in two clusters. Through correlation analysis, ATP was a more important regarding influencing the phosphorylation level of individual protein. The higher ATP content was positively related to phosphorylation level of myofibrillar proteins and restrained the degradation of μ-calpain, troponin T, and desmin. The activity of alkaline phosphatase was lower at −1.5 °C and pH 5.2, so that dephosphorylation of myofibrillar proteins was not significant. The study gave a clear understanding of phosphorylation and dephosphorylation in postmortem muscle, as well as the relationship between ATP content and protein degradation.

Acknowledgments Parts of this chapter are reprinted from Journal of the Science of Food and Agriculture, 100, Ren, C., et al., Effects of temperature on protein phosphorylation in postmortem muscle, 551–559. Copyright (2020), with permission from Elsevier. Parts of this chapter are translated from Food Science, 40(16), REN, C., et al., Effect of temperature and pH on dephosphorylation of myofibrillar protein in vitro (Chinese), 1–7. Copyright (2020), with permission from Journal of Food Science.

References

Aberle, E. D., & Merkel, R. A. (1968). 5′-adenylic acid deaminase in porcine muscle. *Journal of Food Science, 33*, 27–29.

Bi, J. (2015). Research progresses of intestinal alkaline phosphatase in intestinal barrier maintenance. *Parenteral & Enteral Nutrition (Chinese), 22*, 244–247.

Chen, L. J., Li, X., Ni, N., Liu, Y., Chen, L., Wang, Z., et al. (2016). Phosphorylation of myofibrillar proteins in post-mortem ovine muscle with different tenderness. *Journal of the Science of Food and Agriculture, 96*, 1474–1483.

Costa Leite, T., Da Silva, D., Guimaraes Coelho, R., Zancan, P., & Sola-Penna, M. (2007). Lactate favours the dissociation of skeletal muscle 6-phosphofructo-1-kinase tetramers down-regulating the enzyme and muscle glycolysis. *Biochemical Journal, 408*, 123–130.

D'Alessandro, A., Marrocco, C., Rinalducci, S., Mirasole, C., Failla, S., & Zolla, L. (2012b). Chianina beef tenderness investigated through integrated Omics. *Journal of Proteomics, 75*, 4381–4398.

D'Alessandro, A., Rinalducci, S., Marrocco, C., Zolla, V., Napolitano, F., & Zolla, L. (2012a). Love me tender: An Omics window on the bovine meat tenderness network. *Journal of Proteomics, 75*, 4360–4380.

D'Alessandro, A., & Zolla, L. (2013). Meat science: From proteomics to integrated omics towards system biology. *Journal of Proteomics, 78*, 558–577.

Devine, C. E., Graafhuis, A. E., Muir, P. D., & Chrystall, B. B. (1993). The effect of growth rate and ultimate pH on meat quality of lambs. *Meat Science, 35*, 63–77.

England, E. M., Matarneh, S. K., Scheffler, T. L., Wachet, C., & Gerrard, D. E. (2014). pH inactivation of phosphofructokinase arrests postmortem glycolysis. *Meat Science, 98*, 850–857.

Fernandez, X., & Tornberg, E. (1994). The influence of high post-mortem temperature and differing ultimate pH on the course of rigor and ageing in pig longissimus dorsi muscle. *Meat Science, 36*, 345–363.

Gannon, J., Staunton, L., O'Connell, K., Doran, P., & Ohlendieck, K. (2008). Phosphoproteomic analysis of aged skeletal muscle. *International Journal of Molecular Medicine, 22*, 33–42.

Gao, X., Li, X., Li, Z., Du, M., & Zhang, D. (2017). Dephosphorylation of myosin regulatory light chain modulates actin-myosin interaction adverse to meat tenderness. *International Journal of Food Science and Technology, 52*, 1400–1407.

Gratacos-Cubarsi, M., & Lametsch, R. (2008). Determination of changes in protein conformation caused by pH and temperature. *Meat Science, 80*, 545–549.

Guo, J., Liu, K., Yue, B., Zhang, Z., Hou, R., Li, J., et al. (2004). Isolation, purification and properties of alkaline phosphatase from sperm of PIC pigs(PIC344). *Acta Veterinaria et Zoochnic Sinica (Chinese), 35*, 37–41.

Hamm, R. (1977). Postmortem breakdown of ATP and glycogen in ground muscle: A review. *Meat Science, 1*, 15–39.

Huang, H., Larsen, M. R., Karlsson, A. H., Pomponio, L., Costa, L. N., & Lametsch, R. (2011). Gel-based phosphoproteomics analysis of sarcoplasmic proteins in postmortem porcine muscle with pH decline rate and time differences. *Proteomics, 11*, 4063–4076.

Huang, H., Larsen, M. R., & Lametsch, R. (2012). Changes in phosphorylation of myofibrillar proteins during postmortem development of porcine muscle. *Food Chemistry, 134*, 1999–2006.

Hudson, N. J. (2012). Mitochondrial treason: A driver of pH decline rate in post-mortem muscle? *Animal Production Science, 52*, 1107–1110.

Huff-Lonergan, E., Mitsuhashi, T., Beekman, D. D., Parrish, F. C., Olson, D. G., & Robson, R. M. (1996). Proteolysis of specific muscle structural proteins by mu-calpain at low pH and temperature is similar to degradation in postmortem bovine muscle. *Journal of Animal Science, 74*, 993–1008.

Huff-Lonergan, E., Zhang, W., & Lonergan, S. M. (2010). Biochemistry of postmortem muscle - lessons on mechanisms of meat tenderization. *Meat Science, 86*, 184–195.

Hunter, T. (1995). Protein kinases and phosphatases-the yin and Yang of protein phosphorylation and signaling. *Cell, 80*, 225–236.

Kim, Y. H. B., Warner, R. D., & Rosenvold, K. (2014). Influence of high pre-rigor temperature and fast pH fall on muscle proteins and meat quality: A review. *Animal Production Science, 54*, 375–395.

Lametsch, R., Larsen, M. R., Essen-Gustavsson, B., Jensen-Waern, M., Lundstrom, K., & Lindahl, G. (2011). Postmortem changes in pork muscle protein phosphorylation in relation to the RN genotype. *Journal of Agricultural and Food Chemistry, 59*, 11608–11615.

Lametsch, R., Roepstorff, P., & Bendixen, E. (2002). Identification of protein degradation during post-mortem storage of pig meat. *Journal of Agricultural and Food Chemistry, 50*, 5508–5512.

Lehman, W., Rosol, M., Tobacman, L. S., & Craig, R. (2001). Troponin organization on relaxed and activated thin filaments revealed by electron microscopy and three-dimensional reconstruction. *Journal of Molecular Biology, 307*, 739–744.

Li, M., Li, Z., Li, X., Xin, J., Wang, Y., Li, G., et al. (2018). Comparative profiling of sarcoplasmic phosphoproteins in ovine muscle with different color stability. *Food Chemistry, 240*, 104–111.

Li, X., Fang, T., Zong, M., Shi, X., Xu, X., Dai, C., et al. (2015). Phosphorproteome changes of myofibrillar proteins at early post-mortem time in relation to pork quality as affected by season. *Journal of Agricultural and Food Chemistry, 63*, 10287–10294.

Li, Z., Li, X., Gao, X., Shen, Q. W., Du, M., & Zhang, D. (2017). Phosphorylation prevents in vitro myofibrillar proteins degradation by mu-calpain. *Food Chemistry, 218*, 455–462.

Liu, R., Li, Y. P., Zhang, W. G., Fu, Q. Q., Liu, N., & Zhou, G. H. (2015). Activity and expression of nitric oxide synthase in pork skeletal muscles. *Meat Science, 99*, 25–31.

Merkulova, T., Dehaupas, M., Nevers, M. C., Creminon, C., Alameddine, H., & KELLER, A. (2000). Differential modulation of α, β and γ enolase isoforms in regenerating mouse skeletal muscle. *European Journal of Biochemistry, 267*, 3735–3743.

Muroya, S., Ohnishi-Kameyama, M., Oe, M., Nakajima, I., Shibata, M., & Chikuni, K. (2007). Double phosphorylation of the myosin regulatory light chain during rigor mortis of bovine Longissimus muscle. *Journal of Agricultural and Food Chemistry, 55*, 3998–4004.

Ren, C., Hou, C., Li, X., Bai, Y., & Zhang, D. (2019). Effect of temperature and pH on dephosphorylation of myofibrillar protein in vitro. *Food Science (Chinese), 40*, 1–7.

Ren, C., Hou, C., Li, Z., Li, X., Bai, Y., & Zhang, D. (2020). Effects of temperature on protein phosphorylation in postmortem muscle. *Journal of the Science of Food and Agriculture, 100*, 551–559.

Savell, J. W., Mueller, S. L., & Baird, B. E. (2005). The chilling of carcasses. *Meat Science, 70*, 449–459.

Scheffler, T. L., & Gerrard, D. E. (2007). Mechanisms controlling pork quality development: The biochemistry controlling postmortem energy metabolism. *Meat Science, 77*, 7–16.

Schwagele, F., Buesa, P. L. L., & Honikel, K. O. (1996). Enzymological investigations on the causes for the PSE-syndrome, II. Comparative studies on glycogen phosphorylase from pig muscles. *Meat Science, 44*, 41–53.

Shen, Q. W., & Du, M. (2005). Role of AMP-activated protein kinase in the glycolysis of postmortem muscle. *Journal of the Science of Food and Agriculture, 85*, 2401–2406.

Shen, Q. W., Means, W. J., Thompson, S. A., Underwood, K. R., Zhu, M. J., Mccormick, R. J., et al. (2006). Pre-slaughter transport, AMP-activated protein kinase, glycolysis, and quality of pork loin. *Meat Science, 74*, 388–395.

Silverman-Gavrila, L. B., Lu, T. Z., Prashad, R. C., Nejatbakhsh, N., Charlton, M. P., & Feng, Z. P. (2009). Neural phosphoproteomics of a chronic hypoxia model--Lymnaea stagnalis. *Neuroscience, 161*, 621–634.

Sola-Penna, M., Da Silva, D., Coelho, W. S., Marinho-Carvalho, M. M., & Zancan, P. (2010). Regulation of mammalian muscle type 6-phosphofructo-1-kinase and its implication for the control of the metabolism. *IUBMB Life, 62*, 791–796.

Wang, S., Li, C., Xu, X., & Zhou, G. (2013). Effect of fasting on energy metabolism and tenderizing enzymes in chicken breast muscle early postmortem. *Meat Science, 93*, 865–872.

Wang, S., & Yan, S. (2009). Progress in alkaline phosphatase in bone metabolism of animal. *Feed Review (Chinese), 4*, 14–17.

Watabe, S., Kamal, M., & Hashimoto, K. (1991). Postmortem changes in ATP, creatine phosphate, and lactate in sardine muscle. *Journal of Food Science, 56*, 151–153.

Zhang, Y., & Guan, X. (2012). Effects of polar uncharged amino acids on liver alkalinity phosphoric activity in rabbit. *Journal of Hebei Normal University of Science & Technology (Chinese), 26*, 41–46.

Zhao, H., Qian, F., Sun, Y., Huang, L., Zeng, Y., Zhao, J., et al. (2003). Isolation, purification and some properties of alkaline phosphatase from chicken. *Journal of Shanghai Jiaotong University (Agricultural Science) (Chinese), 21*, 194–198.

Chapter 11
Effects of Ionic Strength on Protein Phosphorylation

Abstract Salting is one of the most ancient preservation methods for food, which is widely used to preserve meat and improve the quality of meat products, such as water holding capacity (WHC), texture, tenderness, flavor, taste, and so on. Protein phosphorylation is one of the most ubiquitous post-translational modifications (PTMs). Reversible protein phosphorylation plays an important role in protein structure, function, signaling, and regulation. As one of the most essential processes for meat products, the effects of salting on muscle protein phosphorylation are unclear. Thus, we investigated the phosphorylation of myofibrillar and sarcoplasmic proteins in response to salt content and salting time. We discovered that salting showed significant effect on myofibrillar protein phosphorylation. Four categories of phosphorylated protein were identified by liquid chromatography-tandem mass spectrometry (LC-MS/MS), involved in stress response (heat shock protein), glycometabolism (glycogen phosphorylase, glyceraldehyde-3-phosphate dehydrogenase), oxidation or reduction (superoxide dismutase), and others (myoglobin). The phosphorylation level of the proteins mentioned above was affected by salting. Thus, salting might influence meat quality through protein phosphorylation, which regulated protein degradation and glycolysis. Furthermore, ten different phosphoproteins ($>$1.5 fold) induced by salting were identified by LC-MS/MS and the UniProt database, most of which were involved in glycometabolism, protein function, and protein degradation. It was concluded that salting might influence meat quality through protein phosphorylation, which regulated glycolysis metabolism, protein function, and degradation.

Keywords Protein phosphorylation · Salting · Protein degradation · Glycolysis · Differential phosphorylated proteins · Salting temperature · Sodium chloride · Meat quality

D. Zhang et al., *Protein Phosphorylation and Meat Quality*,
https://doi.org/10.1007/978-981-15-9441-0_11

11.1 Introduction

Salting is commonly used in meat processing to improve the quality of processed meat products, such as flavor, water holding capacity, tenderness, juiciness, firmness, gelation, and appearance (Comaposada et al. 2000; Duranton et al. 2012; Shao et al. 2016). NaCl is the essential salting material used in cured or cooked meat products. It increases the ionic strength within muscle, which causes protein dissociation, denaturation, modification, and myofibril swelling (Kim et al. 2015; Shao et al. 2016; Wang et al. 2016).

Protein phosphorylation is the most common post-translational modification (PTM). The majority of muscle proteins can be phosphorylated in vivo. Many salt-responsive phosphoproteins also have been identified in vivo in plants (Hu et al. 2013). In mice, salt regulates the phosphorylation of STE20/SPS1-related proline/alanine-rich kinase (SPAK) and Na-K-Cl cotransporter isoform 1 (NKCC1) in aorta (Zeniya et al. 2013). Higher salt intake decreases the phosphorylation of with-no-lysine kinase (WNK)-SPAK-NKC1 in the vascular smooth muscle, resulting in disorganized common vascular tone (Zeniya et al. 2013). Protein phosphorylation affects protein stability, enzymatic activity, and muscle contraction, which may then affect muscle development, carcass rigor mortis, and meat quality (Huang et al. 2012; Li et al. 2012). NaCl influences meat quality by regulating myofibrillar protein denaturation, actomyosin dissociation, myosin hydration, and myoglobin redox (Duranton et al. 2012; Kim et al. 2015; Wang et al. 2016), while myosin, actin, and myoglobin could be phosphorylated (Li et al. 2017; Chen et al. 2016).

The standard curing temperature in modern meat production is generally 4 °C. However, in actual production, a few small enterprises adopt room temperature for pickling (25 °C). In addition, ice temperature preservation technology between 0 °C and freezing point temperature has become a research hotspot in recent years. With the application of ice temperature preservation technology, ice temperature curing also becomes a research direction. Garcia-Gil et al. (2014) studied the effect of curing temperature on salt absorption of ham and found out that low-temperature curing was not conducive to salt absorption (Garcia-Gil et al. 2014). Protein phosphorylation was affected by many factors, the endogenous factors were mainly the action of protein kinases and phosphatases (Rubin and Rosen 1975), while the exogenous factors included factors that affected enzymes. Besides, temperature was a very important factor affecting enzyme activity.

Whether NaCl concentration and salting temperature influence protein phosphorylation, and what effects of salting on muscle protein phosphorylation should be elucidated by further research. In this case, systematically studying the response of different phosphorylated proteins to sodium chloride will not only help us understand the relations of the salting and muscle protein phosphorylation, but also help reveal a potential new mechanism of salting changing meat quality.

11.2 Phosphorylation Level and Influencing Pathway of Myofibrillar and Sarcoplasmic Proteins of Muscle in Response to Salting

Phosphorylation level of myofibrillar and sarcoplasmic proteins of muscle in response to salting was discovered recently. In order to explore it clearly, the myofibrillar and sarcoplasmic proteins were extracted from salted meat with 0, 1, 2, 3, 4, and 5% salt and salted for 0, 2, 4, 6, 8, and 16 h. Phosphorylation level was analyzed by dodecyl sulfate sodium salt and polyacrylamide gel electrophoresis (SDS-PAGE), fluorescence staining, and LC-MS/MS. Four categories of phosphorylated protein were identified, which were involved in stress response (heat shock protein), glycometabolism (glycogen phosphorylase, glyceraldehyde-3-phosphate dehydrogenase), oxidation or reduction (superoxide dismutase), and others (myoglobin). Briefly, salting might influence meat quality through protein phosphorylation, which regulated protein degradation and glycolysis (Zhang et al. 2016a).

11.2.1 Global Phosphorylation of Myofibrillar Proteins with Salting Time

The amount of salt added in traditional meat processing is typically 2% (Mora-Gallego et al. 2016). Therefore, the 2% salt was chosen as the experimental dose to profile protein phosphorylation during meat salting. As shown in Fig. 11.1a and b, a total of 23 bands of myofibrillar proteins were detected by Quantity One and selected for the analysis. The global phosphorylation level of proteins was shown in Fig. 11.1c. The phosphorylated proteins divided by total proteins (P/T ratio) of each sample were evaluated by the ratio of the intensity of all bands in Pro-Q Diamond images to the intensity of all bands in SYPRO Ruby images. The P/T ratio of control group increased from 0 to 2 h, then decreased slightly with no significant difference during the whole salting process ($P > 0.05$). Conversely, the P/T ratio of the salted meat with 2% salt significantly decreased from 0 to 16 h ($P < 0.05$).

The results showed that global phosphorylation level of myofibrillar proteins was significantly decreased ($P < 0.05$) during the salting development after adding 2% salt, while it had no significant difference in the control group. The global phosphorylation level of salted group was lower than the control group, which indicated that salt affected protein phosphorylation. The global phosphorylation level at 16 h

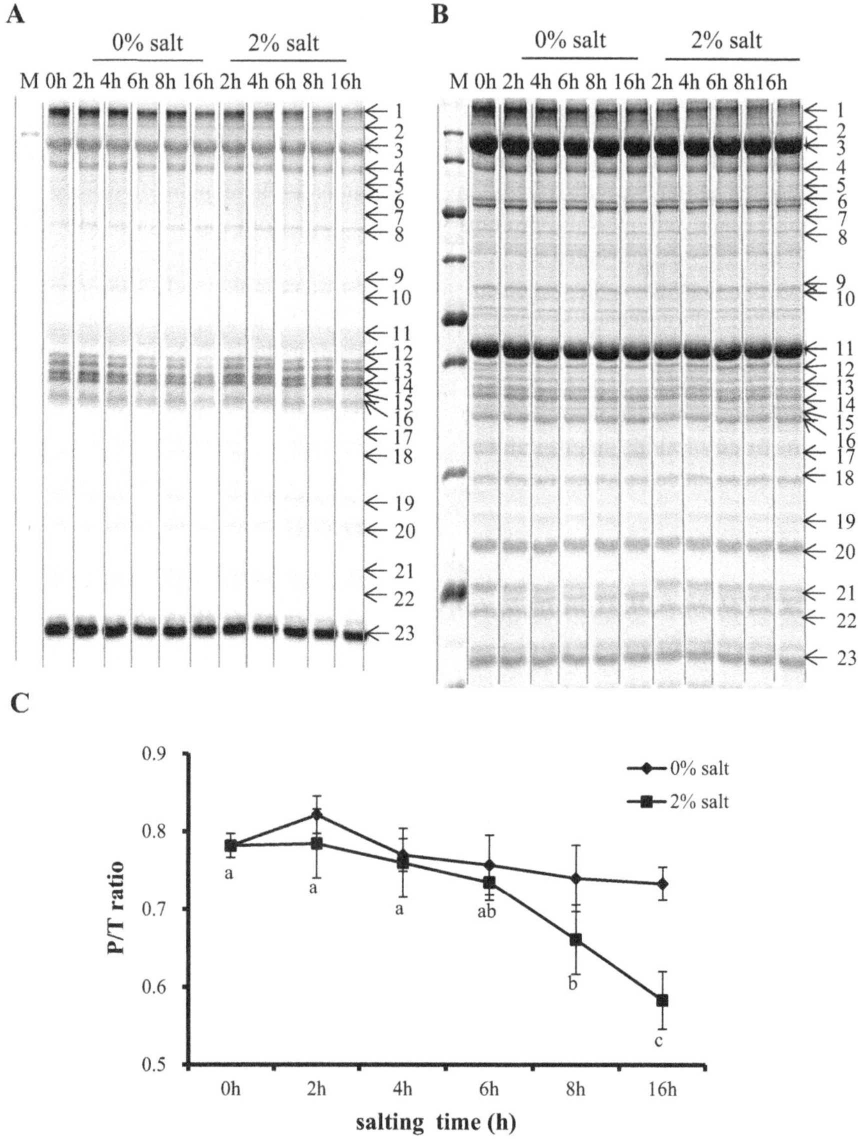

Fig. 11.1 Phosphorylation of myofibrillar proteins during salting. (**a–b**) SDS-PAGE analysis of phosphorylated proteins by Pro-Q Diamond staining (**a**) and total myofibrillar proteins by SYPRO Ruby staining (**b**). (**c**) Quantification of global protein phosphorylation level of myofibrillar proteins at different time points during salting. Different letters indicate significant difference with salting time ($P < 0.05$). Asterisk (*) shows significant differences between the 0% salt and 2% salt groups in the same salting time ($P < 0.05$) (Zhang et al. 2016a)

was lowest during the salting development, so we further studied the influence of salt on muscle protein phosphorylation by adding different concentration of salt.

The phosphorylation level of myofibrillar proteins with different salt concentration was shown in Fig. 11.2. Compared to control group, the phosphorylation of myofibrillar protein of meat salted with 2~5% (m/v) NaCl for 16 h was reduced. Meat salted by 3% salt had the lowest protein phosphorylation (Fig. 11.2c), showing that salt influenced protein phosphorylation during meat processing.

P/T ratio of band 8, 16, 19, and 23 was greater than 1, so these proteins were highly phosphorylated. The P/T ratio of individual protein bands was shown in Fig. 11.2d. In general, the protein phosphorylation level of most bands from salted meat was lower than that of the control group except for band 16 and 20, which showed an increase in protein phosphorylation after salting.

By investigation, it was discovered that salting decreased phosphorylation of myofibrillar proteins. Myofibrillar proteins were structural proteins responsible for sarcomere structure integrity and muscle contraction, both of which were related to meat tenderness. Salting improved meat tenderness (Sheard and Tali 2004), while many researchers found out that protein phosphorylation may affect meat quality (Huang et al. 2012; Li et al. 2015; Chen et al. 2016). Our study showed that the global phosphorylation level of myofibrillar proteins in salted meat was reduced. This was in consistence with previous literature, which reported that protein phosphorylation affected meat tenderness and tender meat had lower protein phosphorylation level (Chen et al. 2016; Zolla 2012). The phosphorylation of myofibrillar proteins reduced the degradation of these proteins by calpains, such as actin binding protein, Troponin T (TnT), and Troponin I (Di Lisa et al. 1995; Toyo-Oka 1982; Zhang et al. 1988). In this study, the phosphorylation of myosin heavy chain (band 3), myosin binding protein C (band 4), and actin (band 11) was reduced by salting, while these bands were related to meat tenderness. According to all the studies, we speculated that relatively low phosphorylation level of myofibrillar proteins in salted meat made them more easily to be degraded than the control group, as a result, improved meat tenderness.

Rigor mortis or muscle contraction shortens sarcomere length and decreases meat tenderness. Many phosphorylated proteins identified in muscle are related to sarcomere function which affects meat quality, so it is speculated that protein phosphorylation may be related to muscle contraction and quality change (Huang et al. 2011). The phosphorylation of myosin light chain (MLC) enhances the intensity of contraction (Stull et al. 2011). Our result showed that the phosphorylation of MLC (Fig. 11.2d, band 23) was reduced by salting, which might inhibit muscle contraction and further affect meat tenderness, providing another mechanism that salting could increase meat tenderness.

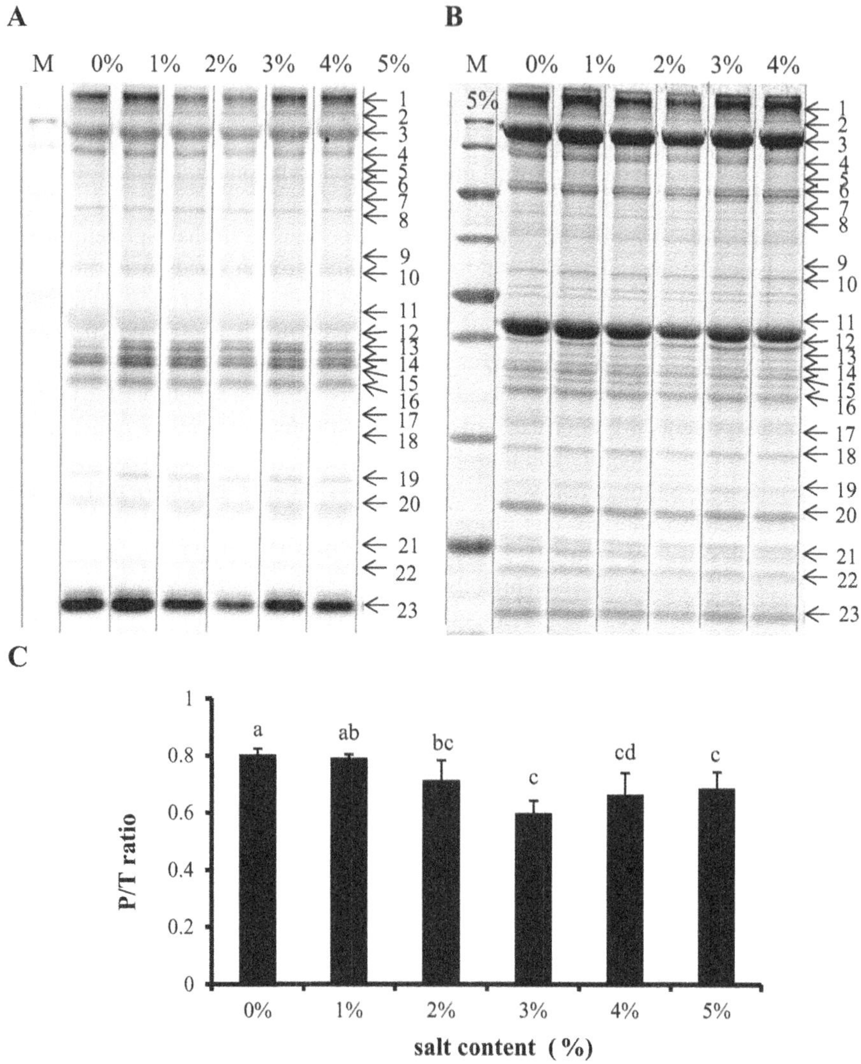

Fig. 11.2 Phosphorylation of myofibrillar proteins in response to salt on concentrations. (**a**–**b**) SDS-PAGE analysis of phosphorylated proteins by Pro-Q Diamond staining (**a**) and total myofibrillar proteins by SYPRO Ruby staining (**b**). (**c**) Quantification of global protein phosphorylation level of myofibrillar proteins from meat salted at different salt concentrations. (**d**) Protein phosphorylation level of nine individual protein bands. Different letters indicate significant differences ($P < 0.05$) (Zhang et al. 2016b)

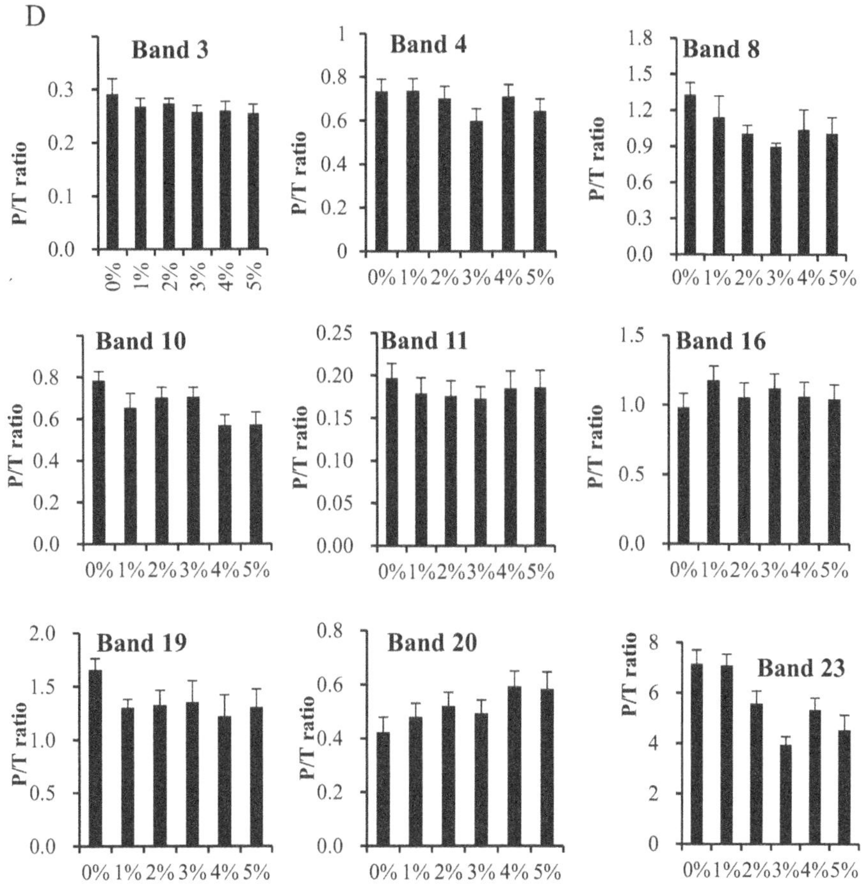

Fig. 11.2 (continued)

11.2.2 Global Phosphorylation of Sarcoplasmic Proteins with Salting Time

In total, 20 sarcoplasmic protein bands were detected to be phosphorylated in meat (Fig. 11.3a). Unlike to myofibrillar proteins, salting showed no significant effect on the phosphorylation of sarcoplasmic proteins. The P/T ratios of both the control group and salted meat with 2% salt showed no significant difference ($P > 0.05$) at different salting time points, though an increasing trend in P/T ratio with salting time might exist in the salted meat (Fig. 11.3c).

To confirm that salt had no effect on the global phosphorylation level of sarcoplasmic proteins, sarcoplasmic proteins were extracted from meat and salted for 16 h with different concentrations of salt. SDS-PAGE and fluorescence staining with

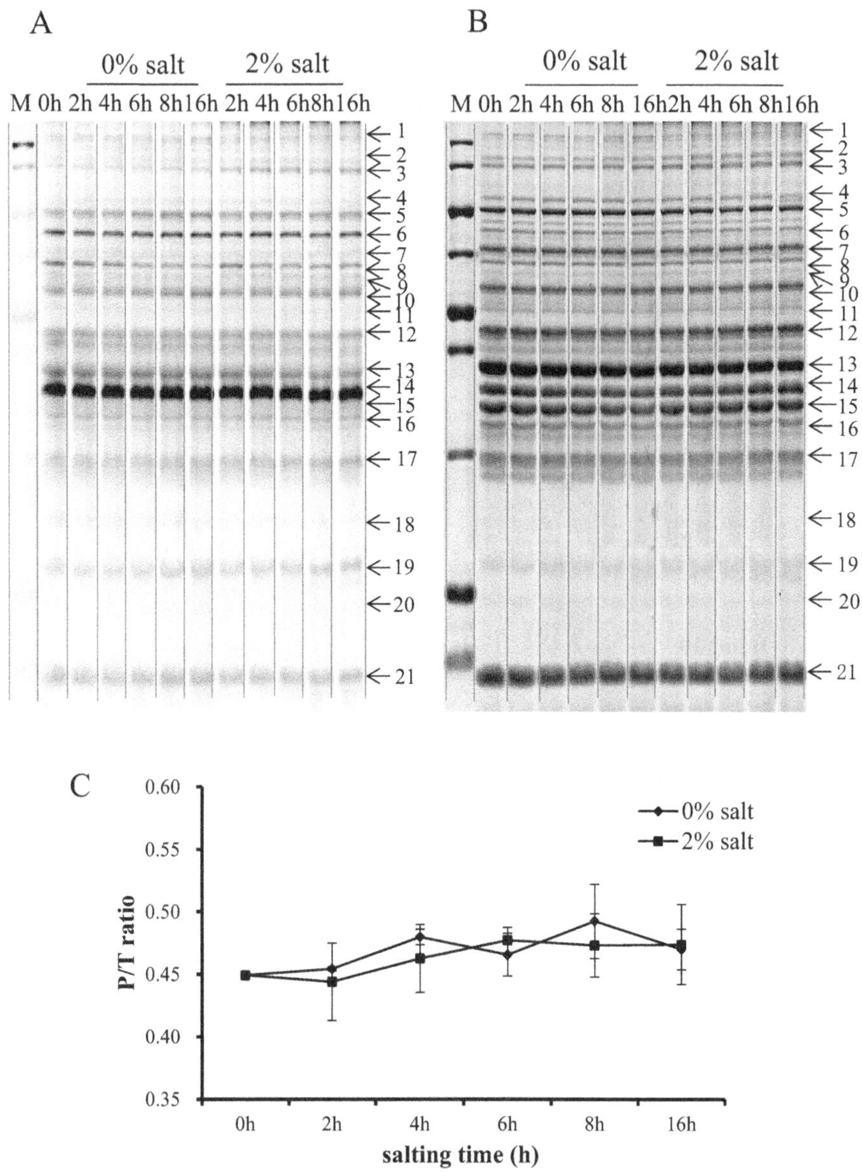

Fig. 11.3 Phosphorylation of sarcoplasmic proteins during salting (**a–b**) SDS-PAGE analysis of phosphorylated proteins by Pro-Q Diamond staining (**a**) and total sarcoplasmic proteins by SYPRO Ruby staining (**b**). (**c**) Quantification of global protein phosphorylation level of sarcoplasmic proteins at different time points during salting (Zhang et al. 2016a)

Pro-Q Diamond dye were conducted to detect phosphoproteins. As shown in Fig. 11.4a–c, no difference was determined ($P > 0.05$) in the P/T ratios among

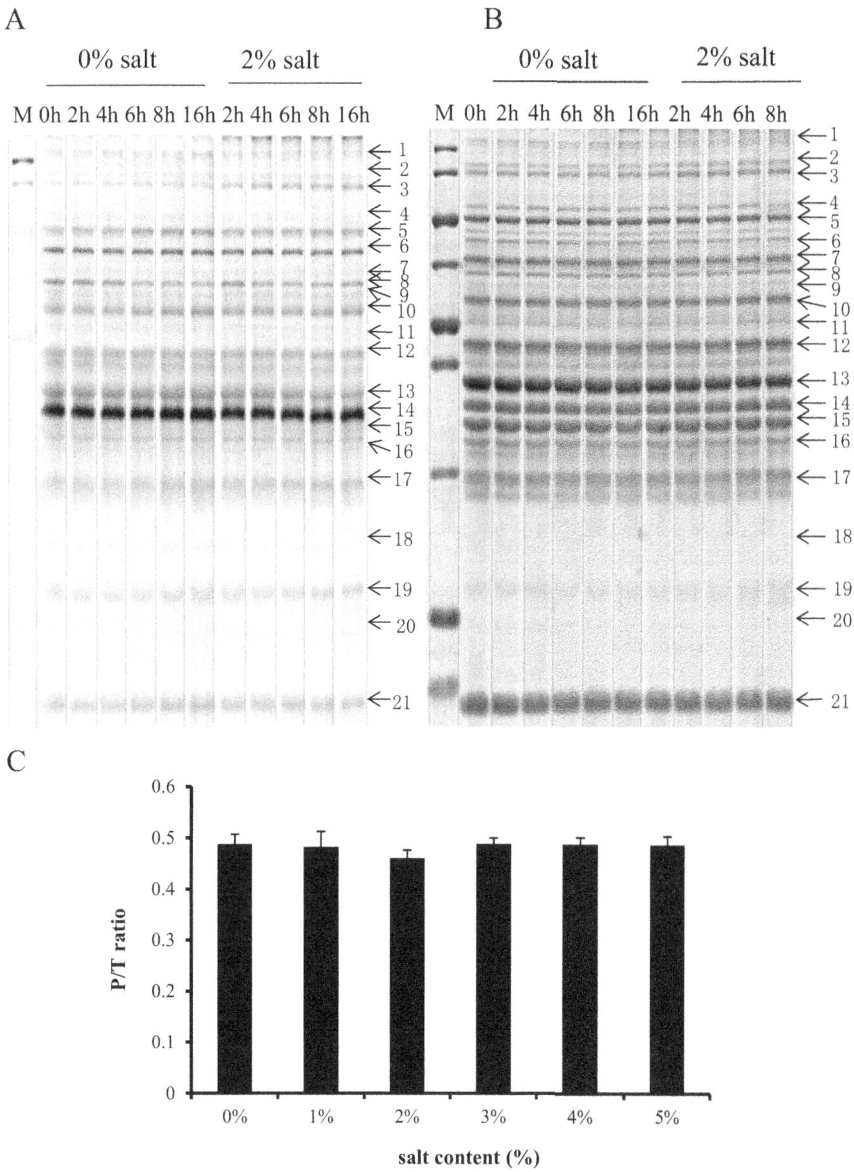

Fig. 11.4 Phosphorylation of sarcoplasmic proteins in response to salt concentrations. (**a–b**) SDS-PAGE analysis of phosphorylated proteins by Pro-Q Diamond staining (**a**) and total sarcoplasmic proteins by SYPRO Ruby staining (**b**). (**c**) Quantification of global protein phosphorylation level of sarcoplasmic proteins from meat salted at different salt concentrations. (**d**) Protein phosphorylation level of 15 individual protein bands (Zhang et al. 2016b)

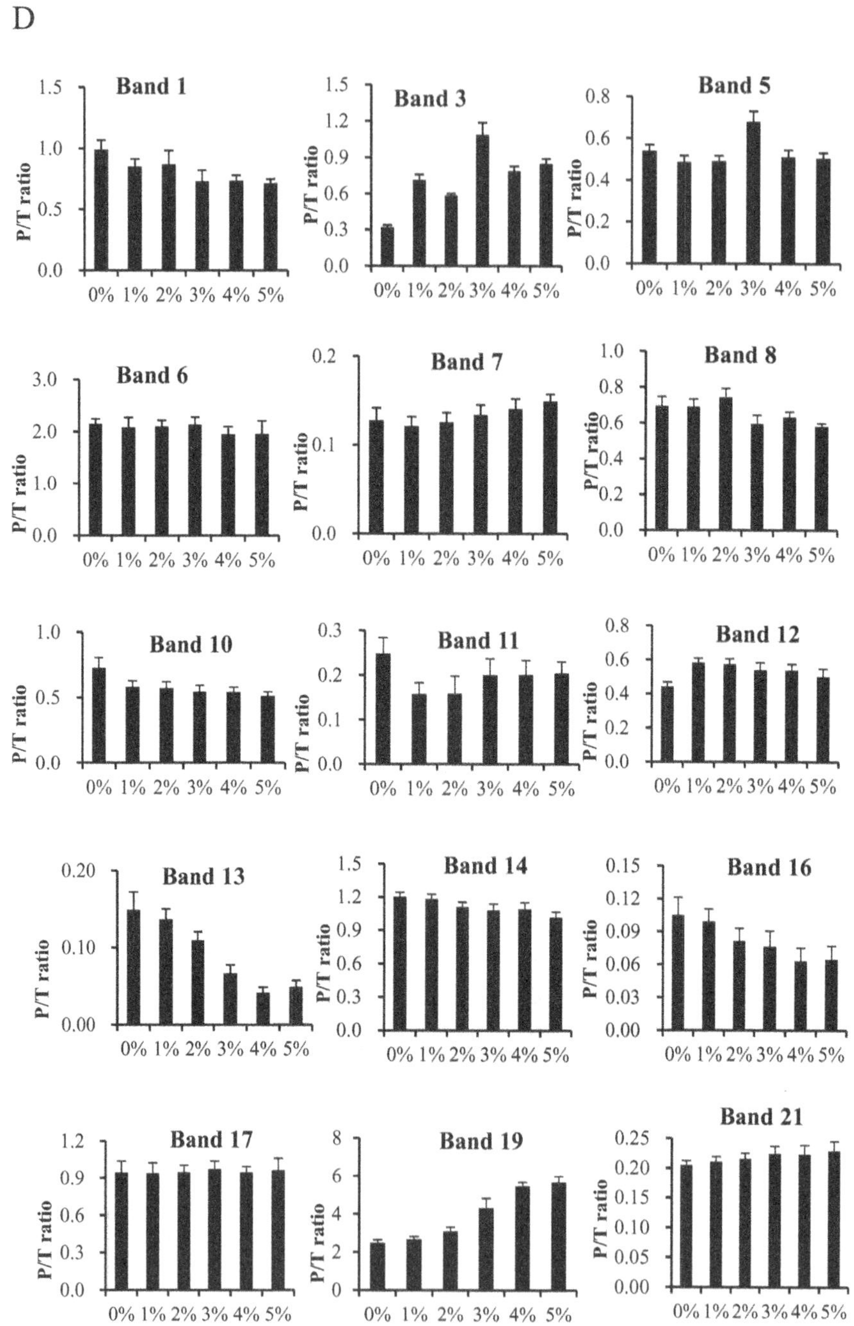

Fig. 11.4 (continued)

different treatments, confirming that salt had no significant effect on the global phosphorylation level of sarcoplasmic proteins of meat. However, the phosphorylation in most protein bands was affected by salt (Fig. 11.4d). The phosphorylation level of eight protein bands was decreased by salt, whereas that of five bands was increased by salt. Although salt had no significant effect on the global phosphorylation of sarcoplasmic proteins, it affected the phosphorylation of individual proteins. The results indicated that salt also had effect on the phosphorylation of sarcoplasmic proteins.

The identified sarcoplasmic proteins by LC-MS/MS are highly abundant. The P/T ratio is a semi-quantitative method for protein phosphorylation (Huang et al. 2011; Schulenberg et al. 2003; Silverman-Gavrila et al. 2009). Several distinctive phosphorylated bands of sarcoplasmic proteins were selected to be identified by LC-MS/MS. The functional categories of the identified proteins were matched manually on the basis of the retrieved information in the Protein Knowledge-base (UniProtKB in www.uniprot.org). According to the score, the theoretical molecular weight, number of matched peptides, and the coverage of the entire amino acid sequence (Chen et al. 2016), proteins were identified from each gel band (Table 11.1). In total, four categories of proteins were identified from these bands, including stress response, glycometabolism, oxidation or reduction, and others. Stress response proteins included heat shock protein alpha (HSPα) and heat shock protein 70 (HSP70). Glycometabolism proteins included glycogen phosphorylase (GP), glyceraldehyde-3-phosphate dehydrogenase (GAPDH), enolase (ENO), AMP-activated protein kinase (AMPK), aldolase (ALD), and phosphoglycerate kinase (PGK). Oxidation or reduction proteins included superoxide dismutase (SOD) and aldehyde dehydrogenase (ALDH). Other proteins included 14-3-3 protein and myoglobin (Mb).

11.2.3 Salting Influences the Phosphorylation of Glycolytic Enzymes

Glycolysis, which is regulated by glycolytic enzymes, is the principal way for energy supply in postmortem muscle and related to meat quality (Donnelly and Finlay 2015). The phosphorylation of many glycolytic enzymes or regulatory proteins (GP, GAPDH, ENO, AMPK, ALD, PGK) has been identified to be influenced by salting. Phosphorylation of glycolytic enzymes affected their activity and stability (Huang et al. 2011; Sale et al. 1987; Müller et al. 2015). GP was an enzyme involved in glycolysis reaction. It was activated into active enzyme form by phosphorylation, catalyzed by phosphorylase kinase, and then further promoted glycolysis (Müller et al. 2015). The phosphorylation of band 5 (GP) in 3% salt group was higher than other groups, so GP had higher activity, promoted glycolysis and then regulated meat tenderness. GAPDH had many post-translation modifications. Phosphorylation of GAPDH decreased its activity and then inhibited the glycolysis (Bell and Storey

Table 11.1 Identification of phosphorylated sarcoplasmic proteins by LC-MS/MS. Adapted from (Zhang et al. 2016b)

Protein names[a]	Band no.[b]	Accession no[a]	Species[c]	MW (Da)[d]	Score[e]	MP[f]	Seq. Cov.[g]	Function
Acetyl-CoA carboxylase 1	1	Q28559	OS	266,736	67	3	2%	Biotin enzyme
Uncharacterized protein	3	F1MHT1	BOS	176,199	19,863	769	69%	
Glycogen phosphorylase	5	O18751	OS	97,702	44,666	1713	77%	Glucose metabolism
Hydroxyacyl-coenzyme A dehydrogenase/3-ketoacyl-coenzyme A thiolase/enoyl-coenzyme A hydratase (Trifunctional protein) alpha subunit	6	B6UV59	OS	83,926	2687	79	50%	Coenzyme Binding
Heat shock protein alpha	6	A8DR93	OS	85,077	1181	37	34%	Stress response
Heat shock protein 70	7	H2DGR2	OS	70,536	5377	207	58%	Stress response
Heat shock protein 70	7	H2DGR0	OS	70,471	5208	199	49%	Stress response
HSP70	7	H9LHX6	OS	69,673	1130	38	8%	Stress response
Malic enzyme	8	B5SZL5	OS	64,601	705	23	61%	Glucose metabolism
AMP-activated protein kinase alpha2 subunit	8	A8D245	OS	62,833	50	1	9%	Glycolysis
Protein disulfide isomerase family A member 3	10	C6JUQ0	OS	57,379	546	24	39%	Rearrangement of -S-S- bonds
Aldehyde dehydrogenase 1 family member L1 (fragment)	10	I1U3B9	OS	56,692	242	9	3%	Oxidation reduction
Enolase (fragment)	11	Q9N113	OS	11,952	14,465	427	81%	Glycolysis
Phosphoglycerate kinase	12	B7TJ13	OS	44,936	13,304	446	73%	Glycolysis
Beta-actin variant 2	12	D7RIF5	OS	42,052	2113	109	51%	Sarcomeric function
Enolase (fragment)	13	Q9N113	OS	11,952	1974	50	79%	Glycolysis
Aldolase A (fragment)	13	Q9N116	OS	15,432	1900	58	61%	Glycolysis
Aldolase A (fragment)	14	Q9N116	OS	17,291	522	9	74%	Glycolysis
Glyceraldehyde-3-phosphate dehydrogenase	15	D7R7V6	OS	36,110	62,155	2031	78%	Glycolysis

Lactate dehydrogenase A (fragment)	16	Q9N109	OS	13,391	15,092	534	73%	Glycolysis
14-3-3 protein epsilon	17	P62262	OS	29,326	4074	140	64%	Cell apoptosis
Superoxide dismutase	18	C8BKD6	OS	24,736	2769	104	69%	Oxidation reduction
RAN	18	C5IJA0	OS	24,579	1113	46	61%	GTP binding
Cytochrome c oxidase subunit 2	18	B0EXQ6	OS	26,134	627	23	23%	Oxidation reduction
AK1	19	C5IJA8	OS	21,750	14,379	623	82%	Creatine energy metabolism
Peroxiredoxin 2	19	C8BKC5	OS	22,059	1768	76	69%	Cell redox homeostasis
Alpha-crystallin B chain	20	Q5ENY9	OS	20,054	6559	341	48%	Aggregation of proteins
S-phase kinase-associated protein 1A	20	B8QGA5	OS	18,817	257	7	49%	Kinase activity
Myoglobin	21	P02190	OS	17,044	32,908	962	97%	Oxygen transport

[a]Protein names and accession numbers were obtained from the UniProt database
[b]Band numbers of the identified proteins
[c]*OS* ovis aries, *BOS* bovine aries
[d]Molecular mass of identified proteins (recorded in the UniProt database)
[e]The MASCOT baseline significant score is 70. For the proteins identified in more than one band, the highest score was presented
[f]Number of matched peptides
[g]Percentage of coverage of the entire amino acid sequence

2014). Finally, although protein oxidation was not analyzed, the influence of salt on the phosphorylation status of SOD, peroxidase, and ALDH suggested that salting might also impact protein oxidation and thus meat quality.

11.2.4 Salting Regulates the Phosphorylation of Heat Shock Proteins

Heat shock proteins (HSPs) were identified in the phosphorylation proteins by LC/MS/MS, such as HSPα and HSP70. HSPs were stress response proteins, and they were divided into multiple subfamilies and had several post-translational modifications which altered their functional roles. HSPα and HSP70 were essential proteins in protection against stress. HSP70 in the myocardia, liver, and soleus muscles was prone to phosphorylation after exercise (González and Manso 2004; Melling et al. 2008). In this study, the phosphorylation of HSPα and HSP70 was detected to be reduced by salting, which may alter their functional roles. Since HSPs had impact on meat tenderness (Carvalho et al. 2014), our study further revealed that salting might alter meat tenderness by regulating the phosphorylation of HSPs.

11.2.5 Research Prospect

In summary, salt influenced phosphorylation and dephosphorylation of muscle proteins during salting. Salt decreased the global phosphorylation level of myofibrillar proteins and affected the phosphorylation level of individual sarcoplasmic proteins. The study may provide a new insight into the mechanism by which salting influences the quality of processed meat products, especially the meat tenderness which is related to protein degradation and muscle contraction. In future, the influencing pathway of salt on myofibrillar and sarcoplasmic protein phosphorylation should be studied more clearly in vivo or in vitro, thus making it possible to explain the protein structure and dynamic function systematically.

11.3 Identification of Specific Phosphorylated Proteins Induced by Ionic Strength

Salting may influence meat quality through protein phosphorylation, which regulates glycolysis metabolism, protein function, and degradation. In order to explore it, ten topside muscles of crossbred sheep were ground, mixed, and divided into two groups, which were cured for 16 h with 0 and 3% NaCl, respectively. Muscle proteins of two cured groups were analyzed by two-dimensional electrophoresis coupled with Pro-Q Diamond and SYPRO Ruby staining. The different phosphorylated proteins were determined using LC-MS/MS and the UniProt database. Ten different phosphoproteins (>1.5 fold) induced by salting were identified, including triosephosphate isomerase, glycogen phosphorylase, creatine kinase M-type, myoglobin, troponin T fast skeletal muscle type, actin, myosin light chain 1/3, tropomyosin beta chain, etc. Most of the different phosphoproteins were involved in glycometabolism, protein function, and protein degradation (Wang et al. 2017).

11.3.1 Identification of Differentially Phosphorylated Proteins After Salting

After electrophoresis, gels were stained with Pro-Q Diamond for phosphoproteins (P) and with SYPRO Ruby for total proteins (T). The relative phosphorylation level of proteins was evaluated by the ratio of relative protein abundance between control and salted group. As shown in Fig. 11.5, 13 protein spots were detected to be differentially phosphorylated (ratio of relative protein abundance <2/3 or >1.5). The isoelectric point (pI) of the differentially phosphorylated proteins was pH 5–8 and the molecular weight was 10–100 kDa. The proteins within the 13 spots were identified by LC-MS/MS, including myoglobin (1, MB), myosin light chain 1/3-skeletal muscle isoform (2, 20, 81, MYL1), tropomyosin beta chain (10, TPM2), actin (18, 62, 84), troponin T fast skeletal muscle type (56, 70, TNT), apolipoprotein A-I preproprotein (124, APOA1), triosephosphate isomerase (125, TPI) and F-actin-capping protein subunit alpha-2 (142, CAPZA2). The relative abundance of phosphoproteins within the 13 spots was quantified by densitometric analysis (Fig. 11.6). The phosphorylated values of spot 2, 81, 84, 124, 125, and 152 in control group were higher than in salt group, and the values of other spots were lower in control group.

The two-dimensional electrophoresis (2-DE) gel images with SYPRO Ruby staining of control group and salt group were showed in Fig. 11.7. Four differential protein spots (>1.5 fold) on the SYPRO Ruby staining gels were excised out with pipette tip, trypsin digestion and LC-MS/MS analysis were conducted. As a result, creatine kinase- M type (95, 100, CK), troponin T fast skeletal muscle type (97, TNT), and glycogen phosphorylase (293, GP) were identified. The gray level value of the four differential proteins was presented in Fig. 11.8.The gray level of spots 97 and 293 was not differential (1.5 > ratio of relative protein abundance >2/3) in Pro-Q Diamond staining (P), but difference was much higher in salt group (ratio of relative protein abundance <2/3 or >1.5) subjected to SYPRO Ruby staining (T), resulting in higher phosphorylation level (P/T) of CK (97) and GP (293) in salt group than that in the control group. Similarly, the value of spot 95 and 100 in control group was lower than the salt group. Conclusively, the 17 spots were identified to be 10 kinds of protein (Tables 11.2 and 11.3), which were classified to glycolysis metabolism enzymes (GP, CK, TPI), function proteins (myoglobin, APOA1), and muscle contractile relative proteins (MYL1, TPM2, actin, TNT, CAPZA2).

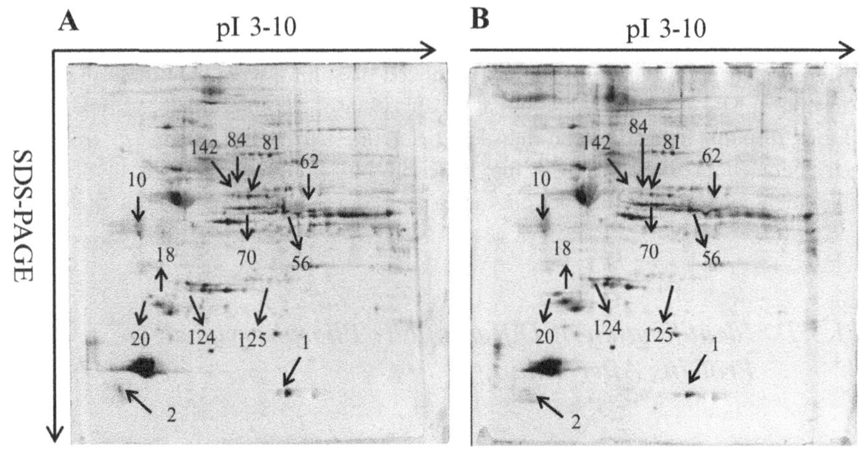

Fig. 11.5 Images of gels stained with Pro-Q Diamond to detect phosphoproteins. (**a**) control; (**b**) meat salted with 3% NaCl (Wang et al. 2017)

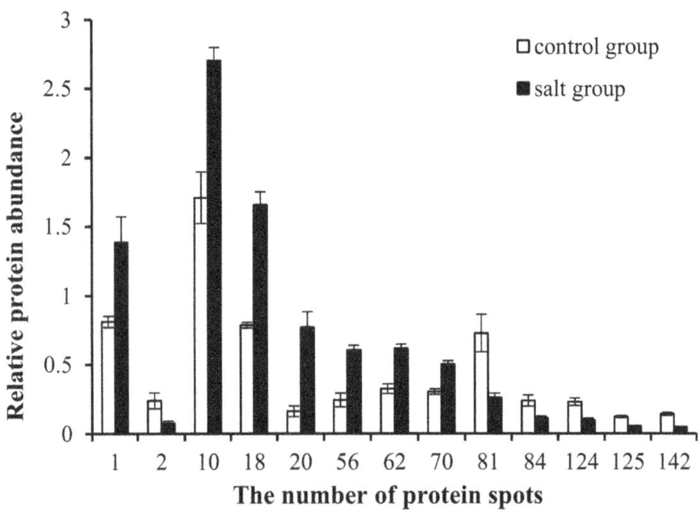

Fig. 11.6 Quantification of phosphoproteins stained by Pro-Q Diamond in selected spots which were different between control and salted meat in protein abundance (the ratio <2/3 or >1.5) (Wang et al. 2017)

11.3.2 Differential Phosphorylated Proteins Regulating Glycolysis Metabolism

The phosphorylation level of some glucometabolism enzymes from salt group was higher than that in the control group, including glycogen phosphorylase (GP),

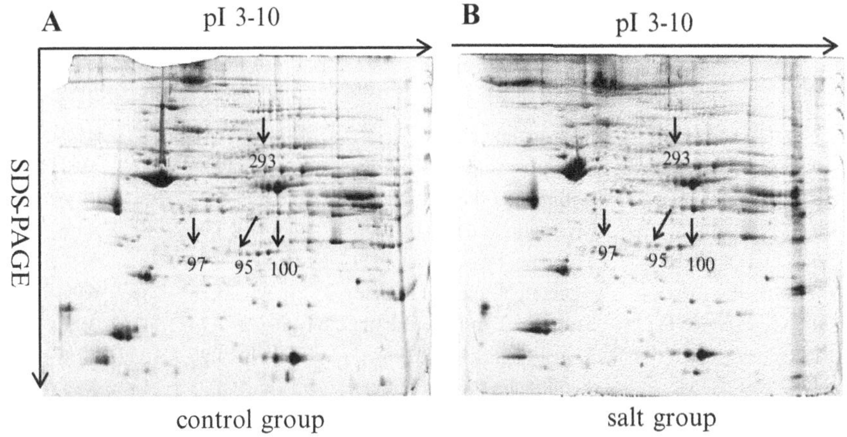

Fig. 11.7 Images of gels stained with SYPRO Ruby for total proteins (**a**) control; (**b**) meat salted with 3% NaCl (Wang et al. 2017)

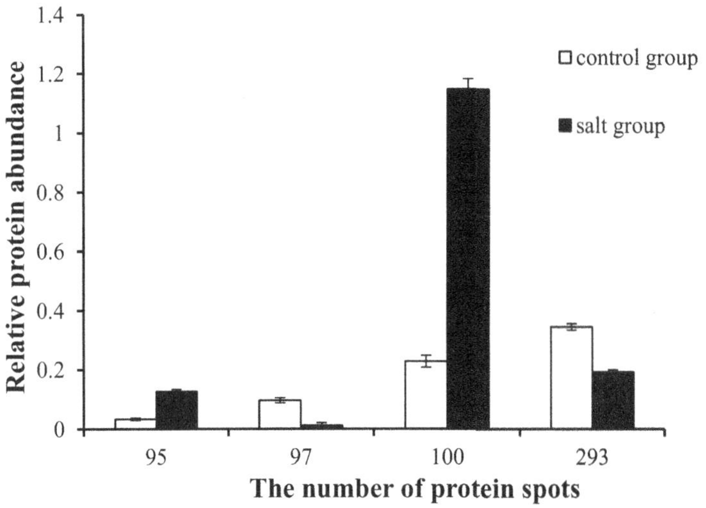

Fig. 11.8 Quantification of proteins stained by SYPRO Ruby in selected spots which were different between control and salted meat in protein abundance (the ratio <2/3 or >1.5) (Wang et al. 2017)

triosephosphate isomerase (TPI), and creatine kinase M-type (CK). The addition of sodium chloride to these enzymes might change glycolysis in muscle and influence the meat color, tenderness, and water holding capacity. GP could be divided into liver type, muscle type, and brain type. GP was one of the glycolytic enzymes which limited the glycolysis rate. Agius et al. found out that phosphorylase kinase could phosphorylate GP on serine 14, which transformed the inactive form (GPb) to the

Table 11.2 Proteins identified to be differentially phosphorylated in control and salted muscles (Wang et al. 2017)

Spot no[a]	Accession no[b]	Protein names[b]	PI[c]	MP[d]	Seq. Cov[e]
1	P02190	Myoglobin	6.9	15	74%
2	A0JNJ5	Myosin light chain 1/3	4.96	11	41%
10	Q5KR48	Tropomyosin beta chain	4.66	43	89%
18	P68138	Actin(alpha skeletal muscle)	5.23	27	71%
20	A0JNJ5	Myosin light chain 1/3	4.96	8	44%
56	Q8MKH8	Troponin T(fast skeletal muscle type)	7.74	25	55%
62	P68138	Actin(alpha skeletal muscle)	5.23	28	61%
70	Q8MKH8	Troponin T(fast skeletal muscle type)	7.74	11	30%
81	A0JNJ5	Myosin light chain 1/3	4.96	12	47%
84	P68138	Actin(alpha skeletal muscle)	5.23	24	59%
124	V6F9A2	Apolipoprotein A-I preproprotein	5.71	12	36%
125	Q5E956	Triosephosphate isomerase	6.45	21	87%
142	Q5E997	F-actin-capping protein subunit alpha-2	5.57	8	24%

[a]Spot number of the identified proteins
[b]Protein names and accession numbers were obtained from the UniProt database
[c]The isoelectric points of identified proteins
[d]Number of matched peptides
[e]Percentage of coverage of the entire amino acid sequence

Table 11.3 Proteins identified to be different in total protein abundance between control and salted meat (Wang et al. 2017)

Spot no[a]	Accession no[b]	Protein names[b]	PI[c]	MP[d]	Seq. Cov[e]
95	Q9XSC6	Creatine kinase (M-type)	6.63	30	58%
97	Q8MKH8	Troponin T(fast skeletal muscle type)	7.74	12	34%
100	Q9XSC6	Creatine kinase M-type	6.63	30	75%
293	O18751	Glycogen phosphorylase	6.65	25	79%

[a]Spot number of the identified proteins
[b]Protein names and accession numbers were obtained from the UniProt database
[c]The isoelectric points of identified proteins
[d]Number of matched peptides
[e]Percentage of coverage of the entire amino acid sequence

active form (GPa) (Agius 2015). The protein phosphorylation level of GP of salt group was higher than the control group, so the enzyme activity of salt group was higher than the control group. Then glycolysis process was accelerated and produced acidic components, indirectly influencing the meat color and tenderness (Cottrell et al. 2008; Huang et al. 2014).

TPI exists in multiple tissues, which is an important isomerase involved in glycolysis. TPI catalyzes the reversible interconversion of dihydroxyacetone phosphate (DHAP) and glucose aldehyde-3-phosphate (G3P) (Hipkiss 2011). Only G3P can proceed further down the glycolytic pathway (Fonvielle et al. 2005). Phosphorylated TPI was a direct substrate of cyclin-dependent protein kinase 2 (Cdk2). Lee et al. (2010) discovered that phosphorylation of TPI decreased its enzyme activity, which was not conducive to the conversion of G3P to DHAP, and then decreased the rate of glycolysis. In the study, salt decreased the phosphorylation level of TPI, which was likely to increase the enzyme activity, probably resulting in accelerating glycolysis process and indirectly improving meat tenderness (Cottrell et al. 2008).

CK ubiquitously exists in skeletal muscles, brain tissues, and other tissues. It is well known that CK is one of the most critical enzymes regulating ATP generation and energy metabolism (Quest et al. 1990; Dieni and Storey 2009). Brain-type creatine kinase can be phosphorylated by AMP-activated protein kinase (AMPK) on serine 6 (Ramírez Ríos et al. 2014). Muscle-type creatine kinase can be phosphorylated by protein kinase A, protein kinase G (PKG), protein kinase C (PKC), and AMPK. Phosphorylation of CK increased its activity, whereas dephosphorylation decreased its activity (Dieni and Storey 2009). The relative protein abundance of CK stained by SYPRO Ruby in salt group was much higher than without salt addition, indicating the phosphorylation of CK might decrease and retard enzyme activity. The reduction of CK activity probably lowered the ATP generation from phosphocreatine and ADP, with the result that ATP was only produced from glycolysis metabolism (Ishida et al. 1994; Grehl et al. 1998).

11.3.3 Differential Phosphorylated Proteins Changing the Protein Function

Myoglobin was also the differential phosphoprotein between the two groups, whose phosphorylation level was higher in the salt group. On the other hand, it was well known that sodium chloride caused the muscle or meat paste discoloration, hence, the phosphorylation level of myoglobin was negative correlated with meat color (Kanner et al. 1991). Huang also found out that myoglobin could be modified by phosphorylation (Huang et al. 2011). Canto et al. further discovered that there were 4 differential phosphorylated myoglobin isoforms by 2-DE gels (Canto et al. 2015), which was due to phosphorylation modification would change myoglobin pI (Zhu et al. 2005). Thus, the phosphorylation modification of myoglobin might actually influence meat color stability and was more likely to make the meat dark or brown, which might be due to phosphorylation accelerated the iron release from myoglobin and tended to be oxidized (Devatkal and Naveena 2010).

Apolipoprotein A-I preproprotein (APOA1) is part of high-density lipoprotein (HDL), which carries cholesterol from tissues to liver, and associated with protection against cardiovascular disease. APOA1 reduces cardiovascular disease significantly, even though it causes a reduction in HDL level and an increase in triglyceride level (Chistiakov et al. 2016; Midtgaard et al. 2015). APOA1 exists in muscle but there is no well understanding of its function related to phosphorylation modification. The salting treatment decreased APOA1's phosphorylation in muscle, but its function to meat quality (lipid oxidation, lipolysis) and meat nutrition should be further investigated.

11.3.4 Differential Phosphorylated Proteins Regulating the Muscle Contractile and Protein Dissociation

The differential phosphorylated proteins identified between salt group and control group included TNT (56, 70, 97), actin (18, 62, 84), TPM2 (10), CAPZA2 (142), and MLC1 (2, 20, 81), while these proteins were all belonged to myofibrillar proteins. Actin, TPM2, and MLC1 might affect the muscle contractile, and TNT, TPM2, and CAPZA2 might influence actomyosin dissociation. The phosphorylation of these myofibrillar proteins reduced their degradation by calpains, and maintained the sarcomere texture (Zhang et al. 1988; Di Lisa et al. 1995). According to the relative phosphorylation (gray level of 56 and 70 in Fig. 11.5, divided by gray level of 97 in Fig. 11.7), it indicated that salt curing increased the phosphorylation level of TNT significantly. TNT contained several phosphorylation sites, while serine 2 was highly phosphorylated (Katrukha and Gusev 2013). The phosphorylation of TNT induced by protein kinase C made it more easier to be degraded by calpains (Toyooka 1982). The relative phosphorylation level of TNT in salt group was much higher than in no salt group, resulting in TNT being degraded easily by calpain.

Tropomyosin (TPM2) is an important protein for regulating muscle contraction and the most important phosphorylated protein in the muscle filament. Lehman found out that tropomyosin was phosphorylated at serine 283 position, then enhanced the binding of the head and tail of tropomyosin (Lehman et al. 2015). The phosphorylation of TPM2 was necessary for long-range cooperative activation of myosin binding (Rao et al. 2009). The phosphorylation of TPM2 in 3% salt group was higher than 0% control group, which might strengthen the binding with actin or myosin, resulting in slowing the actomyosin dissociation.

Myosin and actin were the main components of myofibrillar proteins, while myosin regulatory light chain could change myosin conformation and regulate the binding of myosin with actin. Myosin light chain 1/3 was myosin regulatory light chain. Some authors reported that MLC1's phosphorylation would inhibit actomyosin dissociation, which quickened muscle contraction and restrained the myofibril structure (Miller et al. 2011; Alamo et al. 2008). Other authors reported that actin could be phosphorylated and dephosphorylated on tyrosine residues (Gauthier et al. 1997). Phosphorylation of actin made it difficult to be degraded by μ-calpain (Li et al. 2017), resulting in good texture of meat during the salting. Similar to actin, F-actin-capping protein subunit alpha-2 (CAPZA2) was also a kind of myofibrillar proteins, which bound in a calcium-independent manner to the fast growing ends of actin filaments (barbed end), thereby restricting its growth or dissociation (Hart et al. 1997). The phosphorylation level of CAPZA2 was lower in salt group, which might reveal that salt weakened the binding of CAPZA2 and actin through phosphorylation modification.

Salt curing influences phosphorylation and dephosphorylation of muscle proteins, which accelerates glycolysis metabolism, weakens the protein positive function to meat quality, and regulates the muscle contractile to cause the actomyosin dissociation. Probably, phosphorylation modification altered by salt curing might be a new mechanism to regulate meat quality, especially the meat color and tenderness. In the future, the regulation mechanism of muscle protein function by phosphorylation under sodium chloride should be expounded, moreover, some essential methods were supposed to be invented.

11.4 Effect of Salting Temperature on the Protein Phosphorylation of Muscle

The effect of different salting temperature on the phosphorylation level of myofibrillar and sarcoplasmic proteins in mutton was investigated. The topside muscles from both hind legs were removed immediately after slaughter and aged at 4 °C for 24 h, and then three salting treatment groups were set up at −1 °C, 4 °C, and 25 °C. Samples from each group were collected at 0, 8, 16, and 24 h, and sodium dodecyl sulfate polyacrylamide gel electrophoresis (SDS-PAGE) combined with fluorescence staining were employed to analyze phosphorylation level. The results showed that the effect of salting temperature on the phosphorylation level of individual protein bands was different, while salting temperature could regulate the meat quality by influencing the phosphorylation of muscle proteins.

11.4.1 Effect of Salting Temperature on the Myofibrillar Protein Phosphorylation

The SDS-PAGE electrophoresis of myofibril protein after curing for 0, 8, 16, and 24 h at freezing temperature (−1 °C), 4 °C and room temperature (25 °C) was shown in Fig. 11.9. Figure 11.9a shows the phosphorylated myofibril protein with Pro-Q Diamond staining, and Fig. 11.9b shows the whole myofibril protein with SYPRO Ruby staining. As shown in Fig. 11.9, the separation effect of protein bands was better. The relative optical density of each band was analyzed by the Quantity One 4.6.2 software to obtain the P/T ratio, so as to obtain the changes in the total phosphorylation level of myofibril protein during the curing process at different temperature.

Changes in the total phosphorylation level of myofibril protein during the curing process at different temperatures were shown in Fig. 11.10. As can be seen from the figure, the total phosphorylation level of myofibril protein in the ice temperature treatment group was higher than that in the 4 °C group, while that in the 4 °C group was higher than that in the room temperature treatment group. However, the ice temperature treatment group was significantly higher than the room temperature treatment group at 8 h of curing ($P < 0.05$), and there was no significant difference among the three treatment groups at 16 and 24 h of curing. The results indicated that low curing temperature was conducive to maintaining the total phosphorylation level

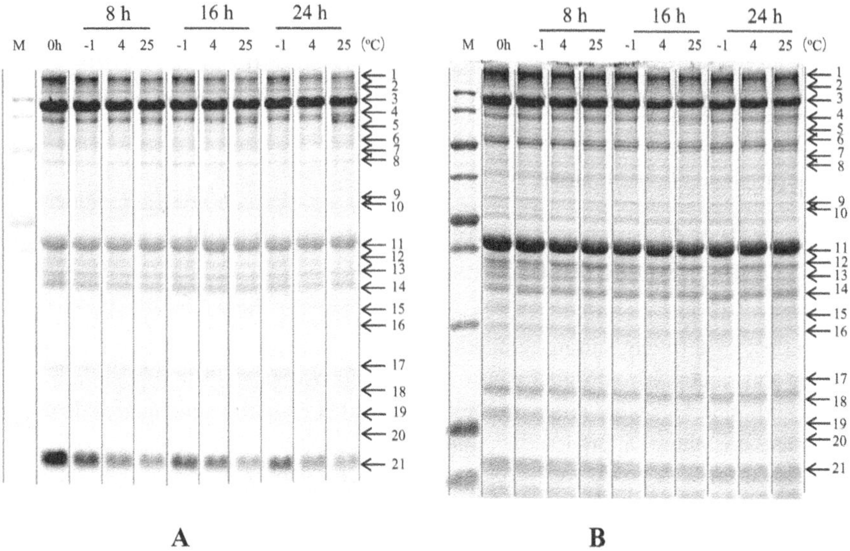

Fig. 11.9 Gel images of phosphoproteins and total myofibrillar proteins of different salting temperature separated by SDS-PAGE. (**a**) Image of phosphorylated myofibrillar proteins by Pro-Q Diamond staining; (**b**) Image of total myofibrillar proteins by SYPRO Ruby staining (Zhang et al. 2016b)

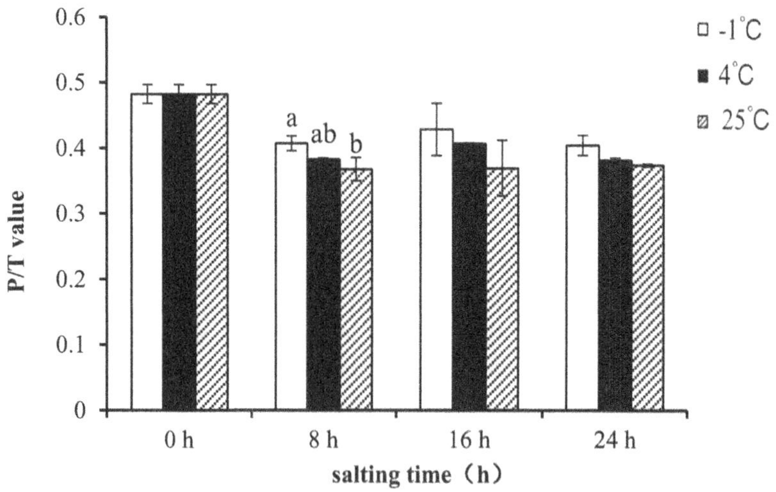

Fig. 11.10 Phosphorylation of myofibrillar proteins during different salting temperature. Different letters at the top of the bar indicate significant difference ($P < 0.05$). Adapted from (Zhang et al. 2016b)

of myofibril protein. After curing, the total phosphorylation level of myofibril protein was significantly lower than that of raw meat at 0 h, and the protein phosphorylation level of all three groups showed a downward trend after curing, indicating that salt reduced the total phosphorylation level of myofibril protein, which was consistent with the experimental results in Chap. 2.

Four important myofibrillar protein bands (band 3, 4, 11, and 21) were selected to analyze the effect of curing temperature on its phosphorylation level, as shown in Fig. 11.11.The phosphorylation level of band 3 in the ice temperature treatment group was higher than that in the 4 °C treatment group at 8, 16, and 24 h after curing, and that in the 4 °C treatment group was higher than that in the room temperature treatment group, but there was no significant difference between the different treatment groups. However, the phosphorylation level of band 21 at 8, 16, and 24 h after curing was significantly higher in the ice temperature treatment group than in the 4 °C treatment group, as well as in the 4 °C treatment group than in the room temperature treatment group. This indicated that the curing temperature was within the range of ice temperature to room temperature, and the lower the curing temperature was, the better the phosphorylation level of band 3 and 21 would be. The phosphorylation level of band 4 and 11 was opposite to that of band 3 and 21. The phosphorylation level of band 11 in the room temperature treatment group was higher than that in the ice temperature treatment group, but there was no significant difference. The phosphorylation level of band 4 in the room temperature group was significantly higher than that in the ice temperature group and 4 °C treatment group ($P < 0.05$). It was indicated that the higher the curing temperature, the more beneficial the improvement of the phosphorylation level of band 4 and 11.

Overall, the whole phosphorylation level of myofibrillar protein in the ice temperature group was significantly higher than that in the other two groups, and the effect of curing temperature on the phosphorylation level of different protein bands was different. According to the mass spectrometry identification results of Chen and Huang et al. (Chen et al. 2016; Huang et al. 2012), the phosphorylation level of some important myofibrillar protein bands was selected and analyzed, as shown in Fig. 11.11. Band 4 was mainly myosin binding protein C (MYBPC), and MYBPC phosphorylation affected muscle contraction through specific sites. Wang et al. (2014) found out that MYBPC was an allosteric activator of myosin, thus enhancing the formation of transverse bridging force. The phosphorylation of MYBPC could regulate this transverse bridging force and accelerate the formation of contraction force (Wang et al. 2014). Band 21 was mainly myosin light chain. The phosphorylation of myosin light chain was regulated by many enzymes, such as myosin regulated light chain kinase (MLCK), protein kinase A, and protein kinase C (Ikebe et al. 1999). Many studies have proved that the phosphorylation of myosin light chain was related to muscle contraction (Farman et al. 2009; Taylor et al. 2014). Phosphorylation can change the structure of myosin light chain, and phosphorylation of myosin light chain 2 can increase the number of strong binding transverse bridges of actin (Miller et al. 2011). The degree of contractility enhancement was positively correlated with the degree of myosin light chain 2 phosphorylation (Stull et al. 2011). In this experiment, the phosphorylation level of myosin light chain in the ice temperature group was significantly higher than that in the other two groups, and the muscle contraction was the strongest, which was not conducive to the formation

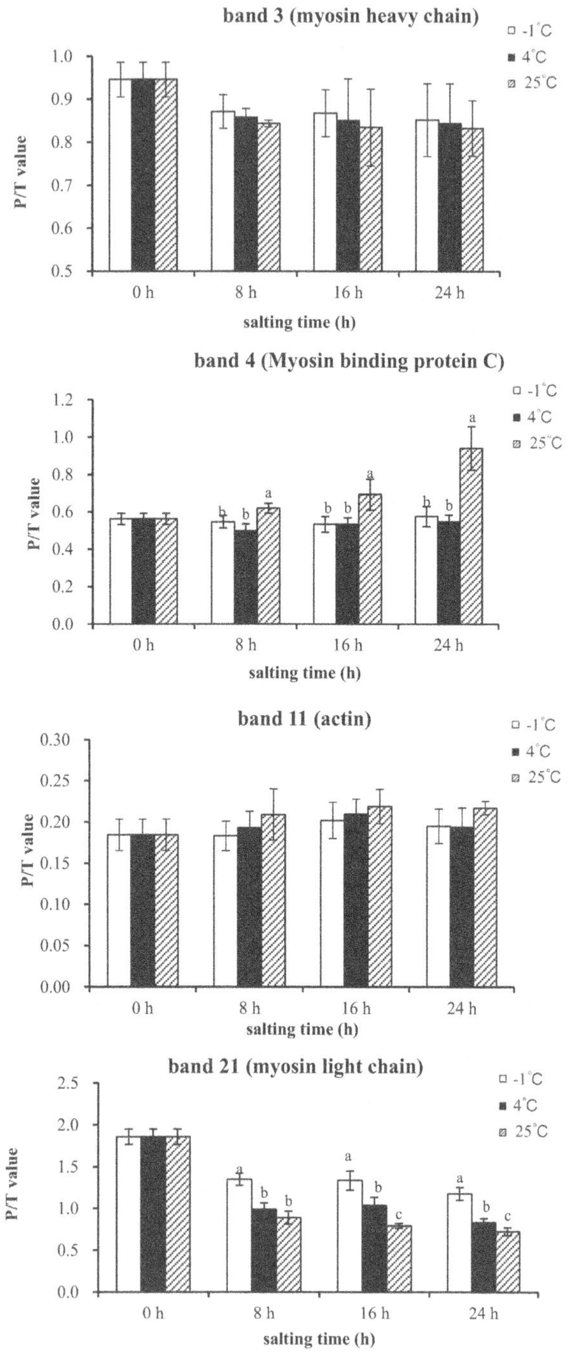

Fig. 11.11 The phosphorylation level of important myofibrillar proteins in different temperature groups. Different letters at the top of the bar indicate significant difference ($P<0.05$). Adapted from (Zhang et al. 2016b)

of tenderness. In addition to the above proteins, band 3 was mainly myosin heavy chain and band 11 was mainly actin, and the phosphorylation level of the above two bands was not significantly different among the three treatment groups ($P > 0.05$).

Studies have found that protein phosphorylation was not conducive to protein degradation by calpain (Zhang et al. 1988, 2012). Calpain in muscle can promote meat tenderization by degrading myofibrillar protein. However, the whole phosphorylation level of myofibrillar protein in the ice temperature salting group was high, which was not conducive to the degradation of myofibrillar protein and the tenderization of meat. The whole phosphorylation level of myofibrillar protein was the lowest at room temperature. Compared with the other two treatment groups, the protein was degraded to a large extent, which was conducive to the tenderization of meat. Therefore, it was of great significance to study the effect of different temperature on protein phosphorylation in muscle to regulate meat quality.

11.4.2 Effect of Salting Temperature on the Sarcoplasmic Protein Phosphorylation

The SDS-PAGE electrophoresis patterns of sarcoplasmic proteins salted at ice temperature (−1 °C), 4 °C and room temperature (25 °C) for 0, 8, 16, and 24 h were shown in Fig. 11.12. Figure 11.12a shows phosphorylated sarcoplasmic proteins stained with Pro-Q Diamond, and Fig. 11.12b shows total sarcoplasmic proteins stained with SYPRO Ruby. Twenty clear protein bands were selected from the gel for total phosphorylation level analysis, so did the myofibril protein.

The total phosphorylation level of sarcoplasmic proteins during salting at different temperature was shown in Fig. 11.13. The total phosphorylation level of sarcoplasmic protein in the three temperature treatment groups was different only at 8 h, but there was no significant difference at 16 and 24 h. After salting for 8 h, the total phosphorylation level of sarcoplasmic protein in ice temperature group was significantly lower than that of 4 °C group and room temperature group ($P < 0.05$), but there was no significant difference between 4 °C group and room temperature group ($P < 0.05$).

According to the results of electropherogram, four protein bands (band 14, 15, 16, and 19) with dark color and obvious gray value were selected to analyze the effect of different salting temperature on their phosphorylation level. The results were shown in Fig. 11.14. The total phosphorylation level of the four bands was affected by salting temperature. The P/T values of band 14 and 19 showed the same overall trend. The ice temperature group was significantly higher than the 4 °C group ($P < 0.05$), and the 4 °C treatment group was significantly higher than the room temperature group ($P < 0.05$). The results showed that when the salting temperature was in the range of ice temperature to room temperature, the lower the salting temperature was, the more favorable it was to maintain the phosphorylation level of band 14 and 19. The P/T value of band 15 in the room temperature group was

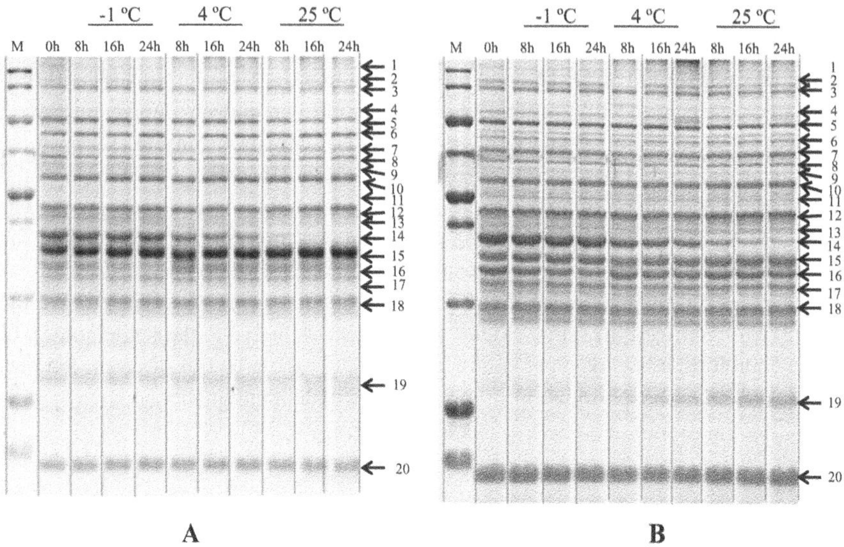

Fig. 11.12 Gel images of phosphoproteins and total sarcoplasmic proteins of different salting temperature separated by SDS-PAGE. (**a**) Image of phosphorylated sarcoplasmic proteins by Pro-Q Diamond staining; (**b**) Image of total sarcoplasmic proteins by SYPRO Ruby staining (Zhang et al. 2016b)

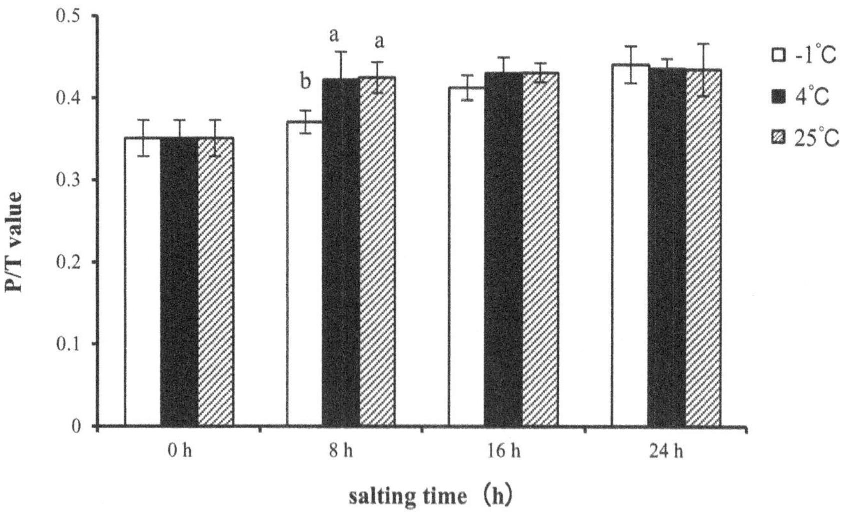

Fig. 11.13 Phosphorylation of sarcoplasmic proteins during different salting temperature. Different letters at the top of the bar indicate significant difference ($P<0.05$). Adapted from (Zhang et al. 2016b)

Fig. 11.14 The phosphorylation level of sarcoplasmic myofibrillar proteins in different temperature groups. Adapted from (Zhang et al. 2016b)

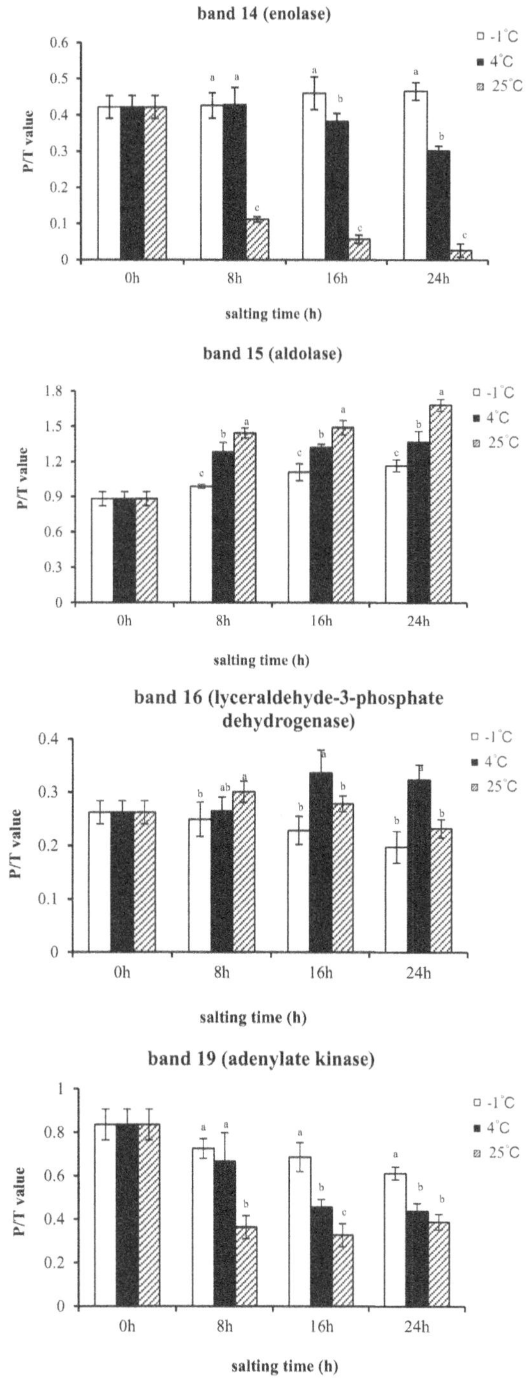

significantly higher than the 4 °C group ($P < 0.05$), and in the 4 °C group was significantly higher than the ice temperature group ($P < 0.05$), which was opposite to that of band 14 and 19. When being salted for 16 h or 24 h, the band 16 in the 4 °C group was significantly higher than that in the other two groups ($P < 0.05$).

The results showed that the total phosphorylation level of sarcoplasmic protein in ice temperature treatment group was lower, but the effect of salting temperature on the phosphorylation of different sarcoplasmic protein bands was significantly different. According to the mass spectrometric results of sarcoplasmic proteins in the second chapter, the protein names of the four selected protein bands were determined. Band 14 may be enolase. Enolase was a rate-limiting enzyme in the pathway of postmortem muscle glycolysis, which catalyzed the transformation between 2-phosphoglyceric acid and phosphoenolpyruvate. The level of phosphorylation decreased significantly with the increase of salting temperature ($P < 0.05$). The results showed that high temperature promoted the dephosphorylation of enolase. Band 15 may be aldolase, and its phosphorylation level increased with the increase of salting temperature. Aldolase was also an important enzyme in the process of glycolysis, and the change of its phosphorylation level may in turn regulate the process of glycolysis by affecting the activity of the enzyme. Band 16 may be glyceraldehyde 3-phosphate dehydrogenase (GAPDH), which was an intermediate enzyme in glycolysis. The activity of GAPDH decreased after phosphorylation, which was not conducive to glycolysis (Bell and Storey 2014). As can be seen from the electropherogram Fig. 11.12b, the gray value of band 14 decreased with the increase of salting time and salting temperature, indicating that the degradation degree of band 14 increased. The band 19 showed the opposite phenomenon, and its gray value increased with the increase of salting time and salting temperature, so it was speculated that band 19 was probably the degradation product of band 14.

The optimum temperature of intestinal alkaline phosphatase (AP) of silver carp was 25 °C (Xu 2007). Therefore, the activity of AP was the highest at room temperature, and dephosphorylation was more likely to occur. Band 14 had the lowest level of phosphorylation at room temperature, indicating that alkaline phosphatase played a full role in the dephosphorylation of enolase, resulting in a low level of phosphorylation. The highest phosphorylation level of band 15 at room temperature indicated that aldolase might not be the substrate of alkaline phosphatase. Other studies have also discovered that temperature also affected the activity of protein kinases. For example, the activity of AMP-activated protein kinase (AMPK) increased with the increase of temperature at 20–32 °C, but decreased at 32–36 °C (Jost et al. 2015). Therefore, the effect of salting temperature on protein phosphorylation was mainly due to the effect of salting temperature on the activities of protein kinase and phosphatase.

Enzyme phosphorylation, especially glycolytic enzyme phosphorylation, can change the activity and stability of the enzyme itself, but the effects of phosphorylation on the activity of different enzymes are different (Huang et al. 2011). The results showed that the lower the salting temperature was, the lower the total phosphorylation level of sarcoplasmic protein was, but the effect on the phosphorylation level of different enzymes was different. It was speculated that the phosphorylation of proteins was regulated by different kinases. The tolerance of these kinases to temperature was not the same. Combined with the results of previous studies, temperature played an important role in meat quality by affecting protein phosphorylation and regulating enzyme activity, and then affecting glycolysis and protein degradation.

In summary, salting temperature will affect the phosphorylation level of myofibrillar protein and sarcoplasmic protein. Protein phosphorylation is regulated by a variety of protein kinases and phosphatases, and there are many kinds of protein kinases (about 300), and phosphatase also has a variety of isozymes. Temperature is an important factor in enzyme activity. Different enzymes have different optimum temperatures (Daniel and Danson 2013), and protein phosphorylation is regulated by many kinds of protein kinases and phosphatases; therefore, the phosphorylation level of different protein bands is affected by temperature. Salting temperature affects the level of protein phosphorylation by altering the activity of enzymes related to protein phosphorylation, thus further regulating the quality of salted meat.

The phosphorylation level of muscle protein was affected by curing temperature. The whole phosphorylation level of myofibrillar protein was the highest under ice temperature. The lower the curing temperature, the higher the phosphorylation level. The global phosphorylation level of myofibrillar protein was the highest under room temperature. The phosphorylation level decreased with the descent of curing temperature. The effect of temperature on the phosphorylation level of individual protein bands was different. The mechanism of protein phosphorylation at different curing temperatures on meat quality needs to be further investigated, and on this basis, a new curing technology based on muscle protein post-modification should be studied.

11.5 Conclusions

Salt concentration and environmental temperature influenced phosphorylation and dephosphorylation of muscle proteins during salting. Higher ionic strength of salt and higher salting temperature decreased the global phosphorylation level of myofibrillar proteins and influenced the phosphorylation level of individual sarcoplasmic protein, which accelerated glycolysis metabolism, weaken the protein positive function of meat quality, and regulated the muscle contractile to cause the actomyosin dissociation. Therefore, salt curing altering protein phosphorylation might be a new mechanism to regulate meat quality, especially the color and tenderness of meat. The study may provide a new insight into the mechanism by which salting influences the quality of processed meat products, especially the meat tenderness which is related to protein degradation and muscle contraction.

Acknowledgments Parts of this chapter are reprinted from Meat Science, 133, Wang Z. et al., Comparative analysis of muscle phospho-proteomics induced by salt curing, 19–25; Food Science and Biotechnology, 25, Zhang C. et al., Phospho-proteomic profiling of myofibrillar and sarcoplasmic proteins of muscle in response to salting, 993–1001, Copyright (2020), with permission

References

Agius, L. (2015). Role of glycogen phosphorylase in liver glycogen metabolism. *Molecular Aspects of Medicine, 46*, 34–45.

Alamo, L., Wriggers, W., Pinto, A., Bártoli, F., Salazar, L., Zhao, F.-Q., et al. (2008). Three-dimensional reconstruction of tarantula myosin filaments suggests how phosphorylation may regulate myosin activity. *Journal of Molecular Biology, 384*, 780–797.

Bell, R. A., & Storey, K. B. (2014). P06: Posttranslational modification of glyceraldehyde-3-phosphate dehydrogenase from a hibernating mammal: Insight into cold-adaptation and structural diversity of a housekeeping enzyme. *Cryobiology, 69*, 196–197.

Canto, A. C. V. C. S., Suman, S. P., Nair, M. N., Li, S., Rentfrow, G., Beach, C. M., et al. (2015). Differential abundance of sarcoplasmic proteome explains animal effect on beef Longissimus lumborum color stability. *Meat Science, 102*, 90–98.

Carvalho, M. E., Gasparin, G., Poleti, M. D., Rosa, A. F., Balieiro, J. C. C., Labate, C. A., et al. (2014). Heat shock and structural proteins associated with meat tenderness in Nellore beef cattle, a Bos indicus breed. *Meat Science, 96*, 1318–1324.

Chen, L., Li, X., Ni, N., Liu, Y., Chen, L., Wang, Z., et al. (2016). Phosphorylation of myofibrillar proteins in post-mortem ovine muscle with different tenderness. *Journal of the Science of Food and Agriculture, 96*, 1474–1483.

Chen, Y., Liu, L., Fu, H., Wei, C., & Jin, Q. (2014). Comparative proteomic analysis of outer membrane vesicles from Shigella flexneri under different culture conditions. *Biochemical and Biophysical Research Communications, 453*, 696–702.

Chistiakov, D. A., Orekhov, A. N., & Bobryshev, Y. V. (2016). ApoA1 and ApoA1-specific self-antibodies in cardiovascular disease. *Laboratory Investigation, 96*, 708–718.

Comaposada, J., Gou, P., & Arnau, J. (2000). The effect of sodium chloride content and temperature on pork meat isotherms. *Meat Science, 55*, 291–295.

Cottrell, J. J., Mcdonagh, M. B., Dunshea, F. R., & Warner, R. D. (2008). Inhibition of nitric oxide release pre-slaughter increases post-mortem glycolysis and improves tenderness in ovine muscles. *Meat Science, 80*, 511–521.

Daniel, R. M., & Danson, M. J. (2013). Temperature and the catalytic activity of enzymes: A fresh understanding. *FEBS Letters, 587*, 2738–2743.

Devatkal, S. K., & Naveena, B. M. (2010). Effect of salt, kinnow and pomegranate fruit by-product powders on color and oxidative stability of raw ground goat meat during refrigerated storage. *Meat Science, 85*, 306–311.

Di Lisa, F., De Tullio, R., Salamino, F., Barbato, R., Melloni, E., Siliprandi, N., et al. (1995). Specific degradation of troponin T and I by mu-calpain and its modulation by substrate phosphorylation. *The Biochemical Journal, 308*(Pt 1), 57–61.

Dieni, C. A., & Storey, K. B. (2009). Creatine kinase regulation by reversible phosphorylation in frog muscle. *Comparative Biochemistry and Physiology. Part B, Biochemistry & Molecular Biology, 152*, 405–412.

Donnelly, R. P., & Finlay, D. K. (2015). Glucose, glycolysis and lymphocyte responses. *Molecular Immunology*, S0161589015300420.

Duranton, F., Simonin, H., Cheret, R., Guillou, S., & De Lamballerie, M. (2012). Effect of high pressure and salt on pork meat quality and microstructure. *Journal of Food Science, 77*, E188–E194.

Farman, G. P., Miller, M. S., Reedy, M. C., Soto-Adames, F. N., Vigoreaux, J. O., Maughan, D. W., . . .Irvinga, T. C. (2009). Phosphorylation and the N-terminal extension of the regulatory light chain help orient and align the myosin heads in Drosophila flight muscle. *Journal of Structural Biology, 168*(2), 240–249.

Fonvielle, M., Mariano, S., & Therisod, M. (2005). New inhibitors of rabbit muscle triose-phosphate isomerase. *Bioorganic & Medicinal Chemistry Letters, 15*, 2906–2909.

Garcia-Gil, N., Muñoz, I., Santos-Garcés, E., Arnau, J., & Gou P. (2014). Salt uptake and water loss in hams with different water contents at the lean surface and at different salting temperatures. *Meat Science, 96*, 65–72.

Gauthier, M. L., Lydan, M. A., O'day, D. H., & Cotter, D. A. (1997). Endogenous autoinhibitors regulate changes in actin tyrosine phosphorylation during Dictyostelium spore germination. *Cellular Signalling, 9*, 79–83.

González, B., & Manso, R. (2004). Induction, modification and accumulation of HSP70s in the rat liver after acute exercise: Early and late responses. *Journal of Physiology, 556*.

Grehl, T., Muller, K., Vorgerd, M., Tegenthoff, M., Malin, J. P., & Zange, J. (1998). Impaired aerobic glycolysis in muscle phosphofructokinase deficiency results in biphasic post-exercise phosphocreatine recovery in 31P magnetic resonance spectroscopy. *Neuromuscular Disorders, 8*, 480–488.

Hart, M. C., Korshunova, Y. O., & Cooper, J. A. (1997). Vertebrates have conserved capping protein α isoforms with specific expression patterns. *Cytoskeleton, 38*, 120–132.

Hipkiss, A. R. (2011). Energy metabolism and ageing regulation: Metabolically driven deamidation of triosephosphate isomerase may contribute to proteostatic dysfunction. *Ageing Research Reviews, 10*, 498–502.

Hu, Y., Guo, S., Li, X., & Ren, X. (2013). Comparative analysis of salt-responsive phosphoproteins in maize leaves using Ti(4+)--IMAC enrichment and ESI-Q-TOF MS. *Electrophoresis, 34*, 485–492.

Huang, H., Larsen, M. R., Karlsson, A. H., Pomponio, L., Costa, L. N., & Lametsch, R. (2011). Gel-based phosphoproteomics analysis of sarcoplasmic proteins in postmortem porcine muscle with pH decline rate and time differences. *Proteomics, 11*, 4063–4076.

Huang, H., Larsen, M. R., & Lametsch, R. (2012). Changes in phosphorylation of myofibrillar proteins during postmortem development of porcine muscle. *Food Chemistry, 134*, 1999–2006.

Huang, J. C., Huang, M., Yang, J., Wang, P., Xu, X. L., & Zhou, G. H. (2014). The effects of electrical stunning methods on broiler meat quality: Effect on stress, glycolysis, water distribution, and myofibrillar ultrastructures. *Poultry Science, 93*, 2087–2095.

Ikebe, R., Reardon, S., Mitsui, T., & Ikebe, M. (1999). Role of the N-terminal region of the regulatory light chain in the Dephosphorylation of myosin by myosin light chain phosphatase. *Journal of Biological Chemistry, 274*, 30122–30126.

Ishida, Y., Riesinger, I., Wallimann, T., & Paul, R. J. (1994). Compartmentation of ATP synthesis and utilization in smooth muscle: Roles of aerobic glycolysis and creatine kinase. *Molecular and Cellular Biochemistry, 133–134*, 39–50.

Jost, J. A., Keshwani, S. S., & Abou-Hanna, J. J. (2015). Activation of AMP-activated protein kinase in response to temperature elevation shows seasonal variation in the zebra mussel, Dreissena polymorpha. *Comparative Biochemistry and Physiology Part A: Molecular & Integrative Physiology, 182*, 75–83.

Kanner, J., Harel, S., & Jaffee, R. (1991). Lipid peroxidation of muscle food as affected by NaCl. *Journal of Agricultural and Food Chemistry, 39*, 1017–1021.

Katrukha, I. A., & Gusev, N. B. (2013). Enigmas of cardiac troponin T phosphorylation. *Journal of Molecular and Cellular Cardiology, 65*, 156–158.

Kim, C. J., Hwang, K. E., Song, D. H., Jeong, T. J., Kim, H. W., Kim, Y. B., et al. (2015). Optimization for reduced-fat/low-NaCl meat emulsion systems with sea mustard (Undaria pinnatifida) and phosphate. *Korean Journal for Food Science of Animal Resources, 35*, 515–523.

Lee, W. H., Choi, J. S., Byun, M. R., Koo, K. T., Shin, S., Lee, S. K., et al. (2010). Functional inactivation of triosephosphate isomerase through phosphorylation during etoposide-induced apoptosis in HeLa cells: Potential role of Cdk2. *Toxicology, 278*, 224–228.

Lehman, W., Medlock, G., Li, X., Suphamungmee, W., Tu, A.-Y., Schmidtmann, A., et al. (2015). Phosphorylation of Ser283 enhances the stiffness of the tropomyosin head-to-tail overlap domain. *Archives of Biochemistry and Biophysics, 571*, 10–15.

Li, C., Zhou, G., Xu, X., Lundström, K., Karlsson, A., & Lametsch, R. (2015). Phosphoproteome analysis of sarcoplasmic and myofibrillar proteins in bovine longissimus muscle in response to postmortem electrical stimulation. *Food Chemistry, 175*, 197–202.

Li, C. B., Li, J., Zhou, G. H., Lametsch, R., Ertbjerg, P., Bruggemann, D. A., et al. (2012). Electrical stimulation affects metabolic enzyme phosphorylation, protease activation, and meat tenderization in beef. *Journal of Animal Science, 90*, 1638–1649.

Li, M., Li, X., Xin, J., Li, Z., Li, G., Zhang, Y., et al. (2017). Effects of protein phosphorylation on color stability of ground meat. *Food Chemistry, 219*, 304–310.

Melling, C. W. J., Thorp, D. B., Milne, K. J., & Noble, E. G. (2008). Myocardial Hsp70 phosphorylation and PKC-mediated cardioprotection following exercise. *Cell Stress and Chaperones, 14*, 10.

Midtgaard, S. R., Pedersen, M. C., & Arleth, L. (2015). Small-angle X-ray scattering of the cholesterol incorporation into human ApoA1-POPC discoidal particles. *Biophysical Journal, 109*, 308–318.

Miller, M. S., Farman, G. P., Braddock, J. M., Soto-Adames, F. N., Irving, T. C., Vigoreaux, J. O., et al. (2011). Regulatory light chain phosphorylation and N-terminal extension increase cross-bridge binding and power output in drosophila at in vivo Myofilament lattice spacing. *Biophysical Journal, 100*, 1737–1746.

Mora-Gallego, H., Guàrdia, M. D., Serra, X., Gou, P., & Arnau, J. (2016). Sensory characterisation and consumer acceptability of potassium chloride and sunflower oil addition in small-caliber non-acid fermented sausages with a reduced content of sodium chloride and fat. *Meat Science, 112*, 9–15.

Müller, M. S., Pedersen, S. E., Walls, A. B., Waagepetersen, H. S., & Bak, L. K. (2015). Isoform-selective regulation of glycogen Phosphorylase by energy deprivation and phosphorylation in astrocytes. *Glia, 63*, 154–162.

Quest, A. F. G., Soldati, T., Hemmer, W., Perriard, J.-C., Eppenberger, H. M., & Wallimann, T. (1990). Phosphorylation of chicken brain-type creatine kinase affects a physiologically important kinetic parameter and gives rise to protein microheterogeneity in vivo. *FEBS Letters, 269*, 457–464.

Ramírez Ríos, S., Lamarche, F., Cottet-Rousselle, C., Klaus, A., Tuerk, R., Thali, R., et al. (2014). Regulation of brain-type creatine kinase by AMP-activated protein kinase: Interaction, phosphorylation and ER localization. *Biochimica et Biophysica Acta (BBA) - Bioenergetics, 1837*, 1271–1283.

Rao, V. S., Marongelli, E. N., & Guilford, W. H. (2009). Phosphorylation of tropomyosin extends cooperative binding of myosin beyond a single regulatory unit. *Cell Motility and the Cytoskeleton, 66*, 10–23.

Rubin, C. S., & Rosen, O. M. (1975). Protein phosphorylation. *Annual Review of Biochemistry, 44*, 831–887.

Sale, E. M., White, M. F., & Kahn, C. R. (1987). Phosphorylation of glycolytic and Gluconeogenic enzymes by the insulin receptor Kinase. *Journal of Cellular Biochemistry, 33*, 12.

Schulenberg, B., Aggeler, R., Beechem, J. M., Capaldi, R. A., & Patton, W. F. (2003). Analysis of steady-state protein phosphorylation in mitochondria using a novel fluorescent Phosphosensor dye. *Journal of Biological Chemistry, 278*, 27251–27255.

Shao, J. H., Deng, Y. M., Jia, N., Li, R. R., Cao, J. X., Liu, D. Y., et al. (2016). Low-field NMR determination of water distribution in meat batters with NaCl and polyphosphate addition. *Food Chemistry, 200*, 308–314.

Sheard, P. R., & Tali, A. (2004). Injection of salt, tripolyphosphate and bicarbonate marinade solutions to improve the yield and tenderness of cooked pork loin. *Meat Science, 68*, 305–311.

Silverman-Gavrila, L. B., Lu, T. Z., Prashad, R. C., Nejatbakhsh, N., & Feng, Z. P. (2009). Neural phosphoproteomics of a chronic hypoxia model—Lymnaea stagnalis. *Neuroscience, 161*, 621–634.

Stull, J. T., Kamm, K. E., & Vandenboom, R. (2011). Myosin light chain kinase and the role of myosin light chain phosphorylation in skeletal muscle. *Archives of Biochemistry and Biophysics, 510*, 0–128.

Taylor, K. A., Feig, M., Brooks III, C. L., Fagnant, P. M., Lowey, M., & Trybus, K. M. (2014). Role of the essential light chain in the activation of smooth muscle myosin by regulatory light chain phosphorylation. *Journal of Structural Biology, 185*(3), 375–382.

Toyo-Oka, T. (1982). Phosphorylation with cyclic adenosine 3′:5′ monophosphate-dependent protein kinase renders bovine cardiac troponin sensitive to the degradation by calcium-activated neutral protease. *Biochemical and Biophysical Research Communications, 107*, 44–50.

Wang, L., Sadayappan, S., & Kawai, M. (2014). Cardiac myosin binding protein C phosphorylation affects cross-bridge cycle's elementary steps in a site-specific manner. *PLoS One, 9*, e113417.

Wang, Z., Xu, W., Kang, N., Shen, Q., & Zhang, D. (2016). Microstructural, protein denaturation and water holding properties of lamb under pulse vacuum brining. *Meat Science, 113*, 132–138.

Wang, Z., Zhang, C., Li, Z., Shen, Q., & Zhang, D. (2017). Comparative analysis of muscle phosphoproteome induced by salt curing. *Meat Science, 133*, 19–25.

Xu, M. (2007) Isolation, purifieation and some properties of intestinal alkaline phosphatase from hypophthalmichthys molitrix [D]. Chongqing: Southwest University.

Zeniya, M., Sohara, E., Kita, S., Iwamoto, T., Susa, K., Mori, T., et al. (2013). Dietary salt intake regulates WNK3-SPAK-NKCC1 phosphorylation cascade in mouse aorta through angiotensin II. *Hypertension, 62*, 872–878.

Zhang, C., Wang, Z., Li, Z., Du, M., & Zhang, D. (2016a). Effect of salting temperature on the protein phosphorylation of mutton. *Modern Food Science and Technology, 32*(11), 215–221. (In Chinese).

Zhang, C., Wang, Z., Li, Z., Shen, Q., Chen, L., Gao, L., et al. (2016b). Phosphoproteomic profiling of myofibrillar and sarcoplasmic proteins of muscle in response to salting. *Food Science and Biotechnology, 25*, 993–1001.

Zhang, Y. Y., Li, S.-F., Qian, S.-W., Zhang, Y.-Y., Liu, Y., Tang, Q.-Q., et al. (2012). Phosphorylation prevents C/EBPβ from the calpain-dependent degradation. *Biochemical and Biophysical Research Communications, 419*, 0–555.

Zhang, Z., Lawrence, J., & Stracher, A. (1988). Phosphorylation of platelet actin binding protein protects against proteolysis by calcium dependent sulfhydryl protease. *Biochemical and Biophysical Research Communications, 151*, 355–360.

Zhu, K., Zhao, J., Lubman, D. M., Miller, F. R., & Barder, T. J. (2005). Protein pI shifts due to posttranslational modifications in the separation and characterization of proteins. *Analytical Chemistry, 77*, 2745–2755.

Zolla, A. D. A. S. R. C. M. V. Z. F. N. L. (2012). Love me tender: An Omics window on the bovine meat tenderness network. *Journal of Proteomics, 75*(14), 4360–4380.

Chapter 12
Improvement of Meat Quality by Novel Technology

Abstract Developing new techniques to improve meat quality is a popular topic. In this chapter, three novel methods for improving the meat quality are provided, including controlled freezing point storage (CFPS), low-variable temperature and high humidity thawing (LVTHHT), and pulsed vacuum pickling (PVP) technologies. Meat stored in CFPS treatment for 10 days proved to have better color stability in comparison with those in 4 °C storage. The LVTHHT could significantly prevent the deterioration of lamb quality, and it was low cost and easy to operate. PVP could improve salting efficiency and meat quality.

Keywords Controlled freezing point storage · Low-variable temperature and high humidity thawing · Pulsed vacuum pickling · Meat quality · Novel technology

12.1 Introduction

Controlled freezing point storage (CFPS), also known as subzero or super chilling storage, is an alternative and promising technique for preservation of fruit, vegetable, and meat products without freezing. Compared to freezing, it provides a storage condition that can lead to a significant retardation of biochemical and microbial activity of muscle tissue without physically damaging the muscle cells due to ice crystal formation. In contrast with regular 4 °C refrigerated storage, CFPS has the characteristics of fast chilling rate, stemming from the high heat exchange capacity, as well as small temperature variation range because of strict control. Due to production factors such as low product turnover rate, meat products often go through to a long production chain. The CFPS condition offers great flexibility for meat retailers and processors to meet the requirement of this situation in comparison with refrigerated storage and freezing.

Freezing storage is the most important storage method for meat. Frozen meat is also the main form of meat products import and export trade and inter-regional circulation (Huang 2005). At present, the thawing methods of frozen meat mainly include air thawing, water immersion thawing, microwave thawing, ultra-high pressure thawing, and the like. The existing methods have their own advantages

D. Zhang et al., *Protein Phosphorylation and Meat Quality*,
https://doi.org/10.1007/978-981-15-9441-0_12

and disadvantages. The conventional air thawing is currently the most widely used method, which has the advantages of simple operation and low cost and the disadvantages of slow thawing speed, easy oxidation and discoloration on the meat surface, and rapid proliferation of microorganisms. Thawing speed of water immersion thawing is faster than the air thawing, but easy to cause loss of nutrients and make meat gray (Yang 2005). Microwave thawing possesses high speed and efficiency, but due to the difference of microwave penetration and absorption of water and ice, it is easy to generate uneven thawing of meat and local overheating (Xia et al. 2012). Ultra-high pressure thawing method can thaw quickly and has little negative impact on product quality, whereas both the requirement and cost of the equipment are higher than the other methods, so it is still immature in domestic industrialization (Liao et al. 2010). Therefore, it is of great significance for the meat industry to develop new and economical thawing techniques that can maintain the quality of thawed meat and solve the problems of juice loss, oxidation, and texture deterioration during the process.

The pickling processing is an important technology for processing meat products, but the traditional pickling (dry curing, wet curing, mixed curing) has many defects, such as tedious processes, long processing time, and hygiene and quality risk. Modern pickling techniques include brine injecting, vacuum pickling, pulsating pickling, and tumbling pickling. Pulse pressure salting technology can be applied to the processing of meat products completely immersed in the marinade. The main reason for the pulse pressure salting to improve the pickling effect is that the hydrodynamic mechanism (Fito, 1994) promotes the pickling liquid from the pores of the meat during the pressure change (Guamis et al. 1997), and the cyclical change of pressure can improve the pickling effect. Pulse pressure salting is an efficient static pickling method, which can effectively improve the pickling rate and better maintain the meat form (Deumier et al. 2011). In recent years, researches on pulse pressure salting have focused on the diffusion of salt and water, for example, previous studied the law of water change of sardine during pulse pressure salting (Corzo and Bracho 2007; Corzo et al. 2006b; Reyes et al. 2008), and Deumier et al. (2011) studied the mass transfer mechanism of turkey meat during pulse pressure salting. Nevertheless, there has been little research on meat quality using pulse pressure salting technology and no report on mutton pickling.

In this chapter, the CFPS, low-variable temperature and high humidity thawing (LVTHHT), and pulsed vacuum pickling (PVP) technologies were introduced, which can provide technical guidance to produce high quality meat.

12.2 Controlled Freezing Point Storage

This study investigated the mechanism of controlled freezing point storage on muscle color and protein phosphorylation level. The 20 *longissimus dorsi* of 10 crossbred sheep were used. The two *longissimus dorsi* of the same sheep were randomly divided into two groups, and each *longissimus dorsi* was equally divided

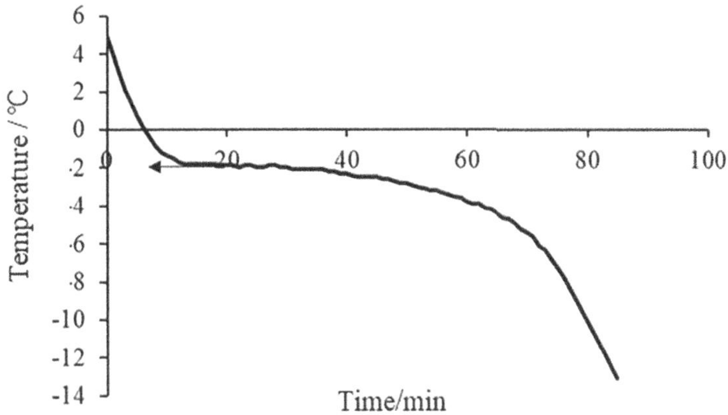

Fig. 12.1 The freezing point of ovine muscle. Translated from Zhang (2016)

and then randomly assigned to each time point. Compared with refrigeration, the effects of controlled freezing point on muscle color and protein phosphorylation level during storage time were studied. In order to estimate the color stability of meat stored at different temperatures, a Konica Minolta colorimeters CM-600d was used to measure instrumental color and reflectance of samples at different wavelengths. The activity of metmyoglobin reductase and concentration of nicotinamide adenine dinucleotide (NADH) were determined by spectrophotometry, then analyzed the mechanism of the effect of controlled freezing point storage on color stability. The activity of protein kinase was tested by enzyme linked immunosorbent assay (ELISA). In order to investigate the phosphorylation level of sarcoplasmic protein and myofibrillar protein in postmortem ovine muscle at different storage temperatures, sodium dodecyl sulfate-polyacrylamide gel electrophoresis (SDS-PAGE) coupled with Pro-Q Diamond and SYPRO Ruby staining was applied. Quantity One software was used to analyze the optical density of protein bands on SDS-PAGE images. By analyzing the changes of color and protein phosphorylation during the CFPS muscle and cold storage conditions, the effects of CFPS on color stability and protein phosphorylation level were investigated, which provided a theoretical basis to study the mechanism of storage temperature controlling meat quality. The research also offered a reference frame to companies and consumers to produce and select ovine muscle with high quality.

As shown in Fig. 12.1, the temperature measurement time of the sample was 85 min, and the maximum ice crystal formation area was around 15–40 min. The temperature fluctuation was small, and the sample stayed in this temperature zone for a long time. According to the temperature recorder test data, the sample temperature continued to decrease before 22 min, a slight rebound occurred at 22 min, from −2 to −1.9 °C, then decreased to −2 °C at 25 min, then again at 27 min. The temperature raised to −1.9 °C and then continued to decrease, so the freezing point of the experimental sample was determined to be −2 °C. Considering the temperature fluctuation range of the incubator with an accuracy of ±1 °C, to prevent the sample from freezing, the temperature of CFPS was set to −0.8 °C.

12.2.1 Effect of Controlled Freezing Point Storage on Meat Color Stability

Figure 12.2 showed the color difference values of samples stored at day 0, 2, 4, 6, 8, and 10, respectively, under CFPS and refrigerated conditions, in which Fig. 12.2a–c were graphs showing changes of brightness value L^*, redness value a^*, and yellowness value b^* with storage time. As shown in Fig. 12.2a, the L^* value of the two treatment groups had the same change in the storage process, which was a trend of first rising, then decreasing and then stabilizing, and the highest brightness values appeared in the day 2–4. During the whole storage process, the L^* value of the group of CFPS fluctuated greatly, with day 2, 4, 6, 8, and 10 significantly higher than day 0 ($P < 0.05$), and day 2 and 4 significantly higher than day 6 and 10. There was no significant difference between day 2 and day 4, 6, and 10. The L^* value of the cold storage group did not change much, and that of day 2, 4, 6, 8, and 10 was significantly higher than day 0 ($P < 0.05$), while there was no significant difference between day 2, 4, 6, 8, and 10. At day 2, 4, and 8, the L^* value of the group of CFPS was significantly higher than that of the refrigerated group ($P < 0.05$).

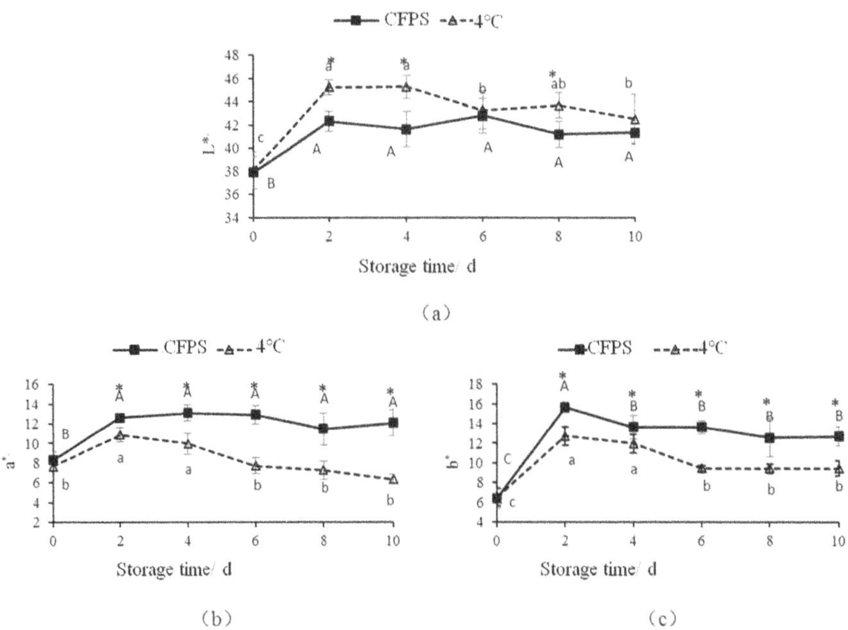

Fig. 12.2 Changes of instrumental meat color at different storage temperature. The lowercases represent the significant difference between different postmortem time of refrigeration ($P < 0.05$), while the capital letters represent the significant difference between different postmortem time of controlled freezing point ($P < 0.05$). * represents the significant difference between controlled freezing point and refrigeration at the same time of storage ($P < 0.05$). (**a**) Changes of brightness value L* with storage time, (**b**) changes of redness value a* with storage time, (**c**) changes of yellowness value b* with storage time. Translated from Zhang (2016)

As shown in Fig. 12.2b, at the initial stage of storage (0–2 days), the a^* values of the two treatment groups showed an upward trend, and then the a^* value of the CFPS group was stable, while the a^* value of the refrigerated group gradually decreased from the day 6 of storage. Both the maximum a^* values of CFPS and refrigerated groups appeared at day 0. Compared with the refrigerated group, the a^* value of the CFPS group fluctuated greatly during the whole storage process, and only that of day 2, 4, 6, 8, and 10 was significantly higher than day 0, while there was no significant difference at day 2, 4, 6, 8, and 10 ($P < 0.05$). The a^* value of the cold storage group was significantly higher than day 0, 6, 8, and 10 at day 2 and day 4 during storage. From the day 2 of storage to the end of the storage period, the a^* value of the CFPS group was always significantly higher than that of the refrigerated group ($P < 0.05$).

As shown in Fig. 12.2c, during the storage process, the b^* values of the CFPS and refrigerated treatment samples showed the same changing pattern as the a^* values, that is, first increased and then decreased, the maximum value of CFPS group appeared on the day 2, decreased on the fourth day, then stabilized. The maximum value of the refrigerated storage group appeared on the day 2 and 4, decreased on the sixth day, then showed a steady state. During the whole storage process, the b^* value of the CFPS group was significantly higher than day 0 at day 2, 4, 6, 8, and 10, day 2 was significantly higher than day 4, 6, 8, and 10, and there was no significant difference on day 4, 6, 8, and 10 ($P < 0.05$). The b^* value of the control group was also significantly higher than day 0 at day 2, 4, 6, 8, and 10, day 2 and 4 was significantly higher than day 6, 8, and 10, and there was no significant difference between day 6, 8, and 10 ($P < 0.05$). From day 2 till the end of the storage period, the b^* value of the CFPS group was always significantly higher than that control group ($P < 0.05$).

As shown in Fig. 12.3, compared with the refrigerated group, the R630/580 value in the CFPS group was relatively stable, with the trend of first increasing and then decreasing. R630/580 value of the control group was first lowered and then increased. The R630/580 value of CFPS group was significantly higher on day 2, 4, and 6 than day 0, 8, and 10, and there was no significant difference between day 2, 4, 6 and day 0, 8 ($P < 0.05$). In the control group, the R630/580 value of day 2 was significantly higher than day 4, 6, 8, 10, day 0 was significantly higher than day 6, 8, 10. But there was no significant difference of the R630/580 value between day 0 and day 2, day 0 and day 4, day 6 and day 8, 10, t ($P < 0.05$). During the whole storage process, from day 2 to the end of storage, the R630/580 value in the CFPS group was significantly higher than that in the control group ($P < 0.05$).

According to Fig. 12.4, during the storage process, the CFPS group was consistent with the changing trend of Chroma value in the control group, both of which increased first and then slowly decreased. The CFPS group was significantly higher than 4 °C for day 2, 4, 6, 8, and 10, and day 2 was significantly higher than day 8 and 10. There was no significant difference between day 2, 4, and 6, and also had no significant difference in day 8 and 10 ($P < 0.05$). The day 2 and 4 of the control group was significantly higher than day 0, 6, 8, and 10. No difference was observed between day 2 and 4, and day 6 was significantly higher than day 0. There was no significant difference between day 6, 8, and 10 ($P < 0.05$). From day 4 to the end of

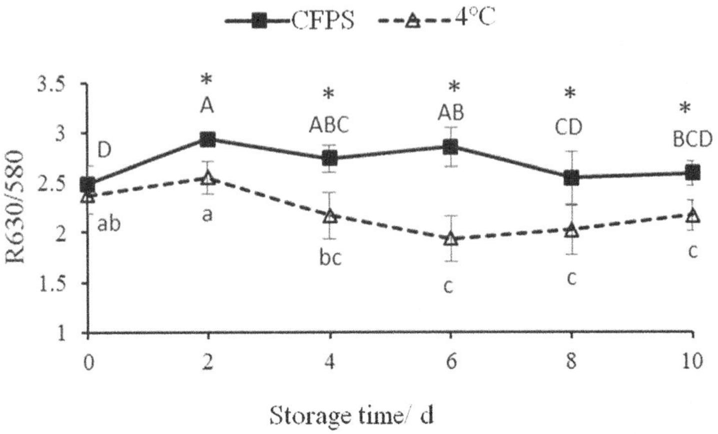

Fig. 12.3 Changes of R630/580 value at different storage temperature. The lowercases represent the significant difference between different postmortem time of refrigeration ($P < 0.05$), while the capital letters represent the significant difference between different postmortem time of controlled freezing point ($P < 0.05$). * represents the significant difference between the controlled freezing point and refrigeration at the same time of storage ($P < 0.05$). Translated from Zhang (2016)

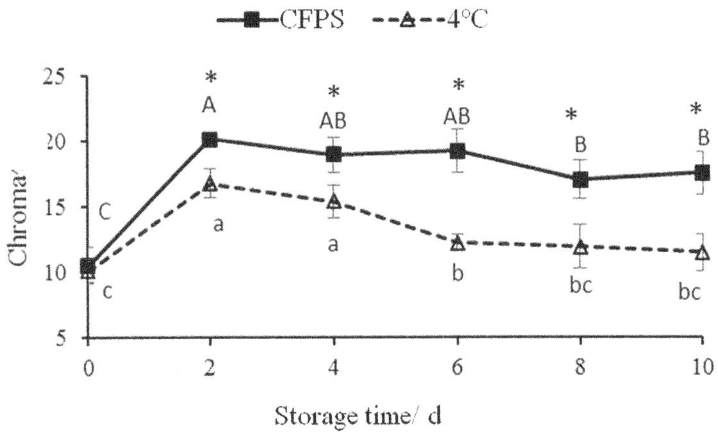

Fig. 12.4 Changes of Chroma value at different storage temperature. The lowercases represent the significant difference between different postmortem time of refrigeration ($P < 0.05$), while the capital letters represent the significant difference between different postmortem time of controlled freezing point ($P < 0.05$). * represents the significant difference between the controlled freezing point and refrigeration at the same time of storage ($P < 0.05$). Translated from Zhang (2016)

the storage period, the Chroma value of CFPS group was significantly higher than the control group ($P < 0.05$).

As shown in Fig. 12.5, Hue angle value of the refrigerated group increased gradually during the whole storage process. As for the CFPS group, Hue angle value first increased, then decreased and then stabilized. The maximum value of the

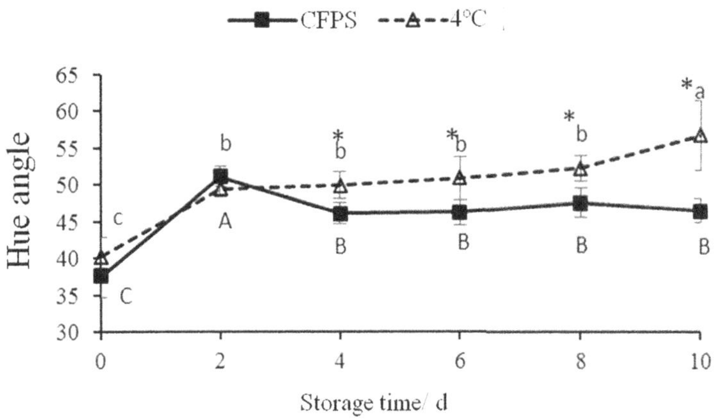

Fig. 12.5 Changes of Hue angle value at different storage temperature. The lowercases represent the significant difference between different postmortem time of refrigeration ($P < 0.05$), while the capital letters represent the significant difference between different postmortem time of controlled freezing point ($P < 0.05$). * represents the significant difference between the controlled freezing point and refrigeration at the same time of storage ($P < 0.05$). Translated from Zhang (2016)

CFPS group appeared on the day 2 of storage, significantly higher than the others. In addition, the day 4, 6, 8, and 10 was significantly higher than day 0, and there was no significant difference between day the 4, 6, 8, and 10. The day 10 of the control group was significantly higher than the rest of the time points, and day 2, 4, 6, and 8 was significantly higher than day 0, and day 2, 4, 6, 8 had no significant difference. Hue angle values of the refrigerated group from the day 4 to the day 10 were significantly higher than those in the CFPS group ($P < 0.05$).

12.2.2 Effect of Controlled Freezing Point Storage on Protein Kinases' Activities

As shown in Fig. 12.6, protein kinase activity increased significantly ($P < 0.05$) at 0.5–12 h after controlled freezing point storage, and then gradually decreased. The protein kinase activity in the refrigerated storage process decreased significantly with the storage time ($P < 0.05$), and the reduction rate in the refrigerated storage group was higher than that in the controlled freezing point storage. The protein kinase activity in the CFPS group was significantly higher than the control group during the storage period of 12 h–9 days ($P < 0.05$). At the end of storage, the protein kinase activity in the control group was close to inactivation and the CFPS group was still at a higher level.

Protein kinases are a class of enzymes that catalyze the phosphorylation of proteins, and they play an important role in almost every cell function (Gonz Lez-Mart Nez et al. 2002). All protein kinases are in the ground state and can be

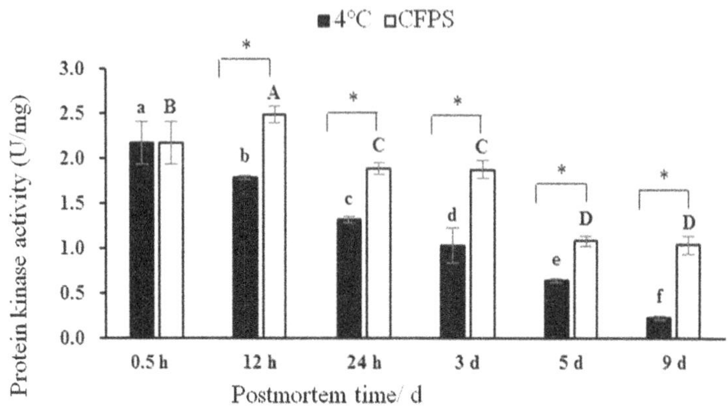

Fig. 12.6 Analysis of the protein kinases' activities at different storage temperature. The lowercases represent the significant difference between different postmortem time of refrigeration ($P < 0.05$), while the capital letters represent the significant difference between different postmortem time of controlled freezing point ($P < 0.05$). * represents the significant difference between the controlled freezing point and refrigeration at the same time of storage ($P < 0.05$). Translated from Zhang et al. (2016)

activated by different stimuli when needed, catalyzing the phosphorylation of certain proteins and activating phosphatases (Roskoski 2015). Kemp et al. (2006) have shown that AMP-activated protein kinase (AMPK) in pale, soft, and exudative (PSE) meat is activated in the early postmortem period and has the highest enzyme activity at 0.5 h, then gradually decreases, while AMPK enzyme activity in normal pork has the highest enzyme activity at 1 h, indicating that AMPK activity has an effect on the quality of postmortem meat. AMPK plays a key role in the postmortem cell glycolysis process and the formation of PSE meat. The consumption of adenosine triphosphate (ATP) and the production of adenosine monophosphate (AMP) in the early postmortem period cause AMPK to be activated, resulting in a rapid decline in pH of muscle, and in pork, it is the formation of PSE meat (Baron et al. 2005; Hwang et al. 2004). Li et al. (2016) found that the muscles stored at controlled freezing point reached their ultimate pH later 4 days than the control. In this study, the activation of protein kinase activity under controlled freezing point conditions was lagged behind the refrigeration, and together with the study of Li, we are able to reveal the reasons for the longer time required for muscles to reach the ultimate pH under CFPS conditions.

12.2.3 *Effect of Controlled Freezing Point Storage on Phosphorylation Levels of Sarcoplasmic Protein*

The SDS-PAGE electrophoresis patterns of sarcoplasmic proteins in the CFPS group and refrigerated group at 0.5 h, 12 h, 24 h, 3 d, 5 d, and 9 d after postmortem were

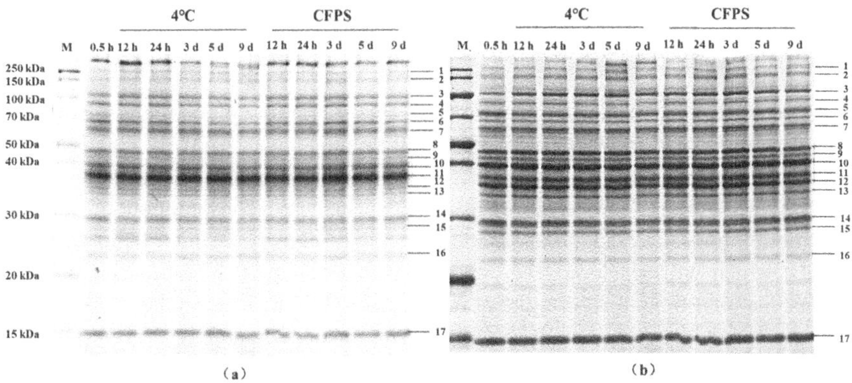

Fig. 12.7 Gel images of phosphoproteins and total proteins of sarcoplasmic protein at different storage temperature separated by SDS-PAGE. (**a**) Image of phosphoproteins stained with Pro-Q Diamond. (**b**) Image of total proteins stained with SYPRO Ruby. Translated from Zhang et al. (2016)

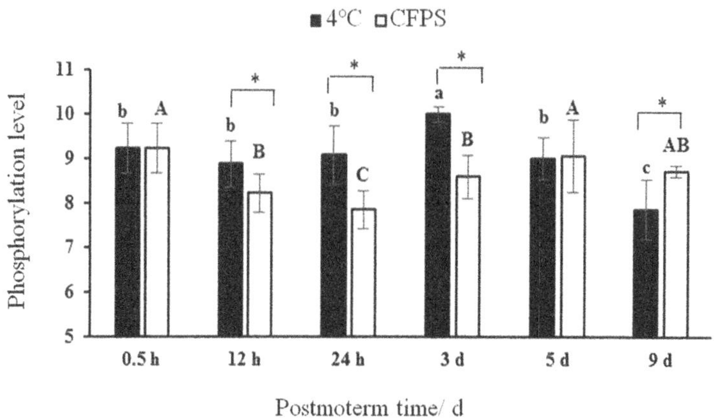

Fig. 12.8 The global phosphorylation level of sarcoplasmic protein at different storage temperature. The lowercases represent the significant difference between different postmortem time of refrigeration ($P < 0.05$), while the capital letters represent the significant difference between different postmortem time of controlled freezing point ($P < 0.05$). * represents the significant difference between the controlled freezing point and refrigeration at the same time of storage ($P < 0.05$). Translated from Zhang et al. (2016)

shown in Fig. 12.7a, b. The gray scale values of the electropherograms in Fig. 12.7 were analyzed, and the overall phosphorylation level of different storage treatment groups was shown in Fig. 12.8. The global phosphorylation level of myofibrillar protein in postmortem lamb increased first and then decreased in ice temperature storage group and cold storage group, the maximum phosphorylation level of the control group was reached at day 3 after postmortem, and the CFPS group reached the maximum phosphorylation level at 12 h postmortem. The myofibrillar protein total level of phosphorylation in the control group was significantly higher than that in the CFPS group, from the 24 h storage period to the end of the storage period ($P < 0.05$).

Glycolysis is the most important mode of energy supply for postmortem muscles. It is directly related to meat color, pH, water holding capacity (WHC), and muscle rigor mortis. Meanwhile, sarcoplasmic protein accounts for more than two-thirds of glycolytic enzymes (Mikkelsen et al. 1999). The large consumption of ATP in postmortem muscles leads to an accelerated rate of glycolysis and a rapid decrease in pH, which will accelerate the appearance of rigor mortis, promote aging, and reduce the total time of rigor mortis (Woods et al. 2005). Other studies have shown that protein phosphorylation plays an important role in regulating glycolytic enzyme activity (Scopes and Stoter 1982). Therefore, the sarcoplasmic protein of postmortem mutton is selected for detecting and analyzing its changes of phosphorylation level under different storage conditions for six time points.

The global phosphorylation level of sarcoplasmic protein in the refrigerated group was higher than that in the CFPS group at 12 h to day 3, and the CFPS group was higher than the refrigerated group on the day 9 (Fig. 12.8). The reason might be that the activity of protein kinase in the early stage of refrigerated storage was higher than that in the CFPS group (Fig. 12.6), and the degree of phosphorylation of sarcoplasmic protein was stronger, with the prolongation of storage time, the rate of glycolysis was faster in the control group (Li et al. 2016). ATP consumption was faster and the content was decreased, resulting in decreased ATP-dependent protein kinase activity (Dickens and Lyon 1995), and the level of reaction of phosphorylation of sarcoplasmic protein was decreased, and it is no longer rising. On the contrary, CFPS was beneficial for maintaining protein kinase activity, thereby promoting the stabilization of the global phosphorylation level of sarcoplasmic proteins. Even though the refrigerated group was good for the global phosphorylation level of sarcoplasmic proteins, not all refrigerated treatment had improved the phosphorylation level of sarcoplasmic protein than CFPS. The global phosphorylation level of sarcoplasmic protein was affected by the storage conditions, storage time, and the interaction between storage conditions and storage time.

12.2.4 Effect of Controlled Freezing Point Storage on Phosphorylation Levels of Myofibrillar Protein

The SDS-PAGE electropherogram of myofibrillar proteins in the CFPS group and refrigerated group at 0.5 h, 12 h, 24 h, 3 d, 5 d, and 9 d after postmortem was shown in Fig. 12.9a, b. Figure 12.9a showed the phosphorylated protein stained by Pro-Q Diamond, and Fig. 12.9b showed the whole myofibrillar protein stained with SYPRO Ruby. The gray scale values of the electropherograms in Fig. 12.9 were analyzed, and the global phosphorylation level of the different storage treatment groups was shown in Fig. 12.10. The global phosphorylation level of myofibrillar

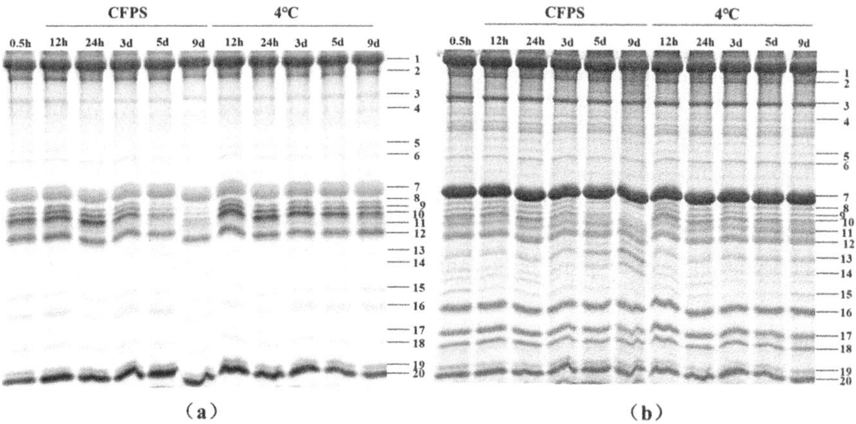

Fig. 12.9 Gel images of phosphoproteins and total proteins of myofibrillar protein at different storage temperature separated by SDS-PAGE. (**a**) Image of phosphoproteins stained with Pro-Q Diamond. (**b**) Image of total proteins stained with SYPRO Ruby. Translated from Zhang et al. (2016)

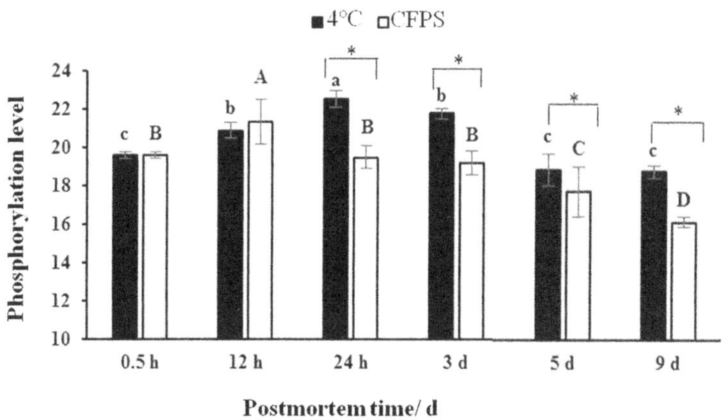

Fig. 12.10 The global phosphorylation level of myofibrillar protein at different storage temperature. The lowercases represent the significant difference between different postmortem time of refrigeration ($P < 0.05$), while the capital letters represent the significant difference between different postmortem time of controlled freezing point ($P < 0.05$). * represents the significant difference between the controlled freezing point and refrigeration at the same time of storage ($P < 0.05$). Translated from Zhang et al. (2016)

protein in postmortem lamb increased first and then decreased in two groups, and the maximum phosphorylation level of the control group was reached at 24 h after postmortem, and the CFPS group reached the maximum phosphorylation level at 12 h, and from the 24 h to the end of the storage period, the global phosphorylation level of myofibrillar protein of the control group was higher than the CFPS group ($P < 0.05$).

Myofibrillar protein is the most abundant protein in skeletal muscle, accounting for 55–60% of the total protein. Myofibrillar protein is associated with muscle contractile properties, and also strongly related to muscle functional properties and raw meat cooking characteristics (Lee et al. 2010). Previous studies have found out that a variety of myofibrillar proteins, such as myosin, troponin, actin, tropomyosin, and myosin light chain, can be phosphorylated (Mazzei and Kuo 1984; Ryder et al. 2007; Toyo-Oka 1982). Combined with the mass spectrometry results from our team (Chen et al. 2016) and Huang et al. (2012), this experiment chose to analyze the phosphorylation level of some important myofibrillar protein bands.

12.3 Low-Variable Temperature and High Humidity Thawing

According to the physicochemical properties and heat exchange characteristics of ice-water during the thawing processes of frozen meat, the frozen meat is thawed at 2–6 °C. Since the density of water is the highest at 4 °C, the temperature fluctuates around 4 °C can change the density of water and promote the migration of water, thus accelerating heat exchange and increasing the thawing speed. At the same time, combined with high humidity thawing conditions, it protects the hydrated surface of the frozen meat protein and inhibits the oxidation of meat during thawing. As presented in Fig. 12.11, the LVTHHT improved based on the 4 °C thawing storage usually used in production. It fluctuates around 4 °C, which has a low requirement of refrigeration equipment and a low energy consumption. The effects of a LVTHHT test group, thawing temperature 2 °C → 6 °C → 2 °C, RH > 90%), and a conventional air thawing method (control group, thawing temperature 4 °C) on the quality of lamb hindquarters were investigated. The indexes including color, cooking loss, thawing loss, protein content of thawing drip, texture profiles analysis (TPA) of thawed hindquarter, and surface hydrophobicity of myofibrillar protein were measured. The microstructures of the transverse section of frozen and thawed samples were observed by scan electric microscopy (SEM). The effects of different thawing methods on the component, aggregation, and degradation of myofibrillar protein were studied by SDS-PAGE gel electrophoresis (Fig. 12.12).

12.3.1 Effect of Low-Variable Temperature and High Humidity Thawing on Meat Quality Color, Thawing Loss and Cooking Loss, Texture

Although the meat color has no significant influence on the nutritional value and flavor of meat, it will affect consumers' preferences. And it is also an apparent manifestation of changes in muscle biology, biochemistry, and physiology, and an

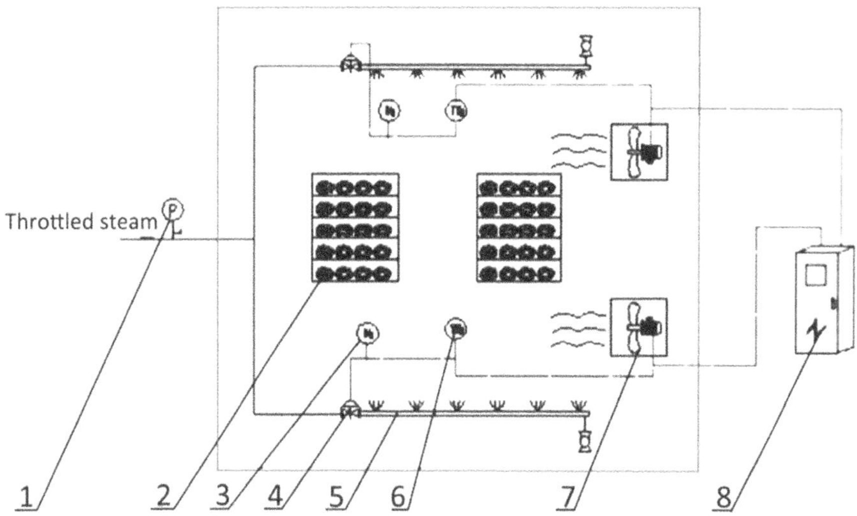

Fig. 12.11 Flow chart of thawing method of low-variable temperature with high humidity. 1-pressure gauge, 2-frozen meat, 3-humidity sensor, 4-steam value, 5-steam pipe, 6-temperature sensor, 7-frequency refrigeration fan, 8-control system. Translated from Zhang et al. (2013)

important indicator of the meat sensory quality. From Table 12.1, we can see that the L^* value and a^* values of the experimental group were both higher than those of the control group. At 48 h, the L^* value of the experimental group was 26.90, significantly higher than that of the control group ($P < 0.05$), and the a^* value was 10.56, which was significantly higher than that of the control group (8.20) ($P < 0.05$). There was no significant difference in the brightness value b^* between the two methods of thawing mutton. L^* value represents the brightness value of the meat sample, the larger the value, the better the gloss of the meat; a^* value represents the redness of the meat sample, the higher the value, the better the color of the meat, the fresher the meat sample; b^* value represents the yellowness of the meat, the higher the value, the less fresh the meat (Xia et al. 2009). Thawing will reduce the freshness of the meat color, but the experimental group had a fresher color than the control group. In the thawing process, L^* and a^* values of both the experimental group and the control group showed a trend of increasing at first and then decreasing, which may be due to the presence of ice crystals on the surface of frozen meat samples in the early thawing stage. The melting of ice crystals and the formation of oxymyoglobin by myoglobin combining with oxygen can result in an increase in brightness and redness values. In the late stage of thawing, the gloss loss occurs due to excessive water loss of the meat sample that can cause a decrease in the brightness value. The increase of the contact time between the meat sample and the air can be the reason to explain the significant decrease in the redness value due to the increase in the oxidation rate of myoglobin. It also reported that the oxidation and degradation of myoglobin during thawing would lead to the deterioration of meat color (Xia et al. 2012).

Table 12.1 Effects of different thawing methods on lamp chromatic aberration. Translated from Zhang et al. (2013)

	Control group			Test group		
Thawing time/h	L*	a*	b*	L*	a*	b*
0	27.32 ± 0.48a	13.16 ± 1.03a	6.84 ± 0.11a	27.39 ± 0.56a	13.06 ± 0.74a	6.86 ± 0.51a
12	34.57 ± 2.8a	15.17 ± 1.46b	7.72 ± 0.89a	34.72 ± 1.22a	18.39 ± 1.89a	7.36 ± 0.84a
24	36.67 ± 0.51b	16.49 ± 1.33b	7.42 ± 1.41a	38.69 ± 1.64a	19.01 ± 1.05a	7.98 ± 1.81a
36	27.35 ± 2.08b	11.10 ± 1.42b	6.81 ± 0.70a	28.36 ± 0.19a	13.43 ± 1.82a	6.94 ± 1.02a
48	24.83 ± 1.03b	8.20 ± 1.03b	5.76 ± 0.55a	26.90 ± 0.96a	10.56 ± 0.96a	5.87 ± 1.40a

Different letters in the same column indicate significant differences between the test group and control group ($P < 0.05$)

Table 12.2 Effects of different thawing methods on thawing loss and cooking loss. Translated from Zhang et al. (2013)

Index	Thawing loss/%	Protein content of drip/%	Cooking loss/%
Control group	7.01 ± 0.06a	13.00 ± 0.04a	39.56 ± 0.01a
Test group	3.01 ± 0.03b	4.02 ± 0.04b	35.97 ± 0.02b

Thawing time was 48 h. Different letters in the same column indicate significant differences between the test group and control group ($P < 0.05$)

Table 12.3 Effects of different thawing methods on texture of lamb. Translated from Zhang et al. (2013)

Treatment	Hardness/N	Springiness	Gumminess	Chewiness/N
Control group	45.8 ± 7.3b	0.62 ± 0.05a	0.59 ± 0.05a	16.6 ± 2.4b
Test group	55.5 ± 13.0a	0.70 ± 0.06a	0.58 ± 0.06a	22.7 ± 3.9a

Thawing time was 48 h. Different letters in the same column indicate significant differences between the test group and control group ($P < 0.05$)

In the thawing process, frozen meat will lose juice, which contains a large amount of soluble protein, resulting in nutrient loss of thawed meat. Compared with the control group, the thawing loss rate of the experimental group was 3.01%, significantly lower than that in the control group (7.01%, $P < 0.05$). The protein content in juice was 4.02%, significantly lower than 13.00% in control group ($P < 0.05$). The cooking loss was 35.97%, which was also significantly lower than the 39.56% of the control group ($P < 0.05$) (Table 12.2). The results showed that, compared with the conventional air thawing, LVTHHT could significantly reduce the nutrient loss and improve the quality of thawed mutton.

Many domestic and international scholars have reported that TPA can be used to evaluate the texture characteristics of meat and meat products (Dai et al. 2008; Johnson et al. 2010; Martinez et al. 2004; Xia et al. 2012). In this experiment, the texture characteristics of thawed mutton under different thawing conditions were analyzed (Table 12.3). The results showed that the hardness and chewiness of the experimental group were 55.5 and 22.7 N, respectively, which were significantly higher than 45.8 and 16.6 N of the control group ($P < 0.05$), but there were no significant differences in the elasticity and cohesion of the two groups. Chewability is a comprehensive manifestation of hardness, elasticity, and cohesiveness, reflecting the energy required for lamb meat from chewable state to swallowing state. Within a certain range, the greater the value is, the better the corresponding "bite sensation" of meat taste will be (Dai et al. 2008; Ruan et al. 2008). Table 12.3 showed that the chewiness of the thawed mutton in the experimental group was significantly better than that in the control group ($P < 0.05$), which indicated that the LVTHHT significantly improved the texture characteristics of the thawed mutton compared with the conventional air thawing method.

12.3.2 *Effect of Low-Variable Temperature and High Humidity Thawing on Myofibril Protein*

Protein surface hydrophobicity is an indicator of protein hydration ability. The larger the value, the smaller the protein hydrophobicity, the weaker the ability to bind with water, the higher the rate of thawing loss. The results showed that the hydrophobicity of proteins increased by two different thawing methods, but the degree of increase was different. For example, the hydrophobicity of the mutton was 213.32 after 48 h of thawing in the control group, which was nearly 5 times higher than the hydrophobicity of the unthawed mutton 44.08, while the hydrophobicity of mutton before and after thawing in the test group was only more than 2 times higher (Table 12.4). Therefore, compared with the conventional air thawing method, the LVTHHT method can delay the reduction of hydration capacity of protein during the thawing process, thus improving the WHC of thawed meat.

Meat proteins can be divided into water-soluble sarcoplasmic proteins, saline-soluble myofibril proteins, and insoluble matrix proteins according to their solubility, and these three types of protein have different physical properties (Ruan et al. 2008). Wagner and Añon (2007) believed that denaturation of salt-soluble protein had the greatest influence on meat quality. In this experiment, SDS-PAGE electrophoresis was used to observe the changes in the composition of the protein of the salt-soluble myofibrillar protein after thawing (Fig. 12.12). As shown in Fig. 12.12, there were 5 major bands of lamb myofibril, and from top to bottom they were: myosin heavy chain (MHC) with molecular weight of 220 kDa, paramyosin with molecular weight of 100 kDa, 43 kDa actin, 36 kDa tropomyosin, and 35 kDa myosin (troponin T) subunit (protoplasts binding subunit). The results of this study were consistent with those reported by Thanonkaew et al. (2006). The SDS-PAGE profiles of unfrozen myofibrillar protein (1,2) were compared with those of air thawing (3,4) and LVTHHT meat (5,6). The comparative analysis showed that thawing resulted in the protein denaturation, degradation and enve new cross-linking, and the thawed mutton electrophoresis map produced new protein bands at about 55 kDa and 30 kDa, and the disappearance of protein bands at about 40 kDa. Similar cases existed in the protein band below 29 kDa. This may be due to the oxidation of myofibril during thawing, which will lead to protein cross-linking and degradation (Park et al. 2007; Xia et al. 2012).

Table 12.4 Surface hydrophobicity of mutton protein during thawing. Translated from Zhang et al. (2013)

	Surface hydrophobicity		
Treatment	0	24 h	48 h
Control group	44.08 ± 2.31a	125.68 ± 4.64a	213.32 ± 5.26a
Test group	44.14 ± 3.21a	72.67 ± 6.77b	115.88 ± 13.55b

Thawing time was 48 h. Different letters in the same column indicate significant differences between the test group and control group ($P < 0.05$)

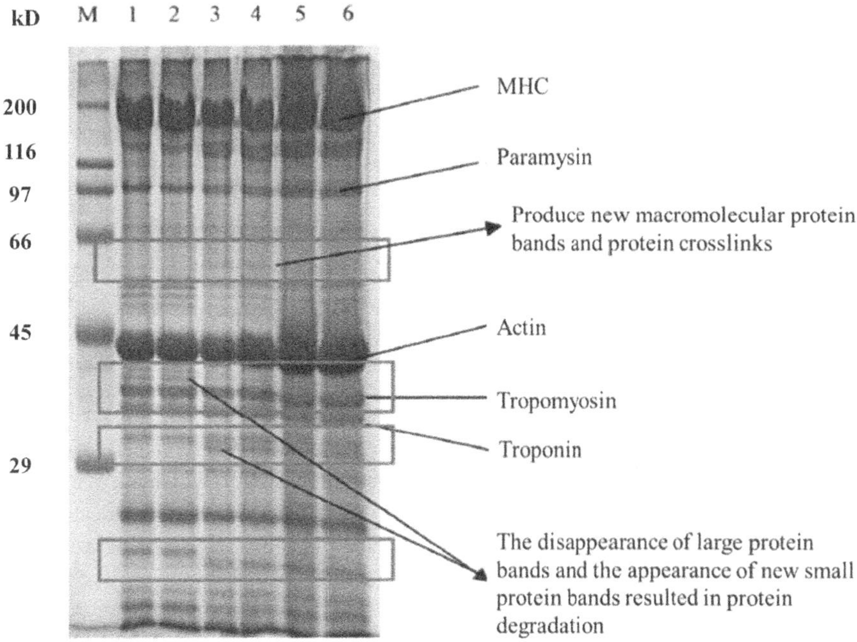

Fig. 12.12 Effects of different thawing methods on SDS-PAGE of myofibril protein. M is standard protein; 1, 2 are non-thawing meat protein; 1, 4 are thawing meat protein in natural condition (in air at 4 °C) for 48 h; 5, 6 are thawing meat protein in low-variable temperature with high humidity for 48 h. Translated from Zhang et al. (2013)

12.3.3 Effect of Low-Variable Temperature and High Humidity Thawing on the Microstructure of Meat

The frozen lamb and lamb thawed by two different methods were observed by scanning electron microscopy to study the integrity of the muscle fiber bundle, the interfiber bundle space and the integrity of the muscle bundle membrane. The results were shown in Fig. 12.13 (magnification 500 times), which showed that the unthawed mutton muscle fiber bundles were parallel to each other, the muscle fiber bundles were closely arranged with small interspaces, and the structure of the perimysium was intact (Figure 12.13a). And thawing can lead to the loss of the integrity of the muscle fiber structure, the increase of the gap between the muscle bundles, the destruction of the dense structure, and the rupture of the myocutaneous membrane. Previous researches have studied the physicochemical changes of the pig muscle *Longissimus thoracis* and the effect on protein oxidation under repeated freezing and thawing conditions (Xia et al. 2012). It was also discovered that freeze-thaw treatment could destroy the microstructure of muscle and increase the gap between muscle fibers. However, the muscle fiber structure of the experimental group was more complete than that of the control group, the gap was smaller than

a. Non-thawing b. Test group c. Control group

Fig. 12.13 Effects of different thawing methods on fiber bundle of mutton for thawing 48 h (500×). Translated from Zhang et al. (2013)

that of the control group, and the muscle bundle membrane was damaged to a lesser extent (Figure 12.13b, c). Muscle fascia has a certain elasticity, which plays a role in maintaining muscle integrity and preventing muscle fiber damage. The destruction of the fascicular membrane will lead to the loss of muscle tissue integrity, the destruction of the dense structure, the enlargement of the space between the muscle fiber bundles, and the decrease of the WHC of the muscle, resulting in more water leakage and serious juice loss. Compared with air thawing, LVTHHT has less damage to the texture of thawed mutton, and the muscle bundle structure of the meat was more complete and dense, thus better WHC, lower juice loss rate, and better chewiness, which also verified the results of this study on the juice loss rate and texture change of thawed mutton.

12.4 Pulse Pressure Pickling

In this part, mutton was used as the research object, and discussed the effects of pulse pressure salting and atmospheric pickling (as shown in Fig. 12.14) on meat quality.

Pulse pressure pickling machine (self-developed): this experimental device simulates the pulse pressure salting process. The material is designed to be placed in the body of the pickling chamber (marked 1 in Fig. 12.15). The pressure regulating device is used to realize the pulse pressure salting process.

12.4.1 Effect of Pulse Pressure Salting on Meat Quality

12.4.1.1 Salt Content of Lamb

As shown in Fig. 12.16, the salt content in the mutton increased with the pickling time, and the pickling efficiency of pulse pressure salting was improved 8–26% than atmospheric pickling. The main reason was that during the pulse pressure salting

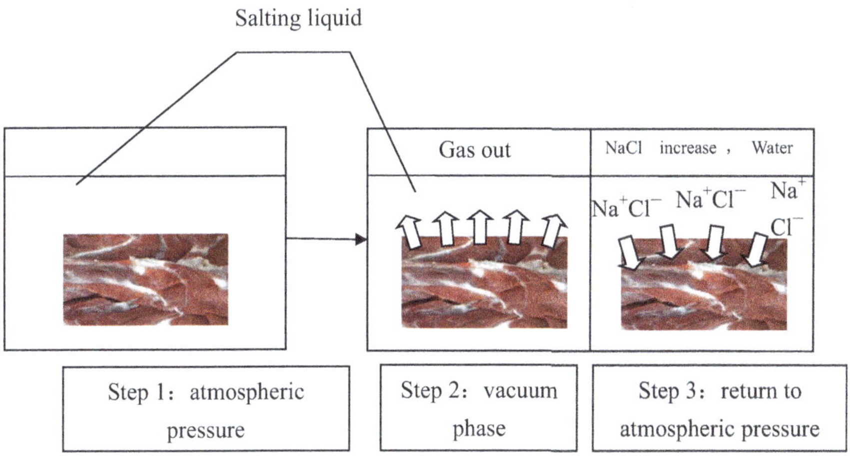

Fig. 12.14 The scheme of the steps in pulsed vacuum pickling. Translated from Xu (2014)

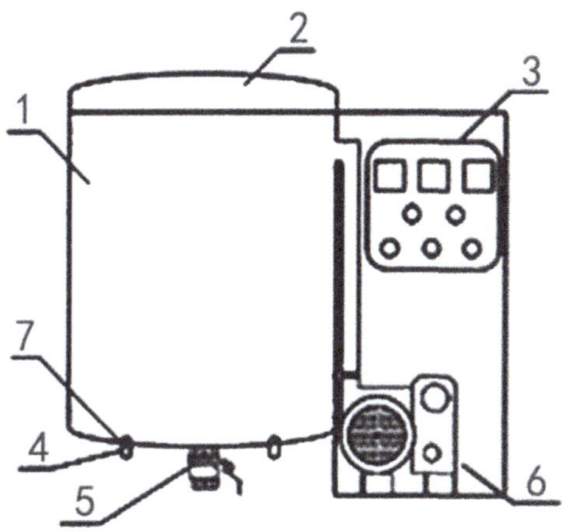

Fig. 12.15 Schematic diagram of pickling machine with pulsed pressure. 1-Salting cavity, 2-Sealing cover, 3-Curing tank and control device, 4-Air inlet, 5-Control valve, 6-Pressure regulator, 7-Drain valve. Translated from Xu (2014)

process, the pressure amplitude changed continuously between vacuum and atmospheric pressure, and the meat structure changed according to the fluid dynamics principle and the deformation relaxation phenomenon, effectively promoting the pickling liquid flow within the meat tissue (Ertekin and Cakaloz, 1996). Compared with the group of atmospheric pickling, when the pulse pressure salting was in a vacuum state, the structure of the meat tissue expanded, and the internal gas and the free state water were discharged, which hindered the entry of the curing liquid. When the pressure returned to normal pressure, the pickling liquid quickly entered the inside of the meat pores, thereby accelerating the migration of the solute material (Dong et al. 2008; Xu et al. 2010; Paes et al. 2006), which promoted the uniform distribution of the salt (Gonz Lez-Mart Nez et al. 2002; Wang et al. 2013).

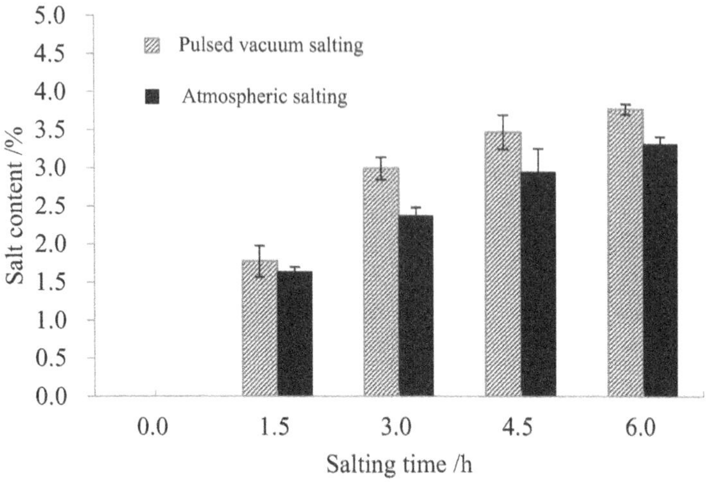

Fig. 12.16 Effect of different brining methods on salt content of lamb. Translated from Xu (2014)

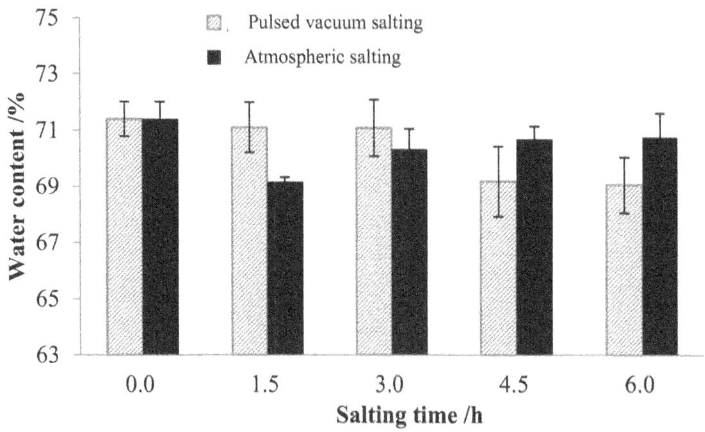

Fig. 12.17 Effect of different brining methods on water content of lamb. Translated from Xu (2014)

12.4.1.2 Water Content of Lamb

It is shown in Fig. 12.17 that the water content of lamb was decreasing during the pulse pressure salting process, and after pickling was significantly lower than before pickling ($P < 0.05$). During the 1.5 h period before pulse pressure salting, the water decreased rapidly and reached equilibrium at around 4.5 h. During the atmospheric pickling process, the water content decreased first and then increased, and the water content in the lamb after pickling was not significantly different from that before pickling ($P > 0.05$). After marinating by two kinds of pickling methods, the water content of lamb after pulsating vacuum pickling was significantly lower than that of atmospheric pickling ($P < 0.05$). The change of water content of lamb during atmospheric pickling was mainly due to the fact that the sodium chloride pickling solution was at stationary statement, causing denaturation and dehydration of the external protein of the lamb, with the advancement of curing, sodium chloride gradually migrates into the interior of the lamb and formed a complex with the protein. This salt complex had a greater driving force, resulting in a small positive increase in water content, and there was no significant change after 4.5 h ($P < 0.05$). The pulsating pressure during the pulse pressure salting process is the main control force of free water in the lamb.

12.4.1.3 Color of Lamb's Surface

It is shown in Table 12.5 that the two pickling methods had a significant effect on the surface color of lamb, and the color indicators L^*, a^*, b^* were significantly lower than the fresh meat. The L^* value represents the brightness of the color, a^* represents the redness value, and b^* represents the yellowness value (Amiryousefi et al. 2014). The decrease of L^* value may be due to changes in the water state and content of the lamb after curing, and the water dispersed in the muscle fibers may affect the color reflectance of the meat (Shi et al. 2011). The redness value decreased ($P < 0.01$), mainly due to the lamb in the marinade, and one was the hemoglobin dissolution in the meat, the other was that the marinade directly contacted the lamb tissue to promote the conversion of myoglobin to high-iron myoglobin and its derivatives. Accompanied by the degeneration of myoglobin (Brewer and Novakofski, 1999; King and Whyte, 2006; Villamonte et al. 2013; Yancey et al. 2011), it made its color lighter. In the pickling process, sodium chloride in the marinade will aggravate the oxidation of fat in the meat, which is one of the causes of the decrease in the yellowness value (Bello and Granados, 1996).

As shown in Table 12.6, the effects of the two pickling methods on the internal color of the lamb were different. The a^* value was significantly lower than that of fresh meat after pulsating pickling, but there was no significant difference after

Table 12.5 Effect of different brining methods on color of lamb' surface Translated from Xu (2014)

		salting time /h		
index	pickling method	0	6	saliency
*L**	Pulse pressure salting	40.56 ± 1.35	37.52 ± 1.41	**
	Atmospheric pickling	40.56 ± 1.35	36.82 ± 2.08	**
*a**	Pulse pressure salting	18.40 ± 0.76	4.50 ± 0.62	**
	Atmospheric pickling	18.40 ± 0.77	5.04 ± 1.55	**
*b**	Pulse pressure salting	12.03 ± 0.40	4.37 ± 1.07	**
	Atmospheric pickling	12.03 ± 0.41	5.54 ± 1.27	**

**Very significant difference ($P < 0.01$)

Table 12.6 Effect of different brining methods on color of lamb's interface Translated from Xu (2014)

index	pickling method	salting time/h		saliency
		0	6	
L^*	Pulse pressure salting	40.06 ± 1.53	34.33 ± 2.12	**
	Atmospheric pickling	40.06 ± 1.53	31.98 ± 1.76	**
a^*	Pulse pressure salting	18.36 ± 0.24	15.86 ± 0.52	**
	Atmospheric pickling	18.36 ± 0.24	18.3 ± 2.85	
b^*	Pulse pressure salting	4.91 ± 0.84	4.33 ± 0.51	
	Atmospheric pickling	4.91 ± 0.84	4.30 ± 0.72	

*Significant difference($P < 0.05$), **very significant difference ($P < 0.01$)

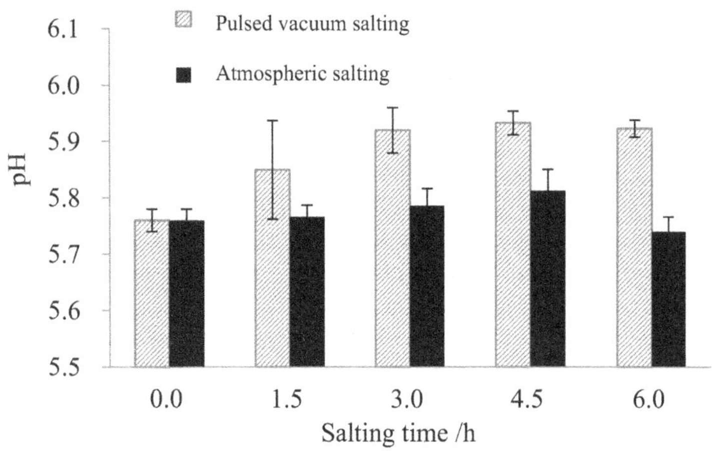

Fig. 12.18 Effect of different brining methods on pH of lamb. Translated from Xu (2014)

atmospheric pickling. The change of the internal color of the lamb was mainly due to the migration of sodium chloride in the marinade. Sodium chloride can promote the conversion of myoglobin to high-iron myoglobin and make the color of the meat brown (Corzo et al. 2006a). Pulse pressure salting can effectively promote the internal penetration of sodium chloride to make its internal salt content significantly higher than that of atmospheric pressure pickling. It was the main reason for the internal color index a^* of the meat sample after pulse pressure salting was significantly lower than 0 h and atmospheric pressure ($P < 0.05$).

12.4.1.4 pH

The change of pH of lamb during pickling was shown in Fig. 12.18. It can be seen that there were some differences in the effects of different pickling methods on the change of pH of lamb. In the 3 h stage before pulse pressure salting, the pH of lamb showed an upward trend, and there was no significant change after 3 h ($P > 0.05$). During the atmospheric pickling process, the pH of lamb increased first and then decreased, and the overall pH was lower than that of the pulsating pickling process. When the curing was finished, the pH of lamb in pulse pressure salting was significantly higher than that of in atmospheric pickling ($P < 0.01$). The main reason for the decrease in the pH of the atmospheric pickling was that as the pickling time prolonged, the liquid in the raw material tissue gradually oozed out due to the influence of the osmotic pressure, resulting in a decrease of pH (Zhang and Xiong, 2013). The pH of lamb during the pulse pressure salting process was mainly because of the action of the pressure during the curing process, which promoted the flow of the marinade in the lamb and accelerated the exudation of the internal gas and free tissue fluid of the lamb. The sodium chloride pickling solution was neutral and the pH rose.

12.4.1.5 Cooking Loss and Water Holding Capacity

The water holding capacity of meat, also known as water power or WHC, refers to the ability of muscles to retain and gain water when subjected to external forces (Yin, 2011). Generally, the indexes to measure the hydraulic retention of muscle include WHC, drip loss, cooking loss, centrifugal loss, purge loss, etc. (Kaale et al. 2014; Sánchez-Valencia et al. 2014; Zhou et al. 2014). Two indicators of cooking loss and WHC were selected for this experiment. As shown in Fig. 12.19, the cooking loss of lamb after pickling decreased significantly ($P < 0.05$), but there was no significant difference between the two pickling methods ($P > 0.01$). As shown in Fig. 12.20, the WHC of lamb increased significantly ($P < 0.05$) before 1.5 h during the pulse pressure salting, and it was stable during the 1.5–6 h ($P > 0.01$). The increase in

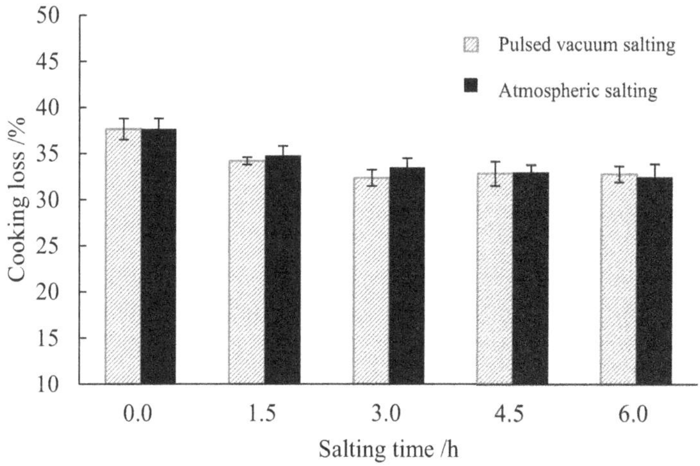

Fig. 12.19 Effect of different brining methods on cooking loss of lamb. Translated from Xu (2014)

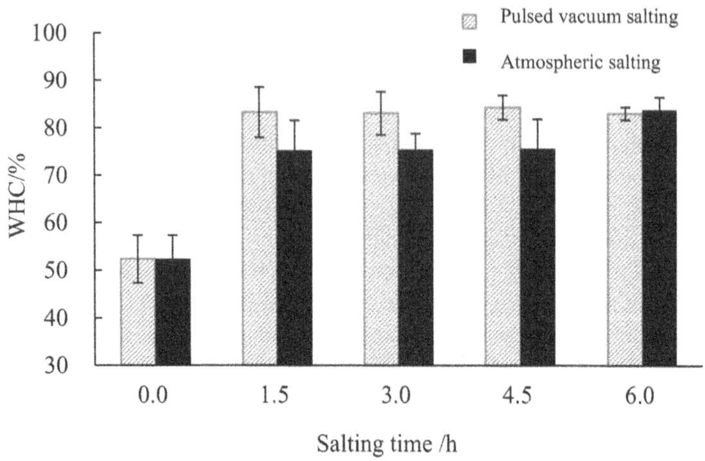

Fig. 12.20 Effect of different brining methods on WHC of lamb. Translated from Xu (2014)

WHC was because when sodium chloride acted directly on muscle fibers, the ionic strength was >0.15, and the muscle fibers would have an electrostatic shielding effect due to the combination of salt ions, resulting in swelling of the muscle fibers, an increase in area, and an increase in WHC (Thorarinsdottir et al. 2011b), and reached a maximum at an ionic strength of 0.8.

The ionic strength of the marinade selected in the experiment was close to 2.56, and the final salt content in the meat sample after curing was 1.5%–2.5% (ionic strength 0.25–0.42), that is, the ionic strength was just in the range of swelling of the muscle fiber. At this time, a large amount of chloride ions were trapped between the myofibrils, increasing the electrostatic repulsion caused by the negative charge, thus leading to the myofibril to swell and increasing the WHC, which was similar to that of Thorarinsdottir reported in the early stage of squid pickling (Thorarinsdottir et al. 2011a). Meanwhile, as the curing time increased, the sarcoplasmic protein and myofibrillar protein were dissolved (Chaijan, 2011). In the case of heat denaturation, the water and fat were wrapped and solidified, and the WHC increased.

12.4.1.6 Tenderness

As shown in Fig. 12.21, with the extension of pickling time, the shear force of lamb presented a downward trend. The shearing force of lamb was significantly decreased ($P < 0.05$) before 1.5 h during pulse pressure salting, and there was no significant change after 1.5 h ($P > 0.01$). The shear force of lamb during pulse pressure salting was lower than the atmospheric pickling during pickling. The shear force is an important indicator for the meat tenderness. The higher the shear force value, the firm the meat is, and vice versa (Devine et al. 2002; Karumendu et al. 2009). The improvement of meat tenderness may be due to the swelling of muscle fibers,

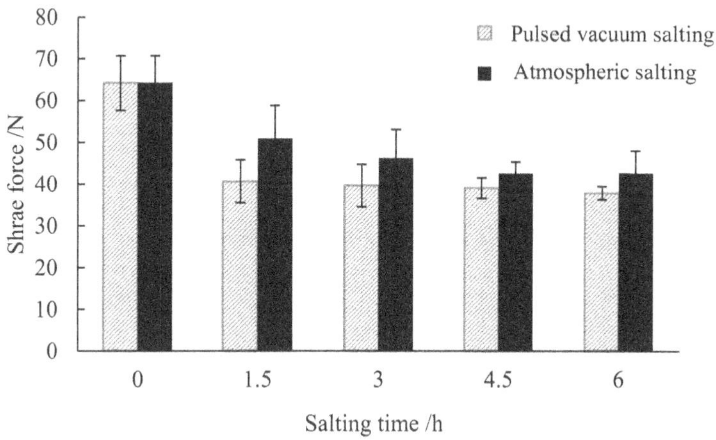

Fig. 12.21 Effect of different brining methods on Warner–Bratzler shear force of lamb. Translated from Xu (2014)

the dissociation of myosin, the increase in WHC, and the increase in tenderness at high ion concentrations. On the other hand, during the pickling process, because of the pulse action, the lamb has a dynamic change of expansion-contraction, which causes the lamb tissue to relax, thereby improving the tenderness (Tobin et al. 2013). It has been reported that pulse pressure salting causes the pickling fluid to flow in the meat due to the hydrodynamic action and the deformation relaxation mechanics due to the cyclical changes in pressure. The mechanical force generated by the flow of the pickling liquid is a kind of massage for lamb, and is also one of the reasons for promoting salt migration and improving tenderness (Koutchma, 2007).

12.4.2 Effect of Pulse Pressure Salting on Myofibrillar and Sarcoplasmic Protein

The results of SDS-PAGE of myofibrillar protein of lamb were shown in Fig. 12.22, as can be seen, the MHC (220 kDa), actin (actin,43 kDa), paramysin (paramysin,100 kDa), and other bands of myofibrillar protein were clearly visible, while tropomyosin was near 36 kDa, and myogenic protein was near 35 kDa, which was consistent with the previous literature (Zhang and Xiong, 2013). The disappearance of the band or the formation of a new band was not observed in Fig. 12.22 compared to 0 h before pickling, the color of electrophoretic pattern of pulse pressure salting and atmospheric pickling became thicker at the bands of the MHC and actin, which was due to the dissolution of myosin and actin in the mutton tissue during the curing process. At the band of 40 kDa, the color of the band gradually darkened as the pickling time increased, which might due to dissociation of the tropomyosin protein.

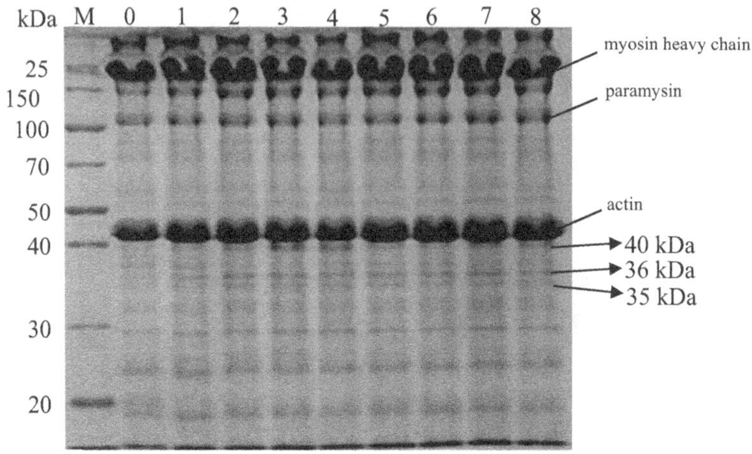

Fig. 12.22 Effects of different pickling methods on SDS-PAGE of myofibril protein. M is standard protein; 0 is fresh lamb protein; 1 ~ 4 are pulsed vacuum pickling meat protein with 1.5, 3, 4.5, 6 h; 5 ~ 8 are atmospheric pressure pickling with 1.5, 3, 4.5, 6 h. Translated from Xu (2014)

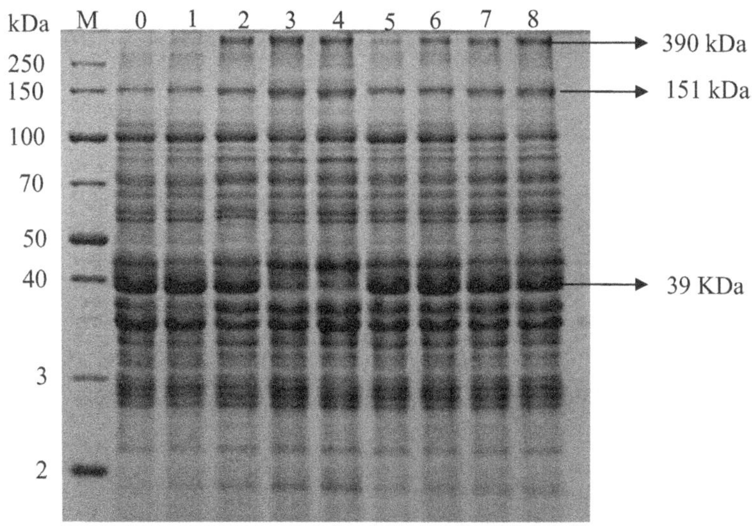

Fig. 12.23 Effects of different pickling methods on SDS-PAGE of sarcoplasmic protein. M is standard protein; 0 is fresh lamb protein;1 ~ 4 are pulsed vacuum pickling meat protein with 1.5, 3, 4.5, 6 h; 5 ~ 8 are atmospheric pressure pickling with 1.5, 3, 4.5, 6 h. Translated from Xu (2014)

Figure 12.23 showed the SDS-PAGE electrophoresis of the sarcoplasmic protein of lamb. The macromolecular bands were generated at 390 kDa; the color of the band at 150 kDa increased with the pickling time; the band at 39 kDa was pulsating. After pickling for 4.5 h, it became lighter, but the change in atmospheric pickling was not obvious.

12.4.3 Effect of Pulse Pressure Salting on Microstructure of Meat

From Fig. 12.24 we could tell that the ultrastructure of lamb has changed after pulse pressure salting. Figure 12.24a shows the raw meat that has not been marinated after thawing. The myofibrils were neat and clearly visible, with regular light and dark stripes, and M line, Z line, A band (dark band), I band (The bright band), and the H zone (the brighter strip in the middle of the dark band) were clearly visible. After 1.5 h of pulse pressure salting, the muscle fibers began to shrink and the Z line began to disappear. After 3 h, the muscle fibers expanded and became thicker. After 6 h of pulse pressure salting, the Z line of the lamb was thinner and fuzzier than the raw meat.

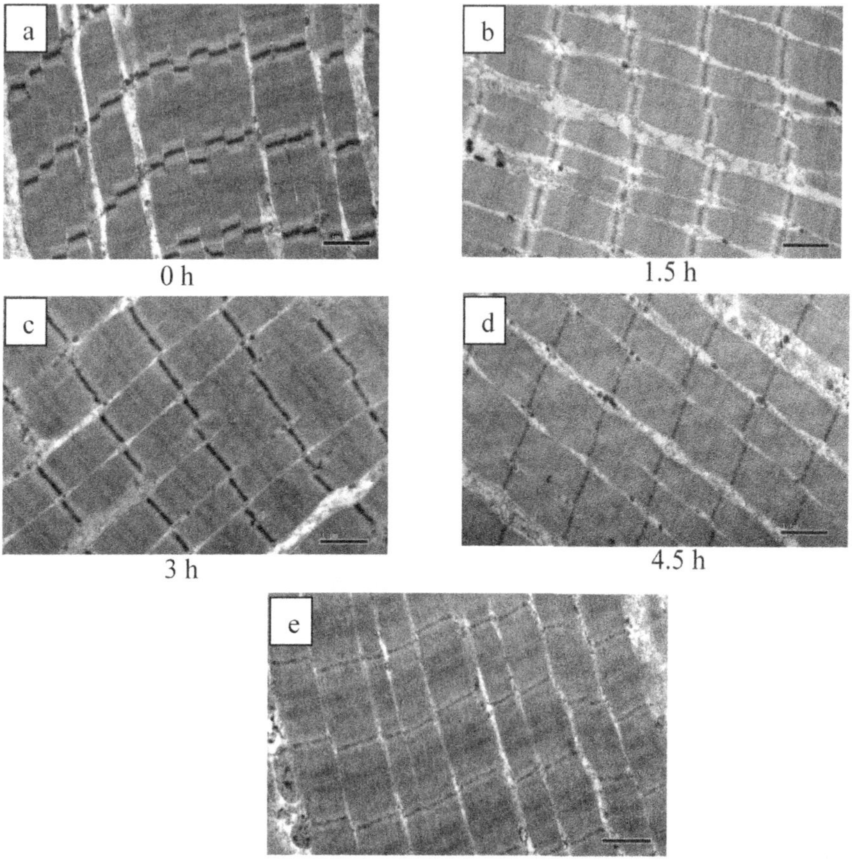

Fig. 12.24 The longitudinal section of lamb muscle at transmission electron microscope under pulsed vacuum pressure. (**a**) Lamb muscle at 0 h; (**b**) lamb muscle under pulsed vacuum pressure for 1.5 h; (**c**) lamb muscle under pulsed vacuum pressure for 3 h; (**d**) lamb muscle under pulsed vacuum pressure for 4.5 h; (**e**) lamb muscle under pulsed vacuum pressure for 6 h. Translated from Xu (2014)

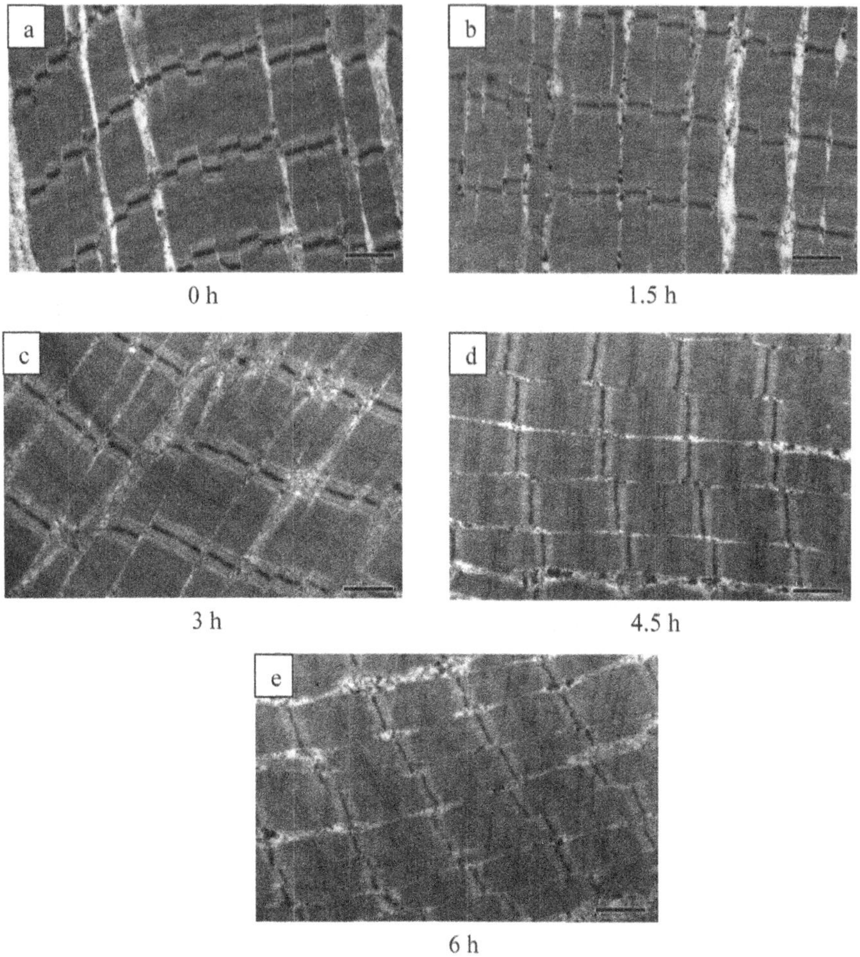

Fig. 12.25 The longitudinal section of lamb muscle at transmission electron microscope under atmospheric pressure. (**a**) Lamb muscle at 0 h; (**b**) lamb muscle under atmospheric pressure for 1.5 h; (**c**) lamb muscle under atmospheric pressure for 3 h; (**d**) lamb muscle under atmospheric pressure for 4.5 h; (**e**) lamb muscle under atmospheric pressure for 6 h. Translated from Xu (2014)

As shown in Fig. 12.25, the atmospheric pickling process has changed the ultrastructure of lamb. After 1.5 h of pickling, the muscle fibers began to shrink and the gap gradually decreased; the Z line became shallower and gradually disappeared; after 6 h of pickling, it can be seen from the figure that the H band of the lamb tended to be blurred.

Compared with pulse pressure salting, the Z-line of lamb tended to be lighter and will be thinner and more blurred after pulse pressure salting, the main component of the Z-line was actin. It indicated that both kinds of pickling methods dispelled the actin of lamb, and the pulse pressure salting caused the dissociation of actin to be

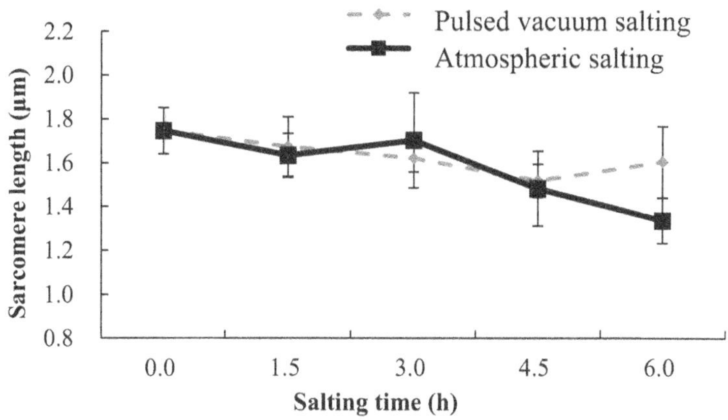

Fig. 12.26 Effect of different brining methods on sarcomere length of lamb. Translated from Xu (2014)

more than that of normal pressure pickling. This was consistent with the results of the effect of pulse pressure salting on myofibrillar protein in Fig. 12.22.

In terms of Fig. 12.26, during the pickling process, the length of the sarcomere of lamb decreased. The length of sarcomere was not significantly changed after pulse pressure salting ($P > 0.05$). The length of sarcomere was significantly lower than that of raw meat after atmospheric pickling ($P < 0.05$), and the length of the sarcomere after pulse pressure salting was significantly higher than the length of the sarcomere after atmospheric pickling ($P < 0.05$). In the present study, the ionic strength is just in the range of muscle fiber expansion, a large amount of chloride ions are trapped between the myofibrils, increasing the electrostatic repulsion caused by the negative charge, causing the expansion of the myofibrils, which is the reason for the increase in WHC. At the same time, the dissolution of sarcoplasmic protein and myofibrillar protein with the increase of pickling time was one of the reasons for myofibrillar structure change during the pickling process. There was no significant change in the length of the sarcomere after pulse pressure salting. It might be that during the pulse pressure salting process, due to the periodic vacuum pulsation process, the vacuum pressure exerted an outward expansion force on the lamb. However, the length of the sarcomere was shortened because of pickling, and the combined effect of the length of the sarcomere was not significantly changed after pulse pressure salting.

12.5 Conclusions

Fresh ovine meat stored in CFPS treatment for 10 days proved to have better color stability in comparison with those in 4 °C storage. CFPS could be used as an option storage method for color improvement of fresh meat. The samples in this study were over wrapped by polyvinyl chloride film during storage and the effects of different packaging conditions could be investigated in the next study as packaging could affect potential effects of storage procedure by interacting with storage treatments.

Compared to the most common thawing method (in air at 4 °C), LVTHHT method had many advantages: it prevented the loss of the surface water, significantly reduced thawing loss, cooking loss, and nutrition loss (protein). Moreover, the formation of water film prevented meat oxidation and delayed the deterioration of meat quality; therefore, the physicochemical characteristics of the thawed meat were closer to fresh meat. LVTHHT method could be used for thawing frozen meat.

Compared to atmospheric pressure pickling, the salt content of lamb with PVP was improved by 8–26%. The cooking loss and Warner–Bratzler shear force decreased notably, and WHC increased significantly. Therefore, the PVP could be used to improve the curing efficiency while maintaining the quality of lamb.

Acknowledgments Parts of this chapter are translated from Society of Agricultural Engineering, 29, Zhang, C. H., et al. Low-variable temperature and high humidity thawing improves lamb quality (Chinese). 267–273; Scientia Agricultura Sinica, 49, Zhang, Y., et al. Effects of controlled freezing point storage on the protein phosphorylation level (Chinese). 4429–4440. Copyright (2020), with permission from two Journals. Xu, W. 2014. Optimization of pulsed vacuum pressure pickling and its effect on the quality of lamb. Ningxia University. Zhang, Y. 2016. Study of controlled freezing point storage regulating lamb color stability and its protein phosphorylation level. Shaanxi Normal University.

References

Amiryousefi, M., Mohebbi, M., & Khodaiyan, F. (2014). Applying an intelligent model and sensitivity analysis to inspect mass transfer kinetics, shrinkage and crust color changes of deep-fat fried ostrich meat cubes. *Meat Science, 96*, 172–178.

Baron, S. J., Li, J., Russell, R. R., Neumann, D., Miller, E. J., Tuerk, R., et al. (2005). Dual mechanisms regulating AMPK kinase action in the ischemic heart. *Circulation Research, 96*, 337.

Bello, R., & Granados, A. (1996). Evaluacion fısico-quımica delpescado seco-salado en Venezuela. *Archivos Latinoamericanos de Nutricion, 46*, 154–158.

Brewer, M. S., & Novakofski, J. (1999). Cooking rate, pH and final end point temperature effects of color and cook loss of a lean ground beef model system. *Meat Science, 52*, 443–451.

Chaijan, M. (2011). Physicochemical changes of tilapia (*Oreochromis niloticus*) muscle during salting. *Food Chemistry, 129*, 1201–1210.

Chen, L., Li, X., Ni, N., Liu, Y., Chen, L., Wang, Z., et al. (2016). Phosphorylation of myofibrillar proteins in post-mortem ovine muscle with different tenderness. *Journal of the Science of Food and Agriculture, 96*, 1474–1483.

Corzo, O., & Bracho, N. (2007). Determination of water effective diffusion coefficient of sardine sheets during vacuum pulse osmotic dehydration. *LWT-Food Science and Technology, 40*, 1452–1458.

Corzo, O., Bracho, N., & Marjal, J. (2006a). Color change kinetics of sardine sheets during vacuum pulse osmotic dehydration. *Journal of Food Engineering, 75*, 21–26.

Corzo, O., Bracho, N., & Marval, J. (2006b). The use of fractional conversion technique on firmness change kinetics of vacuum pulse osmotic dehydration sardine sheets. *Journal of Food Engineering, 73*, 358–363.

Dai, Z., Cui, Y., & Wang, H. (2008). Changes of textural properties of cultured *Pseudosciaena crocea* muscle under different frozen storage conditions. *Food and Fermentation Industries, 34*, 188–191.

Deumier, F., Bohuon, P., Trystram, G., Saber, N. C., & Gnan, A. (2011). Pulsed vacuum brining of poultry meat: Experimental study on the impact of vacuum cycles on mass transfer. *Journal of Food Engineering, 58*, 75–83.

Devine, C. E., Payne, S. R., Peachey, B. M., Lowe, T. E., & Cook, C. J. (2002). High and low rigor temperature effects on sheep meat tenderness and ageing. *Meat Science, 60*, 141–146.

Dickens, J. A., & Lyon, C. E. (1995). The effects of electric stimulation and extended chilling times on the biochemical reactions and texture of cooked broiler breast meat. *Poultry Science, 74*, 2035–2040.

Dong, H., Xian, Y., Wang, S., & Sun, Z. (2008). Transfer rule of osmotic dehydration of carrots under ultrasound treatment. *Journal of Harbin Engineering University, 29*, 83–87.

Ertekin, F., & Cakaloz, T. (1996). Osmotic dehydration of peas. I. Influence of process variables on mass transfer. *Journal of Food Processing and Preservation, 20*, 105–119.

Fito, P. (1994). Modelling of vacuum osmotic dehydration of food. *Journal of Food Engineering, 22*, 313–328.

Reyes, G., Corzo, O., Bracho, N., & Rodriguez, Y. (2008). Optimization of vacuum pulse osmotic dehydration of sardine sheets. *Revista Científica, 18*, 320–328.

Gonz Lez-Mart Nez, C., Chafer, M., Fito, P., & Chiralt, A. (2002). Development of salt profiles on Manchego type cheese during brining. Influence of vacuum pressure. *Journal of Food Engineering, 53*, 67–73.

Guamis, B., Trujillo, A. J., Ferragut, V., Chiralt, A., Andres, A., & Fito, P. (1997). Ripening control of Manchego type cheese salted by brine vacuum impregnation. *International Dairy Journal, 7*, 185–192.

Huang, H. (2005). *The effects of freeze and thaw on ice crystal morphology and physic-chemical quality of pork*. Nanjing: Nanjing Agriculture University.

Huang, H., Larsen, M. R., & Lametsch, R. (2012). Changes in phosphorylation of myofibrillar proteins during postmortem development of porcine muscle. *Food Chemistry, 134*, 1999–2006.

Hwang, J. T., Lee, M., Jung, S. N., Lee, H. J., Kang, I., Kim, S. S., et al. (2004). AMP-activated protein kinase activity is required for vanadate-induced hypoxia-inducible factor 1? Expression in Du145 cells. *Carcinogenesis, 25*, 2497–2507.

Johnson, R. C., Romans, J. R., Muller, T. S., Costello, W. J., & Jones, K. W. (2010). Physical, chemical and sensory characteristics of four types of beef steaks. *Journal of Food Science, 55*, 1264–1267.

Kaale, L. D., Eikevik, T. M., Rustad, T., & Nordtvedt, T. S. L. (2014). Changes in water holding capacity and drip loss of Atlantic salmon (*Salmo salar*) muscle during superchilled storage. *LWT-Food Science and Technology, 55*, 528–535.

Karumendu, L. U., Ven, R. V. D., Kerr, M. J., Lanza, M., & Hopkins, D. L. (2009). Particle size analysis of lamb meat: Effect of homogenization speed, comparison with myofibrillar fragmentation index and its relationship with shear force. *Meat Science, 82*, 425–431.

Kemp, C. M., Bardsley, R. G., & Parr, T. (2006). Changes in caspase activity during the postmortem conditioning period and its relationship to shear force in porcine longissimus muscle. *Journal of Animal Science, 84*, 2841–2846.

King, N. J., & Whyte, R. (2006). Does it look cooked? A review of factors that influence cooked meat color. *Journal of Food Science, 71*, 31–40.

Koutchma, T. (2007). Advanced technologies for meat processing, Leo M.L. Nollet, Fidel Toldra (Eds.). *Food Microbiology, 24*, 801–802.

Lee, S. H., Joo, S. T., & Ryu, Y. C. (2010). Skeletal muscle fiber type and myofibrillar proteins in relation to meat quality. *Meat Science, 86*, 166–170.

Li, P., Li, X., Li, Z., Chen, L., Li, Z., Chen, L., et al. (2016). Effects of controlled freezing point storage on aging from muscle to meat. *Scientia Agricultura Sinica, 49*, 554–562.

Liao, C., Rui, H., Zhang, L., Yin, B., & Chen, Y. (2010). Effects of ultra-high pressure thawing on san Huang chicken quality by different frozen methods. *Transactions of the Chinese Society of Agricultural Engineering, 26*, 331–337.

Martinez, O., Salmer, J., Guill, M. D., & Casas, C. (2004). Texture profile analysis of meat products treated with commercial liquid smoke flavourings. *Food Control, 15*, 457–461.

Mazzei, G. J., & Kuo, J. F. (1984). Phosphorylation of skeletal-muscle troponin I and troponin T by phospholipid-sensitive Ca^{2+}-dependent protein kinase and its inhibition by troponin C and tropomyosin. *Biochemical Journal, 218*, 361–369.

Mikkelsen, A., Juncher, D., & Skibsted, L. H. (1999). Metmyoglobin reductase activity in porcine *M. longissimus* dorsi muscle. *Meat Science, 51*, 155–161.

Paes, S. S., Stringari, G. B., & Laurindo, J. B. (2006). Effect of vacuum impregnation-dehydration on the mechanical properties of apples. *Drying Technology, 24*, 1649–1656.

Park, D., Xiong, Y. L., & Alderton, A. L. (2007). Concentration effects of hydroxyl radical oxidizing systems on biochemical properties of porcine muscle myofibrillar protein. *Food Chemistry, 101*, 1239–1246.

Roskoski, R. (2015). A historical overview of protein kinases and their targeted small molecule inhibitors. *Pharmacological Research, 100*, 1–23.

Ruan, Z., Li, B., Zhu, Z., & Meng, M. (2008). Effects of different freezing rates on the freezing characteristics of Ctenopharyngodon idellus C. et V fillets. *Transactions of the Chinese Society of Agricultural Engineering, 2*, TS254.

Ryder, J. W., Lau, K. S., Kamm, K. E., & Stull, J. T. (2007). Enhanced skeletal muscle contraction with myosin light chain phosphorylation by a calmodulin-sensing kinase. *Journal of Biological Chemistry, 282*, 20447–20454.

Sánchez-Valencia, J., S Nchez-Alonso, I., Martinez, I., & Careche, M. (2014). Estimation of frozen storage time or temperature by kinetic modeling of the Kramer shear resistance and water holding capacity (WHC) of hake (*Merluccius merluccius*, L.) muscle. *Journal of Food Engineering, 120*, 37–43.

Scopes, R. K., & Stoter, A. (1982). [79] Purification of all glycolytic enzymes from one muscle extract. *Methods in Enzymology, 90*, 479–490.

Shi, P. L., Min, H. H., Li, C.-B., Xu, X.-L., & Zhou, G. H. (2011). Changes in meat quality characteristics of goose breast muscle after tumbling. *Food Science, 32*, 88–92.

Thanonkaew, A., Benjakul, S., Visessanguan, W., & Decker, E. A. (2006). The effect of metal ions on lipid oxidation, colour and physicochemical properties of cuttlefish (*Sepia pharaonis*) subjected to multiple freeze-thaw cycles. *Food Chemistry, 95*, 591–599.

Thorarinsdottir, K. A., Arason, S., Sigurgisladottir, S., Gunnlaugsson, V. N., Johannsdottir, J., & Tornberg, E. (2011a). The effects of salt-curing and salting procedures on the microstructure of cod (*Gadus morhua*) muscle. *Food Chemistry, 126*, 109–115.

Thorarinsdottir, K. A., Arason, S., Sigurgisladottir, S., Valsdottir, T., & Tornberg, E. (2011b). Effects of different pre-salting methods on protein aggregation during heavy salting of cod fillets. *Food Chemistry, 124*, 7–14.

Tobin, B. D., O'Sullivan, M. G., Hamill, R. M., & Kerry, J. P. (2013). The impact of salt and fat level variation on the physiochemical properties and sensory quality of pork breakfast sausages. *Meat Science, 93*, 145–152.

Toyo-Oka, T. (1982). Phosphorylation with cyclic adenosine 3′: 5′ monophosphate-dependent protein kinase renders bovine cardiac troponin sensitive to the degradation by calcium-activated neutral protease. *Biochemical and Biophysical Research Communications, 107*, 44–50.

Villamonte, G., Simonin, H., Duranton, F., Ch Ret, R., & De Lamballerie, M. (2013). Functionality of pork meat proteins: Impact of sodium chloride and phosphates under high-pressure processing. *Innovative Food Science and Emerging Technologies, 18*, 15–23.

Wagner, J. R., & Añon, M. C. (2007). Effect of freezing rate on the denaturation of myofibrillar proteins. *International Journal of Food Science & Technology, 20*, 735–744.

Wang, X., Gao, Z., Xiao, H., Wang, Y., & Bai, J. (2013). Enhanced mass transfer of osmotic dehydration and changes in microstructure of pickled salted egg under pulsed pressure. *Journal of Food Engineering, 117*, 141–150.

Woods, A., Dickerson, K., Heath, R., Hong, S.-P., Momcilovic, M., Johnstone, S. R., et al. (2005). Ca^{2+}/calmodulin-dependent protein kinase kinase-β acts upstream of AMP-activated protein kinase in mammalian cells. *Cell Metabolism, 2*, 0–33.

Xia, X., Kong, B., Liu, J., Diao, X., & Liu, Q. (2012). Influence of different thawing methods on physicochemical changes and protein oxidation of porcine longissimus muscle. *LWT-Food Science and Technology, 46*, 280–286.

Xia, X. F., Kong, B. H., Guo, Y. Y., & Liu, Q. (2009). Effects of freeze-thawing cycles on the quality properties and microstructure of pork muscle. *Scientia Agricultura Sinica, 42*, 982–988.

Xu, W. (2014). *Optimization of pulsed vacuum pressure pickling and its effect on the quality of lamb*. Yinchuan: Ningxia University.

Xu, Y., Yuan, Y., Zhang, Y., Li, S., & Dang, X.-A. (2010). Experiments on osmotic dehydration of carrot at vacuum pressure. *Transactions of the Chinese Society of Agricultural Engineering, 26*, 350–355.

Yancey, J. W. S., Wharton, M. D., & Apple, J. K. (2011). Cookery method and end-point temperature can affect the Warner-Bratzler shear force, cooking loss, and internal cooked color of beef longissimus steaks. *Meat Science, 88*, 1–7.

Yang, H. (2005). Research on thawing method for frozen meat. *Meat Research, 7*, 43–44.

Yin, J. (2011). *Animal muscle biology and meat quality*. Beijing: China Agricultural University Press.

Zhang, C. H., Li, X., Li, Y., Sun, H. M., Zhang, D. Q., Jia, W., et al. (2013). Low-variable temperature and high humidity thawing improves lamb quality. *Transactions of the Chinese Society of Agricultural Engineering, 29*, 267–273.

Zhang, L. Y., & Xiong, L. (2013). Effect of vacuum pickling on salt osmotic dehydration and quality changes of pork. *Modern Food Science and Technology, 29*, 2595–2600.

Zhang, Y. (2016). *Study of controlled freezing point storage regulating lamb color stability and its protein phosphorylation level*. Shaanxi: Shaanxi Normal University.

Zhang, Y., Li, X., Li, Z., Li, M., Liu, Y.-F., & Zhang, D.-Q. (2016). Effects of controlled freezing point storage on the protein phosphorylation level. *Scientia Agricultura Sinica, 49*, 4429–4440.

Zhou, F., Zhao, M., Zhao, H., Sun, W., & Cui, C. (2014). Effects of oxidative modification on gel properties of isolated porcine myofibrillar protein by peroxyl radicals. *Meat Science, 96*, 1432–1439.

The manufacturer's authorised representative in the EU is Springer Nature Customer Service Centre GmbH, Europaplatz 3, 69115 Heidelberg, Germany. If you have any concerns regarding our products, please contact ProductSafety@springernature.com

Printed and bound by CPI Group (UK) Ltd, Croydon, CR0 4YY
30/09/2026
02225295-0009